Battery Management System and its Applications

Battery Management System and its Applications

Xiaojun Tan
Sun Yat-sen University, China

Andrea Vezzini
University of Applied Science, Switzerland

Yuqian Fan
Henan Institute of Science and Technology, China

Neeta Khare
Iveco Group, Switzerland

You Xu
Guangdong Polytechnic Normal University, China

Liangliang Wei
Sun Yat-sen University, China

The right of Xiaojun Tan, Andrea Vezzini, Yuqian Fan, Neeta Khare, You Xu, and Liangliang Wei to be identified as the authors of the editorial material in this work has been asserted in accordance with law.

Registered Offices
John Wiley & Sons, Inc., 111 River Street, Hoboken, NJ 07030, USA
John Wiley & Sons Singapore Pte. Ltd, 1 Fusionopolis Walk, #06-01 Solaris South Tower, Singapore 138628

Editorial Office
1 Fusionopolis Walk, #06-01 Solaris South Tower, Singapore 138628

For details of our global editorial offices, customer services, and more information about Wiley products visit us at www.wiley.com.

Wiley also publishes its books in a variety of electronic formats and by print-on-demand. Some content that appears in standard print versions of this book may not be available in other formats.

Limit of Liability/Disclaimer of Warranty
While the publisher and authors have used their best efforts in preparing this work, they make no representations or warranties with respect to the accuracy or completeness of the contents of this work and specifically disclaim all warranties, including without limitation any implied warranties of merchantability or fitness for a particular purpose. No warranty may be created or extended by sales representatives, written sales materials or promotional statements for this work. The fact that an organization, website, or product is referred to in this work as a citation and/or potential source of further information does not mean that the publisher and authors endorse the information or services the organization, website, or product may provide or recommendations it may make. This work is sold with the understanding that the publisher is not engaged in rendering professional services. The advice and strategies contained herein may not be suitable for your situation. You should consult with a specialist where appropriate. Further, readers should be aware that websites listed in this work may have changed or disappeared between when this work was written and when it is read. Neither the publisher nor authors shall be liable for any loss of profit or any other commercial damages, including but not limited to special, incidental, consequential, or other damages.

Library of Congress Cataloging-in-Publication Data
Names: Tan, Xiaojun, author.
Title: Battery management system and its applications / Xiaojun Tan [and five others].
Description: Hoboken, NJ : John Wiley & Sons, 2023. | Includes bibliographical references and index.
Identifiers: LCCN 2022033684 (print) | LCCN 2022033685 (ebook) | ISBN 9781119154006 (hardback) |
    ISBN 9781119154037 (pdf) | ISBN 9781119154020 (epub) | ISBN 9781119154013 (ebook)
Subjects: LCSH: Battery management systems.
Classification: LCC TJ163.9 .T36 2023 (print) | LCC TJ163.9 (ebook) | DDC 621--dc23/eng/20220919
LC record available at https://lccn.loc.gov/2022033684
LC ebook record available at https://lccn.loc.gov/2022033685

Cover image: © petrmalinak/shutterstock
Cover design: Wiley

Set in 9.5/12.5pt STIXTwoText by Integra Software Services Pvt. Ltd, Pondicherry, India

Contents

Preface *xiii*
About the Authors *xv*

Part I   Introduction *1*

1   Why Does a Battery Need a BMS? *3*
1.1   General Introduction to a BMS *3*
1.1.1   Why a Battery Needs a BMS *3*
1.1.2   What Is a BMS? *3*
1.1.3   Why a BMS Is Required in Any Energy Storage System *4*
1.1.4   How a BMS Makes a Storage System Efficient, Safe, and Dependable *4*
1.2   Example of a BMS in a Real System *5*
1.2.1   LabView Based BMS *5*
1.2.2   PLC Based BMS *6*
1.2.3   Microprocessor Based BMS *6*
1.2.4   Microcontroller Based BMS *6*
1.3   System Failures Due to the Absence of a BMS *7*
1.3.1   Dreamline Boeing Fire Incidences *7*
1.3.2   Fire Accident at the Hawaii Grid Connected Energy Storage *8*
1.3.3   Fire Accidents in Electric Vehicles *8*
        References *9*

2   General Requirements (Functions and Features) *11*
2.1   Basic Functions of a BMS *11*
2.1.1   Key Parameter Monitoring *11*
2.1.2   Battery State Analysis *11*
2.1.3   Safety Management *13*
2.1.4   Energy Control Management *14*
2.1.5   Information Management *14*
2.2   Topological Structure of a BMS *16*

2.2.1   Relationship Between a BMC and a Cell   *16*
2.2.2   Relationship Between a BCU and a BMC   *16*
          References   *18*

3   **General Procedure of the BMS Design**   *19*
3.1   Universal Battery Management System and Customized Battery Management System   *19*
3.1.1   Ideal Condition   *19*
3.1.2   Feasible Solution   *19*
3.1.3   Discussion of Universality   *20*
3.2   General Development Flow of the Power Battery Management System   *21*
3.2.1   Applicable Standards for BMS Development   *21*
3.2.2   Boundary of BMS Development   *22*
3.2.3   Battery Characteristic Test Is Essential to BMS Development   *23*
3.3   Core Status of Battery Modeling in the BMS Development Process   *23*
          References   *25*

Part II   Li-Ion Batteries   *27*

4   **Introduction to Li-Ion Batteries**   *29*
4.1   Components of Li-Ion Batteries: Electrodes, Electrolytes, Separators, and Cell Packing   *29*
4.2   Li-Ion Electrode Manufacturing   *31*
4.3   Cell Assembly in an Li-Ion Battery   *32*
4.4   Safety and Cost Prediction   *33*
          References   *35*

5   **Schemes of Battery Testing**   *37*
5.1   Battery Tests for BMS Development   *37*
5.1.1   Test Items and Purpose   *37*
5.1.2   Standardization of Characteristic Tests   *38*
5.1.3   Some Issues on Characteristic Tests   *40*
5.1.4   Contents of Other Sections of This Chapter   *41*
5.2   Capacity and the Charge and Discharge Rate Test   *41*
5.2.1   Test Methods   *42*
5.2.2   Test Report Template   *43*
5.3   Discharge Rate Characteristic Test   *44*
5.3.1   Test Method   *44*
5.3.2   Test Report Template   *45*
5.4   Charge and Discharge Equilibrium Potential Curves and Equivalent Internal Resistance Tests   *46*
5.4.1   Test Method for Discharge Electromotive Force Curve and Equivalent Internal Resistance   *46*
5.4.2   Test Method for Charge Electromotive Force Curve and Equivalent Internal Resistance   *47*
5.4.3   Discussion of the Test Method   *48*
5.4.4   Test Report Template   *49*
5.5   Battery Cycle Test   *49*
5.5.1   Features of Battery Cycle Test   *49*
5.5.2   Fixed Rate Cycle Test Method   *50*

5.5.3 Cycle Test Schemes Based on Standard Working Conditions *52*
5.5.4 Test Report Template *57*
5.6 Phased Evaluation of the Cycle Process *58*
5.6.1 Evaluation Method *59*
5.6.2 Estimation of the Test Time *59*
5.6.3 Test Report Template *63*
  References *65*

6 Test Results and Analysis *67*
6.1 Characteristic Test Results and Their Analysis *67*
6.1.1 Actual Test Arrangement *67*
6.1.2 Characteristic Test Results of the LiFePO$_4$ Battery *68*
6.1.3 Characteristic Test Results of the Li(NiCoMn)O$_2$ Ternary Battery *76*
6.1.4 Characteristics Comparison of the Two Battery Types *79*
6.2 Degradation Test and Analysis *80*
6.2.1 Capacity Change Rule During Battery Degradation *81*
6.2.2 Internal Resistance Spectrum Change Rule During Battery Degradation *87*
6.2.3 Impact of Storage Conditions on Battery Degradation *96*
  References *99*

7 Battery Modeling *101*
7.1 Battery Modeling for BMS *101*
7.1.1 Purpose of Battery Modeling *101*
7.1.2 Battery Modeling Requirement of BMS *102*
7.2 Common Battery Models and Their Deficiencies *102*
7.2.1 Non-circuit Models *102*
7.2.2 Equivalent Circuit Models *103*
7.3 External Characteristics of the Li-Ion Power Battery and Their Analysis *105*
7.3.1 Electromotive Force Characteristic of the Li-Ion Battery *105*
7.3.2 Over-potential Characteristics of the Li-Ion Battery *107*
7.4 A Power Battery Model Based on a Three-Order RC Network *110*
7.4.1 Establishment of a New Power Battery Model *110*
7.4.2 Estimation of Model Parameters *112*
7.5 Model Parameterization and Its Online Identification *117*
7.5.1 Offline Extension Method of Model Parameters *117*
7.5.2 Online Identification Method of Model Parameters *121*
7.6 Battery Cell Simulation Model *124*
7.6.1 Realization of Battery Cell Simulation Model Based on Matlab/Simulink *124*
7.6.2 Model Validation *125*
  References *130*

  Part III Functions of BMS *133*

8 Battery Monitoring *135*
8.1 Discussion on Real Time and Synchronization *135*
8.1.1 Factors Causing Delay *135*

8.1.2   Synchronization *136*
8.1.3   Negative Impact of Non-real-time and Non-synchronous Problems *137*
8.1.4   Proposal on Solution *137*
8.2   Battery Voltage Monitoring *139*
8.2.1   Voltage Monitoring Based on a Photocoupler Relay Switch Array (PhotoMOS) *139*
8.2.2   Voltage Monitoring Based on a Differential Operational Amplifier *140*
8.2.3   Voltage Monitoring Based on a Special Integrated Chip *141*
8.2.4   Comparison of Various Voltage Monitoring Schemes *142*
8.2.5   Significance of Accurate Voltage Monitoring for Effective Capacity Utilization of the Battery Pack *142*
8.3   Battery Current Monitoring *145*
8.3.1   Accuracy *145*
8.3.2   Current Monitoring Based on Series Resistance *146*
8.3.3   Current Monitoring Based on a Hall Sensor *147*
8.3.4   A Compromised Method *148*
8.4   Temperature Monitoring *149*
8.4.1   Importance of Temperature Monitoring *149*
8.4.2   Common Implementation Schemes *150*
8.4.3   Setting of the Temperature Sensor *151*
8.4.4   Accuracy *152*
         References *152*

9   SoC Estimation of a Battery *153*
9.1   Different Understandings of the SoC Definition *153*
9.1.1   Difference on the Understanding of SoC *153*
9.1.2   Difference and Relation Between SoC and SoP as Well as SoE *155*
9.2   Classical Estimation Methods *158*
9.2.1   Coulomb Counting Method *158*
9.2.2   Open Circuit Voltage Method *159*
9.2.3   A Compromised Method *160*
9.2.4   Estimation Methods Not Applicable for the Lithium-Ion Battery *161*
9.3   Difficulty in an SoC Estimation *162*
9.3.1   Difficulty in an Estimation Resulting from Inaccurate Battery State Monitoring *162*
9.3.2   Difficulty in an Estimation Resulting from Battery Difference *164*
9.3.3   Difficulty in an Estimation Resulting from an Uncertain Future Working Condition *165*
9.3.4   Difficulty in an Estimation Resulting from an Uncertain Battery Usage History *165*
9.4   Actual Problems to Be Considered During an SoC Estimation *166*
9.4.1   Safety of the Electric Vehicle *166*
9.4.2   Feasibility *167*
9.4.3   Actual Requirements of Drivers *168*
9.5   Estimation Method Based on the Battery Model and the Extended Kalman Filter *169*
9.5.1   Common Complicated Estimation Method *169*
9.5.2   Advantages of a Kalman Filter in an SoC Estimation *170*
9.5.3   Combination of an EKF and a Lithium-Ion Battery Model *171*
9.5.4   Implementation Rule of the EKF Algorithm *174*
9.5.5   Experimental Verification *176*
9.6   Error Spectrum of the SoC Estimation Based on the EKF *177*
9.6.1   Estimation Error Caused by the Inaccurate Battery Model *177*

9.6.2   Estimation Error Resulting from a Measurement Error of the Sensor   *185*
9.6.3   Factors Affecting SoC Estimation Accuracy   *190*
         References   *191*

10   **Charge Control**   *193*
10.1   Introduction   *193*
10.2   Charging Power Categories   *196*
10.3   Charge Control Methods   *198*
10.3.1   Semi-constant Current   *199*
10.3.2   Constant Current (CC)   *201*
10.3.3   Constant Voltage (CV)   *201*
10.3.4   Constant Power (CP)   *202*
10.3.5   Time-Based Charging   *202*
10.3.6   Pulse Charging   *202*
10.3.7   Trickle Charging   *203*
10.4   Effect of Charge Control on Battery Performance   *203*
10.5   Charging Circuits   *204*
10.5.1   Half-Bridge and Full-Bridge Circuits   *204*
10.5.2   On-Board Charger (Level 1 and Level 2 Chargers)   *205*
10.5.3   Off-Board Charger (Level 3)   *206*
10.5.4   Fast Charger   *206*
10.5.5   Ultra-Fast Charger   *208*
10.6   Infrastructure Development and Challenges   *209*
10.6.1   Home Charging Station   *209*
10.6.2   Workplace Charging Station   *209*
10.6.3   Community and Highways EV Charging Station   *210*
10.6.4   Electrical Infrastructure Upgrades   *210*
10.6.5   Infrastructure Challenges and Issues   *210*
10.6.6   Commercially Available Charges   *210*
10.7   Isolation and Safety Requirement for EC Chargers   *211*
         References   *212*

11   **Balancing/Balancing Control**   *213*
11.1   Balancing Control Management and Its Significance   *213*
11.1.1   Two Expressions of Battery Capacity and SoC Inconsistency   *213*
11.1.2   Significance of Balancing Control Management   *215*
11.2   Classification of Balancing Control Management   *218*
11.2.1   Centralized Balancing and Distributed Balancing   *218*
11.2.2   Discharge Balancing, Charge Balancing, and Bidirectional Balancing   *219*
11.2.3   Passive Balancing and Active Balancing   *220*
11.3   Review and Analysis of Active Balancing Technologies   *221*
11.3.1   Independent-Charge Active Balancing Control   *221*
11.3.2   Energy-Transfer Active Balancing Control   *221*
11.3.3   How to Evaluate the Advantages and Disadvantages of an Active Balancing Control Scheme (an Efficiency Problem of Active Balancing Control)   *223*
11.4   Balancing Strategy Study   *226*
11.4.1   Balancing Time   *227*

11.4.2   Variable for Balancing *229*
11.5   Two Active Balancing Control Strategies *234*
11.5.1   Topologies of Two Active Balancing Schemes *234*
11.5.2   Hierarchical Balancing Control Strategy *238*
11.5.3   Lead-Acid Battery Transfer Balancing Control Strategy *243*
11.6   Evaluation and Comparison of Balancing Control Strategies *245*
11.6.1   Evaluation Indexes of Balancing Control Strategies *245*
11.6.2   Comparison of Flows for Balancing Strategies *247*
11.6.3   Comparison of Balancing Time *249*
11.6.4   Comparison of Energy Consumption *251*
11.6.5   Comparison of the Impact of Balancing on Battery Life *253*
11.6.6   Comparison of the Capacity Utilization Ratio *253*
11.6.7   Analysis of the Optimization Case *253*
          References *255*

12   State of Health (SoH) Estimation of a Battery *257*
12.1   Definition and Indices/Parameters of SoH *257*
12.1.1   Relationship Between Battery Degradation and Battery Life *257*
12.1.2   Relationship Between Battery Degradation and SoH of the Battery *259*
12.1.3   Main Indicators to Describe Battery Degradation *262*
12.2   Modeling of Battery Degradation (Aging) and SoH Estimation *265*
12.2.1   Support Vector Regression *266*
12.2.2   Battery Degradation Model Based on a Support Vector Regression Machine *269*
12.2.3   Steps and Procedures for Evaluating Battery Degradation *276*
12.3   Battery Degradation Diagnosis for EVs *278*
12.3.1   Offline Degradation Diagnosis of the Power Battery *278*
12.3.2   Online Degradation Diagnosis of the Power Battery *281*
          References *289*

13   Communication Interface for BMS *291*
13.1   BMS Communication Bus and Protocols *293*
13.1.1   System Management Bus (SMBus) *294*
13.1.2   BMS: Internal Data Communication *294*
13.1.3   BMS: External Data Communication *295*
13.2   Higher-Layer Communication Protocols *298*
13.3   A Case Study: Universal CiA EnergyBus for a Low-Emission Vehicle (LEV) *299*
          References *300*

14   Battery Lifecycle Information Management *301*
14.1   Data Type of Power Battery *301*
14.2   Vehicle Instrument Data Display *302*
14.2.1   Battery Information Displayed on the Vehicle Instrument *303*
14.2.2   Upgrade Based on a Traditional Instrument Panel *303*
14.2.3   Design of the New Instrument Panel *304*
14.3   Battery Data Transmission Mode *306*
14.3.1   Hardware Implementation of Data Transmission *306*
14.3.2   Control Flow of Data Transmission *307*

14.3.3 Hierarchical Management of Power Battery Data *309*
14.4 Information Concerning a Full-Power Battery Lifecycle *311*
14.4.1 Database Structure of a Power Battery *311*
14.4.2 Power Battery Data Volume Estimation *315*
14.5 Storage and Analysis of Historical Information of a Battery *316*
14.5.1 Necessity for Storage of Historical Information *316*
14.5.2 Achievement of Historical Information Storage *317*
14.5.3 Analysis and Processing of Historical Information *318*
14.6 Battery Detection System Based on a Mobile Terminal *320*
14.6.1 Server Program Design and Implementation *322*
14.6.2 Design and Implementation of the Mobile Terminal *322*
Reference *325*

Part IV Case Studies *327*

15 BMS for an E-Bike *329*
15.1 Balancing *329*
15.1.1 Passive Balancing *330*
15.1.2 Active Charge Compensation *330*
15.2 Battery Pack Design for an E-Bike *331*
15.2.1 E-Bike Battery Pack Design Specifications *332*
15.2.2 Testing *332*
15.3 Methodology *333*
15.4 Active Balancing Solutions *337*
15.4.1 Structure of LTC3300 *338*
15.4.2 Discharging Procedure *338*
15.4.3 Charging Process *339*
15.5 Test Results *341*
15.5.1 Measurements with Different Discharges *341*
15.5.2 Comparison Between the Batteries *346*
15.6 Possibility with Active Balancing *349*
15.7 Results and Evaluation *349*
Reference *351*

16 BMS for a Fork-Lift *353*
16.1 Lithium-Iron-Phosphate Batteries for Fork-Lifts *353*
16.2 Battery Management Systems for Fork-Lifts *355*
16.3 The LIONIC® Battery System for Truck Applications *356*
16.4 Application *357*
16.5 The Usable Energy Li-Ion Traction Batteries *359*
Reference *361*

17 BMS for a Minibus *363*
17.1 Internal Resistance Analysis of a Power Battery System and Discharging Strategy Research of Vehicles *361*
17.1.1 Internal Resistance Change Characteristic Research of a Power Battery *364*

17.1.2   Internal Resistance Characteristic–Based Discharge Strategy   *369*

17.1.3   Research of a Charging Method for a Power Battery System Based on an Internal Resistance Characteristic   *374*

17.2   Consistency Evaluation Research of a Power Battery System   *377*

17.2.1   Analysis of a Battery Pack Maintenance Strategy and Performance Evaluation Index   *377*

17.2.2   Comparison of the Battery Pack Performance Evaluation Methods   *378*

17.2.3   Internal Resistance Characteristic-Based Consistency Evaluation Theory of the Battery Pack   *379*

17.2.4   Internal Resistance Characteristic-Based Consistency Evaluation of the Battery Pack   *380*

17.2.5   Internal Resistance Characteristic-Based Staged Consistency Evaluation Method for the Battery Pack   *381*

17.2.6   Internal Resistance Consistency Evaluation Test of the Battery Pack for a Pure Electric Vehicle   *384*

17.3   Safety Management and Protection of a Power Battery System   *386*

Index   *389*

Preface

We have known each other since the autumn of 2005 and have been engaged in the research of electric vehicles together. We often worked in each other's laboratories in the form of short-term visits. In China and Switzerland, we have a group of good partners. We trusted each other and always shared information together.

In 2015, two of us discussed writing down our knowledge and thoughts in the field of BMS and publishing a book to share with researchers and engineers around the world. However, the work of editing and publishing is huge. Our teams on both sides have added many young researchers to participate in the organization and editing of the manuscript. With everyone's efforts, we had completed the basic compilation of the manuscript by the end of 2019.

Due to the influence of COVID-19, our final editing work has been affected. Finally, we were glad to see that in the winter of 2021, we were able to send the manuscript of this book to the publishing house.

In this book, Chapters 1, 4, 10, 15, and 16 were mainly written by Andrea Vezzini, with the assistance of Neeta Khare. Chapters 2, 3, 5, 6, 7, 8, 9, 11, 12, 13 and 14 were mainly written by Xiaojun Tan, with the assistance of Yuqian Fan. Chapter 17 was written by You Xu. Liangliang Wei helped to revise all the text and mature the figures and tables.

Many thanks are due to the faculties and students from both research teams in China and Switzerland, who gave valuable help to this book.

Xiaojun Tan and Andrea Vezzini

About the Authors

Professor Dr. Xiaojun Tan received his PhD degree from Sun Yat-sen University, China, in 2005. He has worked for Sun Yat-sen University since 2005 and is now a professor at the School of Intelligent Systems Engineering, Sun Yat-sen University. At the same time, he is the Director of the Electric Vehicle Research Center which is an engineering laboratory of Guangdong Province. He has worked on electric vehicles for more than 17 years. His research interests include battery management systems and intelligent driving.

The research fields of the Electric Vehicle Research Center include the safe battery system as well as the lightweight vehicle body. Since 2005, Sun Yat-sen University began to collaborate with BFH on the area of battery management systems and driving motors.

Professor Dr. Andrea Vezzini received his PhD in electrical engineering from ETH Zurich in 1996 and successfully completed the Mastering Technology Enterprises (MTE) programme at IMD Lausanne in 2002. In 1996 he became a professor at the Bern University of Applied Sciences.

Professor Dr. Andrea Vezzini heads the BFH Centre for Energy Storage in Biel and is the head of Innosuisse's flagship project "CircuBAT – Swiss Circular Economy Model for Lithium Car Batteries."

Since 2020, he has also been president of "iBAT," the Swiss innovation platform for companies and research institutions on the topic of batteries.

He has been a member of the Swiss Federal Energy Research Commission (CORE) since 2015 and an official member of the Scientific Advisory Board of AEE Suisse, the umbrella organization of the renewable energy and energy efficiency industry, since 2018.

Professor Dr. Yuqian Fan received the PhD degree in Intelligent Transportation Engineering from Sun Yat-sen University, China, in 2019, an ME degree in Electronics and Communication Engineering from Wuhan University, Wuhan, China, in 2011, and a BE degree in Communication Engineering from Harbin Engineering University, Harbin, China, in 2009.

From 2020 to 2022, he was a postdoc with the School of Intelligent Systems Engineering, Sun Yat-sen University. Since 2022, he has been a Professor at Henan Institute of Science and Technology. Dr. Fan was the author of more than 10 scientific publications published on the Web of science. His research interests include intelligent control and optimization design for power battery systems, battery thermal management and thermal safety, and battery state of health prediction, etc.

Neeta Khare has 19-plus years of experience in developing BMS and battery pack for EV and Energy storage systems. Dr. Khare's broad work experience covers a wide range of battery technologies (Li-ion, LFP, LTO, NMC-G, lead-acid and zinc hybrid), failure analysis, diagnostics, and optimized solutions for energy storage and hybrid power systems. Her core expertise is in aging algorithms of a battery/cell using AI and adaptive algorithms, Battery Pack, Battery Management System (BMS) development, multistring controller, EMS, innovative Battery Health Monitoring, and an optimized power controller for Hybrid Power Systems for Electric Vehicles.

Currently, Dr. Khare is serving as Director (Battery and Fuel Cell) with the Iveco Group. Prior to this position, she served as CTO/CIO in Green Cubes Technology (GCT-EU), Uster and Vice President of Technology in Leclanche, Switzerland.

Dr. Khare acquired her doctoral degree on "Intelligent Battery Monitoring" from Banasthali University in India and served as post doc and in the research faculty in Villanova University in USA. She claims multiple patents and international publications to her credit.

You Xu received his BSc degree and PhD degree from Sun Yat-sen University in 2006 and 2011 respectively. He is now an associate professor in Guangdong Polytechnic Normal University, where he has been engaged in power battery systems and precision reverse equipment.

Dr. Xu was the author of more than 20 scientific publications, among which, 12 papers were published or accepted in Engineering Village or Web of Science. His research interests include battery management for electrical vehicles and accuracy of multi-joint robot and visual measurement systems, etc. Dr. Xu has presided over one general project of Guangdong Provincial Natural Science Foundation, one sub-project of Guangdong Provincial Department of science and technology, one project of Guangdong Provincial Department of education, one research and development project of Guangdong Provincial Key Laboratory, won third prize of Guangdong Provincial Scientific and technological progress in 2018, won second prize of Dongguan municipal scientific and technological progress in 2017, and second prize of Guangdong Provincial Science and technology award of machinery industry.

Liangliang Wei received BS and PhD degrees in Electrical Engineering from the Wuhan University, Wuhan, China, in 2012 and 2017 respectively.

From 2017 to 2018, he was a postdoc in Kyoto University, where he has been engaged in permanent magnet machines and renewable energy generation. From 2018 to 2020, he became an Assistant Professor in Kyoto University. Since 2020, he has been an Associate Professor in control science and engineering at Sun Yat-Sen University. Dr. Wei was the author of more than 20 scientific publications, among which, 10 papers were published or accepted in IEEE Transactions or Web of Science. His research interests include electrical machines and motor control for electrical vehicles and renewable energy generation, battery management systems, and health condition monitoring, etc.

Part I

Introduction

1

Why Does a Battery Need a BMS?

1.1 General Introduction to a BMS

1.1.1 Why a Battery Needs a BMS

A battery management system (BMS) is an essential part of any energy storage system. It controls battery charging and discharging, manages optimum operating conditions, governs the safety limits, runs the battery charge and health algorithms, monitors battery parameters, and communicates with other associated devices [1, 2]. A BMS or similar monitoring and control system is strongly recommended for other electrical energy systems, such as a fuel cell, supercapacitor, superbat capacitor, or other hybrid combinations of electrical energy storage systems. A BMS allows the system to be efficient and to use an application for stored energy up to the safe operating limit [3]. It makes energy storage cost effective for short-term applications such as consumer electronics. With an efficient control over optimum charge and discharge ranges, the BMS adequately extends the life of energy storage. The increased life makes the energy storage economically viable for long-term applications such as grid, automotive, and stationary applications [4].

1.1.2 What Is a BMS?

A BMS is a control system that ensures optimum use of the battery energy in powering any portable or non-portable system. This is achieved by monitoring and controlling the battery's charging and discharging processes along with careful control over the surrounding environment. The BMS becomes essential in all storage systems to prevent the risk of damaging the battery by misuse. The features of a BMS design should include:

- Charge control
- Battery capacity and efficiency calculations
- Remaining run-time information
- Cycle counting
- Battery life prognosis
- Thermal management
- Prediction of battery failure
- Safety and alarm indications for over the limit usage

An effective BMS can protect the battery from damage, ensure safety, predict battery life, and maintain the battery operation in order to keep efficiency high.

A general block diagram of a BMS is shown in Figure 1.1. The battery-charger charges the battery from the mains. A protection integrated circuit (IC) connected to the battery indicates the unsafe condition of the battery.

Battery Management System and its Applications, First Edition. Xiaojun Tan, Andrea Vezzini, Yuqian Fan, Neeta Khare, You Xu, and Liangliang Wei.

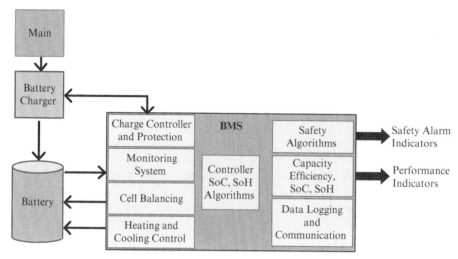

Figure 1.1 Block diagram of a battery management system.

A protection IC specifically deals with the over/under-voltage protection, over current protection, imbalance of cells, and thermal runaway. In addition, protection circuits also include a blocking diode, each of which is outfitted with a series string that prevents parallel strings from discharging through a battery with an unforeseen short circuit [5]. Researchers, such as Kim et al. [6], have also proposed more robust circuits capable of mitigating the electrical impacts of a single cell failure. Manufacturers of large battery systems typically integrate a proprietary control system as well, in order to control issues such as cell balance, cell temperature, and an estimation of the battery life.

The battery state indicates the current state and future prediction of the battery by using the State of Charge (SoC) and State of Health (SoH). The processor runs the battery management algorithms that compute the SoC, SoH, and property parameters [7]. The subsequent parts in the book will discuss prognostic and diagnostic approaches for determination of the SoC and SoH. Finally, to establish communication between the BMS and other devices, most commonly used interfaces are I2C, Modbus, and CAN ports and protocols.

1.1.3 Why a BMS Is Required in Any Energy Storage System

The demand for an energy storage system is increasing day by day with exponential growth in the area of consumer electronics, portable devices, and e-mobility. In addition, a budding urge for clean energy usage in order to address the challenge of reducing carbon footprints makes energy storage more popular than other stationary applications. At present, stationary applications, such as a grid-connected energy storage, are aggressively being tested around the world. Grid-connected electrical energy storage is a potential candidate for load shifting, PV smoothing, stabilizing the grid, etc. The most popular solution for electric energy storage is a battery pack due to its high energy density, long life, and cost-effective features. However, challenges lie with its optimum performance and safety. The requirement for a BMS controller with energy storage is quite obvious when considering the increasing challenge regarding safety and optimum utilization together with high efficiency. A BMS allows energy storage to function within the safety limits and provides high-performance capabilities.

1.1.4 How a BMS Makes a Storage System Efficient, Safe, and Dependable

An important aspect of BMS functions is to control the battery charging and usage within safe limits. A BMS recommends relevant parameters to the battery charger and commands it to use the most effective charging algorithm. A charging algorithm helps to reduce the charging time, offers a long battery life, and maintains high efficiency, while keeping the operation within given safety limits of voltages, temperature, current, and SoC. The BMS monitors real-time electrical parameters such as terminal voltage, charging and discharging current, temperature, impedance, and number of cycles [8]. Further, it calculates compensation factors, estimates the SoC and SoH, and determines other performance characteristic parameters such as energy efficiency, capacity, and remaining life time. The SoC and SoH are the most critical parameters for maintaining the operation under safe conditions [9, 10]. Monitoring battery health is one of the prime factors affecting the system reliability.

A BMS helps energy storage in the following three ways:

1) Increases efficiency by
   - Compensating cut-off voltage with temperature variations, C-rate charge and discharge, and aging.
   - Selecting appropriate charging current to maintain the current density limit at the electrode surfaces.
   - Controlling and compensating the SoC range for charging and discharging over the operating range in order to maintain coulombic efficiency.
   - Keeping all cell voltage and SoC balanced to increase its operating range.
   - Thermal controlling the pack in order to maintain the optimum temperature range.
2) Increases battery life time by
   - Saving the battery from abuses of over-charging. Over-charging causes heating and out-gassing that reduces the life of the battery.
   - Preventing deep discharging by limiting the discharge at the end of the discharge cut-off voltage. Metal plating is a major cause of shortening the battery age when operating below the end of discharge cut-off voltage.
   - Maintaining current density to prevent electrode surfaces from damage.
   - Keeping the SoC within the operating range that provides a balance between capacities in and out at various operating conditions.
   - Cell balancing prevents under-charging of good cells and over-charging of weak cells, which increase the overall age of the pack.
3) Provides safety and reliability by
   - Maintaining and controlling operations within the safety limits.
   - Indicating safety alarms for events beyond the operating condition.
   - Shutting down the operation during a critical safety threat.
   - Employing a thermal controlling system to prevent any thermal runaway conditions
   - Giving an indication of the remaining battery life and thus facilitating timely action taken proactively in alarming conditions, reducing the risk of running into a disaster.

1.2 Example of a BMS in a Real System

1.2.1 LabView Based BMS

A LabView (Laboratory Virtual Instrument Engineering Workbench) based BMS provides easy execution on a PC. A LabView from the National Instruments Corporation is a software development application that uses a graphical programming language to create programs in a block diagram. Since a LabView includes libraries of functions for data acquisition, serial instrument control, data analysis, data presentation, and data storage, it is recommended for the BMS application. A BMS designed using LabView offers higher flexibility and much better graphic tools for data visualization. The central unit of a LabView based BMS consists of the following blocks:

a) Data Processing
b) Parameter Adaptation

c) Monitoring

d) Management

The central unit and input/output interfaces have been recognized as a LabView application.

Due to the flexible design of the LabView BMS, the system is able to perform control and surveillance activities for any kind of battery application and battery technology (e.g. Pb, VRLA, NiCd, NiMH, etc.) [11].

1.2.2 PLC Based BMS

The PLC (programmable logic controller) plays an important role in the field of industrial automation because of its excellent performance. Its multiple functions include logic arithmetic, calculation, communication, noise resistance, and stability. A PLC based BMS design is shown in Figure 1.2. The analog and digital data from the battery were passed to PLC on real time. This system controls the battery charging and discharging.

It can be seen that the PLC controls the action of relays and delivers the signal to the sensor for judging whether charging is being done or not. The development cost of a PLC based BMS is high and offers only limited functions. Thus, it does not match the demand in price.

1.2.3 Microprocessor Based BMS

The microprocessor based BMS consists of (a) a data acquisition unit (DA); (b) an ampere-hours counter unit (AhC) with a battery current measuring unit, a battery voltage measuring unit, and a battery ambient temperature measuring unit; (c) a cell or mono blocks voltage measuring unit (CV); (d) a modem; and (e) a personal computer (PC). In this BMS, AhC and CV units are used to measure the battery parameters. All measured data are read out in a definite time period with the help of a DA unit, and are then stored in an internal memory of the system. Figure 1.3 represents the use of the central unit – a microprocessor 80C535 – in a DA unit to estimate the battery state.

Personal computer (PC) with utility program is used to read saved data from system memory and to transfer it in appropriate database files. Data were transmitted from Battery Monitoring System to personal computer PC via modem.

1.2.4 Microcontroller Based BMS

In this BMS design, an 80C196 KB microcontroller was used for developing the system hardware to estimate the battery SoC and SoH. The general block diagram of the microcontroller based BMS is shown in Figure 1.4.

The five battery parameters, which include the current drawn, terminal voltage, temperature, internal resistance, and time, are supplied as inputs to the microcontroller, which was programmed according to the Neuro Fuzzy process model. The weights, biases, and battery history were stored in EPROM. The microcontroller processed the input parameters and gave the output as SoC and SoH [12]. In the complete setup of the microcontroller real-time battery, the parameters were extracted by the interfacing circuitry. These parameters were applied to a programmed microcontroller as inputs and finally the SoC and SoH were displayed.

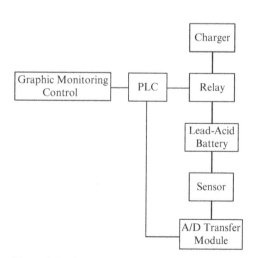

Figure 1.2 Schematic diagram of a PLC based BMS.

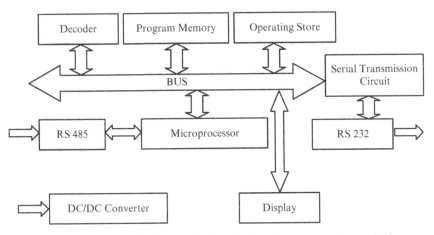

Figure 1.3 Schematic block diagram of a DA unit of a microprocessor based BMS.

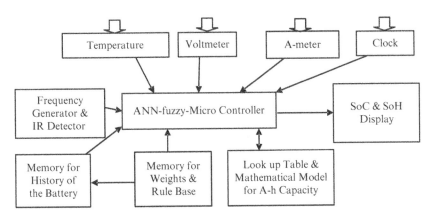

Figure 1.4 Block diagram of a microcontroller based BMS implementation.

1.3 System Failures Due to the Absence of a BMS

The internal state information of the battery is one of the most important factors used to protect the system from failure. There are a large number of examples of system failure due to the absence of a BMS or non-accurate BMS algorithms or malfunctioning in BMS control.

In the recent past, there have been major EV and energy storage failures highlighted in the media. The following incidences are a few examples.

1.3.1 Dreamline Boeing Fire Incidences

During 2013 and 2014, a series of incidents of smoke and fire in the battery pack of the Boeing 787 of Dreamline Airlines were detected, including at Boston airport a jet fire accident and at Narita airport smoke detected by the maintenance crew. The airline used lithium-ion (Li-ion) batteries to deliver power for its energy-hungry electrical systems.

Various agencies including NTSB and US Japanese Joint team carried out investigations and found a few possible causes:

a) It was found that electrolytes, a flammable battery fluid, had leaked from the main Li-ion battery pack and the entire system was damaged.

b) *The Wall Street Journal* reported on February 12, 2013: a possible theory is that the formation of microscopic structures known as dendrites inside the Boeing Co. 787's Li-ion batteries played a role in twin incidents of fire and smoke in the battery pack of the Boeing.

c) NTSB declared that overvoltage was not the cause of the Boston incident, as voltage did not exceed the battery limit of 32 V and the charging unit passed tests. The battery had signs of short-circuiting and thermal runaway.

d) Based on the data and the recorded events, there were several factors that suggest a problem with integration of the battery system into the plane. Two smoke detectors in the electrical/electronic bay, where the APU battery is located, failed to trigger an alarm.

e) These include problems that may arise from poor systems integration between the engine indicating and crew alerting system (EICAS) and the battery management system.

f) Another theory said that it is possible that the fail-safe devices in the battery management system, the charger, and the battery cells functioned properly and prevented the short circuit from becoming a catastrophic failure [13]. However, the reboot of the APU by a different subsystem in the plane could have caused the final surge in the current that led to the fire.

1.3.2 Fire Accident at the Hawaii Grid Connected Energy Storage

Electricity storage using a Li-ion battery with the BMS and system integration caused fire in 2011 and 2012 at Kahuku, Hawaii. This 15 MW battery energy storage pilot project was connected to a 30 MW wind power plant and was among the first of a few utility scale projects. The project was developed by First Wind, the Kahuku facility with a 15 MW battery from VC-funded Xtreme Power and sells power to the island utility, HECO. Dynapower had supplied the power inverter. Although there were disputes about the cause of the fire, one theory claimed that both fires were attributed to ECI capacitors in inverters from Dynapower. Xtreme, who designed the BMS later, sued Dynapower for malfunctioning of the power controller. Besides legal battles between the companies, the BMS design and safety alarms were always questioned. It was found that the BMS should shut down the process in any extreme conditions and disconnect the AC/DC switchgears for safety.

1.3.3 Fire Accidents in Electric Vehicles

1. The Chevrolet Volt

The Chevrolet Volt recorded the first famous fire when obliterated by the NHSTA (National Highway Traffic Safety Administrative). A fire occurred in a parking lot due to a failure to discharge the battery after crash testing. NHTSA found the Volt to meet its five-star crash rating. After the test, they stashed the mangled Volt outside and three weeks later the vehicle's battery pack shorted and caught fire – an incident. As a result, NHTSA has opened an inquiry into not only the Volt's battery pack performance but also the post-crash performance of all hybrid vehicle battery packs [3]. The prevailing theory explaining the battery fire is that the coolant lines serving the battery were probably severed during the crash, leading to a short or eventual overheating condition. However, this has not yet been confirmed. GM contends that the fire happened because the prototype test vehicle's programming was incomplete. All production Volts have programming for depowering the battery after a crash, dissipating any remaining charge and rendering the battery inert [14].

2. Tesla S model battery fire

There has been more than one incidence where Tesla Model S caught fire. A few happened when the driver ran into something due to high speed at turning or after the driver ran over a chunk of metal on the road.

One theory claimed that the fire did not spread quickly to the whole battery pack and, as a result, no one was hurt. In a L-ion battery, fire spreads quickly throughout the battery, as the cells within the battery ignite their neighboring cells. Tesla's CTO, J. B. Straubel, said that the company had engineered the pack to prevent fires from spreading and this prevented the fire from spreading throughout the whole battery, and, according to Tesla, it did not enter the passenger compartment.

There are a couple of schools of thought among battery experts about the causes of fire. In a battery fire, the main thing that burns is the liquid electrolyte, which burns easiest when it is exposed to air. One school of thought is that even in the absence of air there are other oxidants within the battery that can create and sustain a fire. It is thought that the battery electrodes themselves can release oxygen, fueling the fire from within.

Other research suggests that this is not the case. Instead, a possibility is that even once the fire is put out, the cells stay very hot and keep releasing more electrolyte in the form of vapor. Once firefighters turn off the water and oxygen can once more come into contact with the vapor, it can reignite.

The second negative is the placement of the battery in the car. The fire accident initiated due to the road debris raises questions about the packaging and protection of the battery. The Tesla battery spreads out over most of the floor of the car in contrast to the battery on the Chevy Volt. Does packaging of the battery cause the accidents?

The most recent incident, reported in Norway, is of an S model that caught fire at a super charging station on New Year's Day in 2016.

For the five Model S fires that occurred, the true cause of one is still unknown. Different theories claim different answers for the cause. However, safety aspects cannot be ignored and so has the role of the BMS. It is clear that more tests are needed and the best ways to put out battery fires are required, especially as battery-powered cars proliferate.

Certainly, one should keep in mind that mechanical or electrical failures or malfunctions were factors in roughly two-thirds of vehicle fires.

References

1 Zhang, J., Ci, S. and Sharif, H., "An enhanced circuit-based model for single-cell battery," *Applied Power Electronics Conference and Exposition (APEC)*, IEEE, 2010, pp. 672–675.

2 Cheng, K. W. E., Divakar, B. P., Wu, H. J., et al., "Battery-Management System (BMS) and SOC development for electrical vehicles," *Vehicular Technology*, 2011, 60: 76–88.

3 Andrea, D., *Battery Management Systems for Large Lithium-Ion Battery Packs*, Boston, USA: Artech House, 2010, pp. 71–76.

4 Heymans, C., Walker, S. B., Young, S. B., et al., "Economic analysis of second use electric vehicle batteries for residential energy storage and load-levelling," *Energy Policy*, 2014, 71: 22–30.

5 Lee, Y. S. and Cheng, M. W., "Intelligent control battery equalization for series connected lithium-ion battery strings," *IEEE Transactions on Industrial Electronics*, 2005, 52(5): 1297–1307.

6 Park, H. S., Kim, C. E., Kim, C. H., et al., "A modularized charge equalizer for an HEV lithium-ion battery string," *IEEE Transactions on Industrial Electronics*, 2009, 56(5): 1464–1476.

7 Lee, D. T., Shiah, S. J., Lee, C. M., et al., "State-of-charge estimation for electric scooters by using learning mechanisms," *IEEE Transaction on Vehicular Technology*, 2007, 56(2): 544–556.

8 Waag, W., Käbitz, S., and Sauer, D. U., "Experimental investigation of the lithium-ion battery impedance characteristic at various conditions and aging states and its influence on the application," *Applied Energy*, 2013, 102: 885–897.

9 Andre, D., Appel, C., Soczka-Guth, T., and Sauer, D. U., "Advanced mathematical methods of SOC and SOH estimation for lithium-ion batteries," *Journal of Power Sources*, 2013, 224: 20–27.

10 Valer, P., Henk, J. B., Danilov, D., et al., *Battery Management Systems: Accurate State-of-Charge Indication for Battery-Powered Applications* (Philips Research Book Series), Germany: Springer, 2008.

11 IEEE, "*IEEE recommended practice for maintenance, testing, and replacement of valve-regulated lead-acid (VRLA) batteries for stationary applications,*" in IEEE Std 1188-2005 (Revision of IEEE Std 188-1996), pp.1–44, 8 February 2006.

12 Zenati, A., Desprez, P., and Razik, H., "Estimation of the SOC and the SOH of Li-ion batteries, by combining impedance measurements with the fuzzy logic inference," in *IECON 2010 – 36th Annual Conference on IEEE Industrial Electronics Society*, IEEE, 2010, pp. 1773–1778.

13 Fleckenstein, M., Bohlen, O., Roscher, M. A., and Bäker, B., "Current density and state of charge inhomogeneities in Li-ion battery cells with $LiFePO_4$ as cathode material due to temperature gradients," *Journal of Power Sources*, 2011, 196(10): 4769–4778.

14 Zheng, Y., Ouyang, M., Lu, L., et al., "On-line equalization for lithium-ion battery packs based on charging cell voltages: Part 1. Equalization based on remaining charging capacity estimation," *Journal of Power Sources*, 2014, 247: 676–686.

2

General Requirements (Functions and Features)

2.1 Basic Functions of a BMS

Electric vehicles are the most typical representatives of BMS applications at present. In this section, the main functions of a BMS are described by the functional block diagram of the BMS for electric vehicles (Figure 2.1).

The functions of the BMS for the electric vehicle may be different from those shown in Figure 2.1 in different applications, but the BMS should have the basic functions shown in the figure, for example, voltage monitoring, SoC estimation, safety protection, etc. The main functions of a BMS are described as follows.

2.1.1 Key Parameter Monitoring

Key parameter monitoring of a battery refers to the monitoring of voltage, current, and temperature. Unlike voltage and current monitoring, in addition to battery temperature monitoring, the temperature monitoring includes in-car temperature and ambient temperature monitoring, and the battery can be more reasonably used by monitoring multiple temperature parameters.

Key parameter monitoring of a battery is the most basic function of a battery management system and the premise and foundation of other functions. For example, the battery state analysis (SoC, SoF, and SoH evaluations) described in the next section depends on accurate and timely voltage, current, and temperature monitoring.

2.1.2 Battery State Analysis

The battery state analysis mainly includes an SoC estimation, an SoH estimation, and an SOF estimation. With the deepening of the research, a state of power (SoP) evaluation, a state of life (SoL) evaluation, a state of energy (SoE) evaluation, and a state of range (SoR) evaluation have also been gradually considered. The book only discusses the more common SoC, SoH, and SoF evaluations.

1. State of charge evaluation

Just as the driver of a conventional car is often required to keep an eye on the oil level of the car, the driver of an electric vehicle needs to know the SoC, which is the function implemented by the SoC evaluation module of the battery management system. This is the most basic and important function of the battery management system. If the SoC of the electric vehicle is unknown, the danger of running out of power may happen. Hence, it is very important to evaluate the SoC. In recent years, more than half of researches have been conducted for SoC evaluation in the field of battery management systems. The SoC evaluation method is briefly discussed in Chapter 9.

Battery Management System and its Applications, First Edition. Xiaojun Tan, Andrea Vezzini, Yuqian Fan, Neeta Khare, You Xu, and Liangliang Wei.

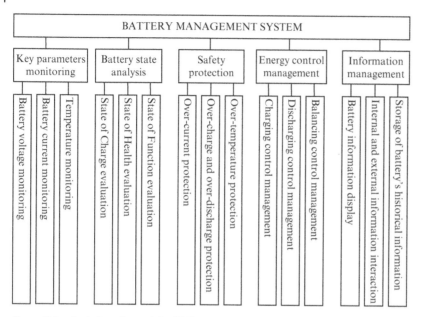

Figure 2.1 Basic functions of the BMS for electric vehicles.

2. State of function evaluation

During the use of the battery, as the energy source for the motor, the vehicle, and other loads, the SoF of the battery is also very important. For the electric vehicle, the SoF can be defined as the power that can be supplied by a battery pack to a variety of electrical loads, such as the motor, at any given moment. The SoF can be simply considered as a function of the SoC and temperature:

$$SoF = f(SoC, T)$$

In fact, for the powertrains of many electric vehicles, the BMS not only evaluates the SoF of the battery pack at a given time, but also provides the maximum power at which it is allowed to charge the battery pack, as SoF_2. On the one hand, SoF_2 should be sent to the motor through the communication bus in order to require the motor not to exceed a certain limit value during braking energy recovery. On the other hand, SoF_2 is sent to the charger together with the charging strategy to avoid damage to the battery resulting from the over-high charging current provided by the charger.

3. State of health evaluation

The battery performance will gradually decline from the beginning of use, which is an irreversible process, so the worse the SoH of the battery, the closer the battery is to the end of its life. The degradation of the battery is a gradual and complex process [1]. Even so, it is hoped that quantifiable indicators will be found to describe the SoH of the battery. For example, capacity loss and a DC internal resistance spectrum can be used as typical indicators to judge the SoH of the battery (Chapter 12 will discuss the SoH evaluation in greater detail) [2]. Information of many aspects is required for an SoH evaluation since the SoH is affected by the operating temperature, the discharge current, and other factors during the use of the power battery. It is necessary to continuously evaluate and upgrade the SoH during use in order to ensure that the driver obtains more accurate information.

2.1.3 Safety Management

Safety management is undoubtedly the most important function of the BMS for electric vehicles. This function is placed in the third place because it is often premised on the above "parameter monitoring" and "state analysis" functions. The most common contents of the safety management are "over-current protection," "over-charge and over-discharge protection," and "over-temperature protection."

1. Over-current protection

Over-current protection means that the corresponding safety protection measures should be taken if the working current exceeds the safe value in the process of charging and discharging. Most lithium-ion (Li-ion) power batteries support short periods of overload discharge, which can provide a higher current to meet the requirements of power performance during starting and accelerating [3]. However, the overload current rate and overload duration of the power battery vary with different manufacturers and models. For example, a certain type of power battery can support 3C overload current for not more than one minute, which must be considered for the over-current protection function of the battery management system.

2. Over-charge and over-discharge protection

Another basic function of the safety protection is over-charge and over-discharge protection. Over-charge protection means that a protection measure, such as disconnection of the charging circuit of the battery, is taken to prevent damage to the battery resulting from continuous charging after 100% SoC [4]. On the other hand, if the battery is continuously discharged in 0% SoC, the battery may be damaged, so a corresponding measure, such as disconnection of the discharging circuit, is taken to protect the battery, which is called over-discharge protection. In the actual operation process, a simple way, namely setting threshold charging and discharging voltages, is available to realize over-charge and over-discharge protection; that is, if it is detected that battery voltage is higher or lower than the set threshold voltage, the current circuit is disconnected in time to protect the battery.

It should be noted that, in the practical application of electric vehicles, battery cells are generally connected in series to constitute a battery pack [5–7]. As long as one of the cells in the battery pack is lower than the discharging voltage threshold, the whole battery pack is protected. At this time, other cells in the battery pack often carry a certain amount of residual charge, resulting in a certain degree of invisible waste, so it is necessary to carry out "balancing management" for the battery, which belongs to the category of "energy control management."

3. Over-temperature protection

Over-temperature protection means that a protective measure is taken to the power battery when the temperature exceeds a certain limit value. The power battery is a chemical product and it is difficult to control the chemical reaction during operation at high temperature, causing damage to the battery, accidents, and casualties depending on its seriousness. For over-temperature protection, consideration should be given to the ambient temperature, battery pack temperature, and battery cell temperature. Since the temperature change needs a process and the temperature control is often hysteretic, some "advance quantities" should be considered for temperature protection. For example, if the ambient temperature or battery box temperature is detected to be too high and close to the temperature threshold for battery damage, corresponding protection measures should be taken. Alternatively, in case of a sudden and rapid temperature rise in a cell, some protective measures should be taken, such as giving an alarm to the driver through the instrument, although the safety threshold is not reached.

2.1.4 Energy Control Management

Energy control management is often classified as "optimal management" of the battery; that is, it does not belong to the basic and essential functions of battery management systems. In the past, many battery management systems did not participate in the charge and discharge management of the battery and nor did they have the balance control management function.

1. Charging control management

Charging control management means that the BMS provides the real-time optimal control for the charging voltage, the charging current, and other parameters during charging. The optimization objectives include charging time, charging efficiency, etc. In early applications of electric vehicles, no communication channel was available between the BMS and the charger. In other words, the BMS could only control the start and stop of the charging procedure, but could not control the charging parameters. However, this situation has been improved in today's mainstream applications. Whether it is the on-board charger or the charging pile on the ground, generally there is a communication interface with the BMS to control the charging voltage and current according to received parameter information.

2. Discharging control management

Discharging control management refers to the control of the discharging current based on the battery state during discharging. This is a feature that has often been overlooked in simple systems where the battery pack is thought to supply only power without safety problems. However, after adding the discharging control management function in a more advanced and perfect system, the power battery pack shows greater efficiency. For example, if the maximum discharging current of the battery pack is properly limited when the SoC of the power battery pack is less than 10%, this is helpful in extending the SoR of the car. More importantly, it is useful in extending the life of the power battery pack, although it may have an impact on the maximum speed of the electric vehicle [8].

In addition, regenerative braking control is often an important part of energy control management. For example, in some hybrid vehicles, the SoC of the battery is required to be about 60%~80% through charging and discharging control management to provide enough charge capacity to receive the regenerative braking energy. Another purpose is to operate the battery within a smaller equivalent internal resistance range in order to improve the charging and discharging efficiency.

3. Balancing control management

The difference between the cells of a battery pack is attributable to two factors. The primary factor is the instability of manufacturing process, while the secondary factor is operational conditions. By balancing control management certain measures can be taken to reduce the negative effects of cell inconsistency as much as possible in order to optimize the whole discharging efficiency of the battery pack and extend its life. As mentioned in the "safety protection," as long as the discharging voltage of one of the cells in the battery pack is lower than the discharging voltage threshold, the whole battery pack is protected while other cells in the battery pack often carry a certain amount of residual charge. The battery balancing management is helpful as it uses the residual charge to improve the discharge efficiency of the battery pack [9, 10].

2.1.5 Information Management

Due to the large number of cells in the power battery pack for an electric vehicle, a large number of data are generated in every second, some of which are provided to the driver by the instruments, some of which are sent to the components other than the battery management system (such as the car controller, motor controller, etc.), and some of which are stored as history data in the system.

1. Battery information display

The battery management system displays the battery status to the driver or the vehicle maintainer through the instruments. The information that needs to be displayed usually includes the following three types.

First, real-time voltage, current, and temperature information. Due to the large number of cells in the electric vehicle, there is no need to display the information of each cell; it is generally required only to display the total voltage, total current, maximum battery voltage, minimum battery voltage, maximum battery temperature, minimum battery temperature, and other information of the battery pack on the instruments [11].

Second, SoC. Like the fuel gauge of orthodox cars, the SoC is displayed as a percentage. In addition, the estimated SoR is generally displayed on the instrument to help the driver get a visual understanding of the SoC.

Third, alarm information. If the battery pack has or will have a safety problem, the driver is informed of that problem by the instrument. In this case, a voice alarm and other methods of alarm are required to catch the timely attention of the driver.

2. Internal and external information interaction

A vehicle-mounted information network is essential for the control of the advanced electric vehicle. The battery management system often consists of both "internal" and "external" networks. The internal network is used to transmit the internal information of the battery management system. For example, in a distributed battery management system for electric vehicles, the whole power battery pack is first divided into several "cells," each of which is managed by a circuit board. The circuit board of each cell then sends the basic information of each cell to the main circuit board of the battery management system through the internal network [12, 13]. The external network is used for the battery management system to perform information interaction with the car controller, the motor controller, and other components. The external network should be duplex (support two-way communication). On one hand, the battery management system is required to send voltage, current, temperature, and other information to other components. On the other hand, the car controller is also required to send information, such as "whether the charger is connected" or "whether charging is allowed" to the battery management system.

3. Historical information storage

Historical information storage is essential in an advanced power battery management system. The information storage is divided into "temporary storage" and "permanent storage" according to the duration of storage. Temporary storage means temporary storage of the battery information by RAM. For example, temporary storage of the estimated SoC and current change in one minute are used to estimate the SoC at this moment. Permanent storage may be realized by EEROM, Flash Memory, and other devices, and the historical information can be stored for a longer time using the permanent storage function.

Historical information storage is of significance in the following functions:

The first is the data buffering and improvement of analysis and estimation accuracy. For example, in case that errors occur in the voltage and current monitored in real-time due to interference, historical data are helpful in filtering error data in order to obtain more accurate data.

The second is for battery state analysis. In particular, the aging state of the battery can be evaluated according to the historical data of the battery over a period of time.

The third is for failure analysis and removal. The historical information storage of the battery is similar to that of the black box of an aircraft, and in the case of failure in the electric vehicle, the cause of failure can be analyzed and identified by the historical data to help to remove the failure.

2.2 Topological Structure of a BMS

In a battery management system, the hardware circuit is typically divided into two functional modules: a battery monitoring circuit (BMC) and a battery control unit (BCU). The topological structure of a battery management system can be studied at two levels: first, the topological relationship between a BMC and each cell; second, the topological relationship between a BCU and a BMC.

2.2.1 Relationship Between a BMC and a Cell

The topological relationship between the BMC and the cells includes the following two types.

1. One BMC corresponding to a single cell

In practical work, each cell can be equipped with a separate monitoring circuit board to monitor the voltage, current, temperature, and other physical properties of the battery, as shown in Figure 2.2.

As shown in Figure 2.2, the BMC circuit board is responsible for monitoring the voltage, temperature, and current of the battery [14]. Communication, balance and control functions can be added in the BMC to report relevant information to the BCU and achieve the energy dissipation balancing management for the controlled cells by a bypass resistance.

Sometimes such a BMC circuit board can be packaged inside the cell to form a "smart battery," in which the cell has some autonomous function. This "one-to-one" topological structure has the following advantages: a shorter distance between the BMC and the cell, which can reduce the length and complexity of an acquisition line to some extent, and improve acquisition accuracy and anti-interference. However, its disadvantage is relatively higher cost of circuit board. At the same time, as the power is often supplied to the battery management system by the monitored power battery, it is possible to cause relatively higher energy consumption of the whole battery management system.

2. One BMC corresponding to multiple cells

Relative to the "one-to-one" structure, another topological structure for battery monitoring is one BMC corresponding to multiple cells, as shown in Figure 2.3.

As shown in the figure, one BMC circuit board is responsible for monitoring the information of multiple cells. Compared with the "one-to-one" structure, this structure is cheaper as the circuit board is shared by multiple power batteries. However, it can be seen that due to a longer acquisition line, it is possible to increase the complexity of wiring and reduce the anti-interference performance. Additionally, the voltage acquisition accuracy may be reduced because of a longer acquisition line, and the actual cost of this structure is increased due to the cost of wires.

2.2.2 Relationship Between a BCU and a BMC

The topological structure relationship between a BCU and a BMC includes three types.

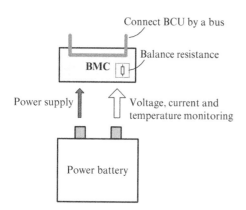

Figure 2.2 Structure of one BMC monitoring one cell.

1. A BCU and a BMC share a board

In some application cases of electric vehicles, a relatively small battery management system is required for less powerful batteries, so the BCU and the BMC can be designed on the same circuit board to uniformly manage all power batteries. Under some special conditions, the functions of the BCU and the BMC can even be combined into the same IC chip. The battery management system with this topological structure is relatively cheaper, but it is not suitable for electric vehicle applications with a large number of cells or a large system.

2. Star type

Relative to type 1, the BMC is separated from the BCU in other topological relationships, so it is necessary to achieve the communication between the BMC and the BCU. Generally, the communication is achieved by a specific communication protocol. However, the physical connections of the communication bus can be realized by combining different topological structures. The first possible connection is a star connection, as shown in Figure 2.4.

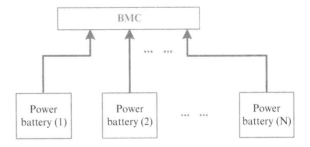

Figure 2.3 Structure of one BMC monitoring multiple cells.

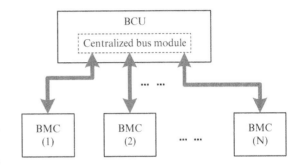

Figure 2.4 Star connection of a BCU and a BMC.

In appearance, in the star connection, the BCU is located centrally and is connected to each BMC module by a wiring harness. The BCU generally comprises a centralized bus module to share a communication channel among multiple BMCs. The star connection has advantages, such as a convenient medium access control and absence of communication impact with other BMCs while certain BMCs fail or stop working. There are some disadvantages, such as the difficulty in maintenance for a longer communication line, poor expandability, or failure of the increase the BMC to the limit of the centralized bus module ports without control.

3. Bus type

Figure 2.5 shows the bus type connection of the BCU and the BMC. As shown in the figure, each circuit board is a part of the communication bus. Compared with the above star connection, the bus connection has advantages such as the lower cost of wires used for the communication channel, a more flexible connection, and strong expandability. If it is required to add the cell in the battery pack and the corresponding number of the BMC, an additional communication line is only required to implement the addition. Conversely, if it is required to remove a certain BMC from the system, it is only required to extend the communication line of the adjacent BMC. The bus connection has a most obvious disadvantage. Due to the interdependency of communication lines, that is, it is required to use (N–1) circuit boards for communication between the circuit board N and the BCU, in case of failure in a certain circuit board, it is possible to immediately cause an impact on the communication between the next BMCs and the BCU.

It is worth mentioning that the physical connection mode, regardless of a star or a bus connection, refers to its topological form. From the perspective of a communication network, there is "media access competition" in both modes. Information interaction between the BCU and the BMC is generally achieved by the bus communication protocols.

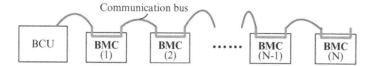

Figure 2.5 Bus type connection of a BCU and a BMC.

References

1 Purewal, J., Wang, J., Graetz, J., et al., "Degradation of lithium ion batteries employing graphite negatives and nickel–cobalt–manganese oxide + spinel manganese oxide positives: Part 2, chemical–mechanical degradation model," *Journal of Power Sources*, 2014, 272(2): 1154–1161.

2 Ramadass, P., Haran, B., White, R., and Popov, B. N., "Mathematical modeling of the capacity fade of Li-ion cells," *Journal of Power Sources*, 2003, 123(2): 230–240.

3 Thomas, E. V., Case, H. L., Doughty, D. H., et al., "Accelerated power degradation of Li-ion cells," *Journal of Power Sources*, 2003, 124: 254–260.

4 Hsieh, Y. C., Moo, C. S., and Tsai, I. S., "Balance charging circuit for charge equalization," in *Power Conversion Conference – Osaka*, 2002, vol. 3, pp. 1138–1143.

5 Zhao, J., Jiang, J., and Niu, L., "A novel charge equalization technique for electric vehicle battery system," *Power Electronics and Drive Systems*, 2003, 2: 853–857.

6 Kutkut, N. H., Wiegman, H. L., Divan, D. M., and Novotny, D. W., "Design considerations for charge equalization of an electric vehicle battery system," *IEEE Transactions on Industry Applications*, 1999, (1): 28–35.

7 Einhorn, M., Roessler, W., and Fleig, J., "Improved performance of serially connected Li-ion batteries with active cell balancing in electric vehicles," *IEEE Transactions on Vehicular Technology*, 2011, 60(6): 2448–2457.

8 Narang, A., Shah, S. L., and Chen, T., "Continuous-time model identification of fractional-order models with time delays," *IET Control Theory and Applications*, 2009, 42(10): 916–921.

9 Gallardo-Lozano, J., Lateef, A., Romero-Cadaval, E., and Milanés-Montero, M. I., "Active battery balancing for battery packs," *Electrical Control and Communication Engineering*, 2013, 2(1): 40–46.

10 Cao, J. C. J., Schofield, N., and Emadi, A., "Battery balancing methods: A comprehensive review," in *IEEE Vehicle Power and Propulsion Conference*, IEEE, 2008.

11 Amine, J., Liu, J., and Belharouak, I., "High-temperature storage and cycling of C-LiFePO$_4$/graphite Li-ion cells," *Electrochemistry Communications*, 2005, 7: 669–673.

12 Morstyn, T., Momayyezan, M., Hredzak, B., et al., "Distributed control for state of charge balancing between the modules of a reconfigurable battery energy storage system," *IEEE Transactions on Power Electronics*, 2015, 31(11): 7986–7995.

13 Gao, F., Zhang, L., Zhou, Q., et al., "State-of-charge balancing control strategy of battery energy storage system based on modular multilevel converter," in *2014 IEEE Energy Conversion Congress and Exposition (ECCE)*, IEEE, 2014.

14 Zhang, J., Ci, S., and Sharif, H., "An enhanced circuit-based model for single-cell battery," in *Applied Power Electronics Conference and Exposition (APEC)*, IEEE, 2010, pp. 672–675.

3

General Procedure of the BMS Design

3.1 Universal Battery Management System and Customized Battery Management System

It is much better for users to buy a "universal" battery management system like the charger. However, the battery management system is generally more complicated; for example, the characteristics, such as the operating environment and the condition of the battery, should be considered. The universality of the battery management system is mainly discussed in this section.

3.1.1 Ideal Condition

Ideal conditions of a BMS include feasibility in all types of the batteries, rapidly self adapt to replaced batteries and wide applicability on different powertrains, such as hybrid electric vehicles, battery electric vehicles, and energy systems other than vehicles.

However, it is difficult to achieve the ideal condition at present, mainly because of the following causes:

First, the operating characteristics vary with the battery type; for example, there are significant differences in the threshold voltage for battery charging and discharging protection, as well as balancing implementation measures among the lead acid battery, the nickel metal hydride battery, the lithium-ion (Li-ion) battery, etc.
Second, there is a difference in the same battery type of different manufacturers or the different batches from the same manufacturer, which cause different SoC estimation algorithms and balancing management strategies [1].
Third, the power battery pack may be used for different environments or under different operating conditions. For example, for the balancing management, even if the same balancing strategy is used, different considerations are required for different environmental conditions. The high-power balancing circuit has advantages, such as a high balancing speed and a good balancing effect, but there are also disadvantages, such as a larger volume and higher heating and production costs [2]. In the case where there is no customized development, there may be resource waste and even failures of the battery management system caused by high heating during balancing. Therefore, the hardware of the balancing circuit is determined according to different environments and conditions used to finalize the final design scheme for the battery management system.

3.1.2 Feasible Solution

Since it is impossible to design a powerful and universal battery management system at present, the battery management system is designed with two tendencies [3].

Battery Management System and its Applications, First Edition. Xiaojun Tan, Andrea Vezzini, Yuqian Fan, Neeta Khare, You Xu, and Liangliang Wei.
© 2023 China Machine Press. All rights reserved. Published 2023 by John Wiley & Sons Singapore Pte. Ltd.

1. Design of a universal and simple protection board

As it is possible for the secondary power battery to cause a safety accident during use in over-charging, over-discharging, and over-current states, a circuit board with a basic protection function can be added to the power battery. The upper and lower limits of charging and discharging voltages may be set on such a circuit board so that the current circuit is disconnected once a voltage comparison shows that the operating voltage of the battery is more or less than the threshold value. Similarly, the corresponding threshold value may be set for the operating current and operating temperature so that when the threshold value is exceeded, the power supply circuit of the battery is disconnected in order to protect the safety of the battery pack [4].

The power battery may be charged and discharged safely with the protection of such a circuit board. However, due to the lack of monitoring of the voltage, current, and other information, the user cannot know the current status of the power battery or assess the SoC and SoR of the battery pack. Just as an ordinary household flashlight will stop when it is out of power, it is difficult to know the SoC and usable time at any moment.

2. Customization of a more complicated solution for a specific battery

Due to the smooth electromotive force (EMF) characteristic curve of an Li-ion power battery and the complex working condition of an electric vehicle, it is necessary to customize the battery management system for different battery types and different applications. The battery management system design discussed in this section is to customize the functions of the battery management system according to the characteristics of the power battery and the application requirements. The battery management system design can be roughly divided into the following steps.

First, the working characteristics of the relevant power battery type are obtained according to the data provided by the manufacturer and the previous battery sample evaluation record. Second, the basic design scheme is determined for the power battery management system, including:

a) Select a suitable topological structure.
b) Determine the battery safety protection strategy.
c) Establish the power battery model and design the SoC estimation algorithm.
d) Determine the balancing management strategy and energy control strategy.

Third, the corresponding software and hardware systems are designed according to the above schemes and verified for their reliability [5, 6].

The power battery management system developed according to the above steps is customized based on the battery type and its application and has high precision. However, the early-stage workload is large, including a full battery performance test and evaluation, before a design is made of the basic scheme.

3.1.3 Discussion of Universality

There has been a hot discussion in the electric vehicle industry in recent years: "Who should develop the battery management system? Is it the car manufacturer, power battery manufacturer, or a third party?" It is generally acknowledged that, if developed by the battery manufacturer, the battery management system is universal, that is, it can be used for different electric vehicles, but the battery management system is not necessarily optimized according to the use conditions of the electric vehicle. If developed by the electric vehicle manufacturer, the battery management system may not be universal, but only applicable to that electric vehicle manufacturer's own cars, and the production costs may be very heavy due to the amount of time it took to understand their particular power battery characteristics. If it is developed by a third party, the two above-mentioned will be burdened by the third party.

However, whoever develops the battery management system, consideration should be given to the operating environment and working conditions (which seems to be the responsibility of the electric vehicle manufacturers) and the working characteristics of the battery (which seems to be the responsibility of the battery manufacturers). A preliminary conclusion may be drawn according to the previous two sections: the simple protection circuit has certain universality, but its function is limited; the customized battery management system can have relatively complete functions, but the workload is large while requiring more human intervention. A battery management system R&D engineer can consider the following aspects in the near future:

1. **Optimization of hardware design**

The hardware facilities that are necessary for the power battery pack of the electric vehicle, such as a protection circuit for the hardware (contactor, current breaker, etc.), balance circuit, heater, cooler, etc., should be optimized according to the car features including volume, weight, cost, and other factors.

2. **Self-adaptation of the software system**

On the one hand, according to the batteries of different manufacturers and different batches of the battery of the same manufacturer, the software system can be adaptively adjusted to acquire the characteristic parameters during normal use of the car and reduce the workload required for many characteristic tests of the battery. At the same time, the software system should have more intelligent algorithms to accurately evaluate the SoC, SoH, etc.

3. **Low-power design**

Energy consumption is required for operation of the battery management system. Since it is often difficult to forecast the idle time of an electric vehicle, a power-saving design must be adopted for the battery management system. The low-power design should be considered from software and hardware aspects.

3.2 General Development Flow of the Power Battery Management System

Figure 3.1 shows the design and implementation flow of the BMS for electric vehicles.

3.2.1 Applicable Standards for BMS Development

Relevant standards should be collected before BMS development. Many standards related to BMS development have been issued in the past decade. Specific standards are not listed here due to their continuous perfection and upgrade. In simple terms, the applicable standards can be divided into the following types:

First, standards directly related to BMS quality, performance, and function, which directly specify the functional and performance indicators of the BMS, such as voltage and current sensor test requirements of BMS, accuracy requirement of main functions of BMS such as SoC estimation, etc.

Second, BMS requirements deduced according to battery system standards. For example, some standards specify that the battery system should have a self-protection function during charging and discharging in order to prevent over-charge and over-discharge. According to these standards, the BMS shall have charging and discharging protection functions. Therefore, relative electronic and electrical standards and battery system (battery pack) standards shall be considered during collection of the applicable standards for BMS development.

Third, BMS requirements deduced according to vehicle standards. For example, the Electro Magnetic Compatibility (EMC) is specified for the vehicle electronic system in the standards of many countries and regions, so the external radiation intensity of the BMS shall not be higher than the specified threshold value during operation. Furthermore, BMS shall resist a certain intensity of external radiation in order to avoid an operation stop resulting from any degree of electro-magnetic interference during operation.

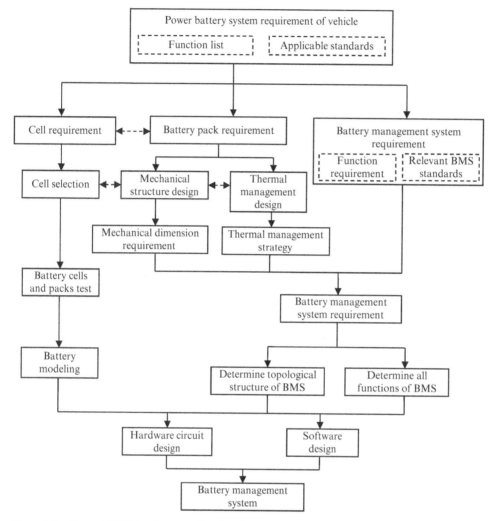

Figure 3.1 General BMS development flow for electric vehicles.

For the above-mentioned standards, more considerations were given to hardware standards in the past. In recent years, more attention has been paid to software standards, such as the representative standards AUTOSAR Software Architecture Specification and ISO-26262.

3.2.2 Boundary of BMS Development

In the past, the software and hardware were respectively considered during BMS development. In fact, the practice in recent years shows that some related fields must be considered during BMS development. In other words, the BMS engineers should not only understand software and hardware development, but also consider the professional knowledge of related fields, including:

First, understanding of electrochemical characteristics of cells and establishment of a battery model according to the cell characteristics. The battery model can reflect the characteristics of the battery in the working process [7]. In order to obtain a more comprehensive battery model, a cell test is required. It can be seen from Figure 3.1 that the cell test is very important and is the precondition for battery management system development.

Second, combination of BMS and battery pack designs. For quite a long time, the BMS has been separately designed from the battery pack. In other words, the battery pack designer and manufacturer only consider the BMS manufacturer as a regular supplier who only provides standardized BMS products. As a result, the designed BMS fails to exactly match the battery system performance and the battery system therefore cannot show its best performance. Similarly, many electric vehicle manufacturers only consider the battery system manufacturer as a regular supplier, give less consideration to the battery pack layout at the beginning of the electric vehicle design, and require the battery pack to fit the vehicle during a later period of design. As a result, "freak" battery systems are installed in many vehicles.

As shown in Figure 3.1, the BMS design is combined with the battery pack design, which is specifically embodied in "mechanical structure design" and "thermal management design." The former involves the BMS's hardware arrangement, acquisition line wiring design, and high-voltage insulation design. The latter involves the thermal management model, the thermal management algorithm, etc.

3.2.3 Battery Characteristic Test Is Essential to BMS Development

In fact, the battery is indispensable for the BMS development process, mainly reflected by two aspects. First, the BMS software is based on the battery characteristic; that is, the BMS functions are developed according to the battery performance characteristic. Second, the BMS is often tested with the battery to ensure its performance and reliability.

For the relationship between the performance and BMS software, the following examples can be cited: before development of the BMS hardware, there is often a requirement to determine the voltage acquisition accuracy, but this accuracy is determined according to the battery characteristics. Compared with the $Li(NiCoMn)O_2$ ternary battery, the EMF curve of the $LiFePO_4$ battery is significantly smoother. For the $Li(NiCoMn)O_2$ ternary battery, 1% SoC is approximately 5 mV, while the 1% SoC of the $LiFePO_4$ battery is 1 mV. For the same SoC estimation accuracy requirement, the voltage sensor accuracy of the BMS for the $LiFePO_4$ battery is five times the voltage sensor accuracy of the BMS for the $Li(NiCoMn)O_2$ ternary battery (in other words, the allowable error of the voltage sensor of the BMS for the $LiFePO_4$ battery is 20% that of the BMS for the $Li(NiCoMn)O_2$ ternary battery) [8].

As another example, attention is given to the battery balance control strategy, which belongs to the software category. In fact, when the balance function test is performed for the BMS, the management object, namely the battery pack, must be tested together with the BMS to obtain the balance effect.

3.3 Core Status of Battery Modeling in the BMS Development Process

According to the above analysis, it can be seen that for the boundary of BMS development, there is a requirement to take into account the electrical function of the entire automotive system, involve the force and thermal design of the battery pack, and thoroughly understand the characteristics of the cell. The characteristics of the cells can be deeply understood only by establishing the battery model based on the cell and battery pack tests.

It can be seen from Figure 3.2 that, as the basic theoretical part during the development of the battery management system, power battery modeling can show the BMS designer's understanding of the battery performance and degradation characteristics, and provide a basis for development of other battery management

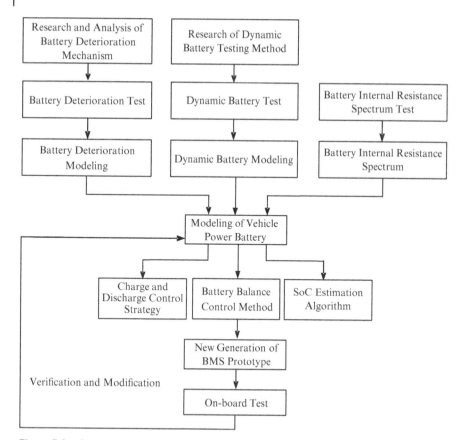

Figure 3.2 Core status of battery modeling in the BMS development process of vehicles.

functions (such as SoC estimation, battery balance, charge and discharge control, etc.) [9]. It is mainly shown in the following aspects. First, the accurate battery model is the precondition for SoC estimation. Second, the battery model is the basis for the battery balancing control strategy to avoid the situation where "the battery is OK during imbalance, but may be damaged after balancing." Third, it is beneficial to develop an advanced battery charge and discharge control method.

Generally, the following specific problems should be solved to establish a battery model.

1. Modeling for dynamic operating characteristics of the battery

Due to the variability and uncertainty of the operating condition of electric vehicles, the dynamic characteristics of the vehicle power battery are one of the important factors considered for the battery management system [10]. The dynamic operating characteristics of the battery may be simulated by the equivalent circuit model, including electromotive force hysteresis characteristics, open-circuit voltage rebound characteristics, dynamic temperature characteristics, etc. [11–13]. The following important issues should be considered in the modeling process:

First, the impact of the battery operation history on the current battery state.

Second, the impact of the battery operating temperature on the battery performance.

Third, the practicability analysis of the battery model, namely the algorithm complexity of the established battery model during implementation, to provide a basis for selection of the master chip for the battery management system.

2. Research and modeling of the battery deterioration mechanism

The "battery degradation" refers to the performance deterioration of the battery. If the person responsible for developing the battery management system is not clear about the aging and deterioration mechanism of the battery, the software of the battery management system cannot adapt to the performance deterioration of the battery [14, 15]. When the deterioration model is established for the battery, the major factors causing the battery performance deterioration, such as temperature, charging and discharging rates, charging and discharging degrees, etc., are identified by plenty of samples and experimental data, and the effect of all factors on the deterioration is quantitatively described by mathematical methods [16].

3. Real-time correction of battery model parameters

Since the operation of a power battery is dynamic and battery deterioration is constantly occurring in the operating process, the battery model parameters should be corrected in real time. In general, the battery model parameters are corrected in real time using a Kalman filter as the most common method [17].

Nevertheless, the cell and battery pack tests are essential for battery modeling. The battery tests will be discussed in the next chapter.

References

1 El Lakkis, M., Sename, O., Corno, M., and Bresch Pietri, D., "Combined battery SOC/SOH estimation using a nonlinear adaptive observer," in *2015 European Control Conference (ECC)*, 2015.

2 Kim, M. Y., Kim, C. H., Kim, J. H., and Moon, G. W., "A chain structure of switched capacitor for improved cell balancing speed of lithium-ion batteries," *IEEE Transactions on Industrial Electronics*, 2013, 61(8): 3989–3999.

3 Ahmadi, L., Yip, A., Fowler, M., et al., "Environmental feasibility of re-use of electric vehicle batteries," *Sustainable Energy Technologies and Assessments*, 2014, 6: 64–74.

4 Maccario, M., Croguennec, L., Cras, F. L., et al., "Electrochemical performances in temperature for a C-containing LiFePO$_4$ composite synthesized at high temperature," *Journal of Power Sources*, 2008, 183: 411–417.

5 Chen, P., "The entity-relationship model: Toward a unified view of data," *Readings in Artificial Intelligence and Databases*, 1989, 10(3): 98–111.

6 Chen, P. P.-S., "The entity relationship model – Toward a unified view of data," in M. Broy and E. Denert (eds), *Software Pioneers*, Berlin, Heidelberg: Springer, 2002, pp. 311–339.

7 Chen, M. and Rincon-Mora, G. A., "Accurate electrical battery model capable of predicting runtime and i-v performance," *IEEE Transactions on Energy Conversion*, 2006, 21(2): 504–511.

8 He, H., Xiong, R., and Guo, H., "Online estimation of model parameters and state-of-charge of LiFePO$_4$ batteries in electric vehicles," *Applied Energy*, 2012, 89(1): 413–420.

9 Zhang, Y., Wang, C. Y., and Tang, X., "Cycling degradation of an automotive LiFePO$_4$ lithium-ion battery," *Journal of Power Sources*, 2011, 196: 1513–1520.

10 Kutkut, N. H. and Divan, D. M., "Dynamic equalization techniques for series battery stacks," in *Proceedings of Intelec'96-International Telecommunications Energy Conference*, IEEE, 1996, pp. 514–521.

11 Snihir, I., Rey, W., Verbitskiy, E., et al., "Battery open-circuit voltage estimation by a method of statistical analysis," *Power Sources*, 2006, 159(2): 1484–1487.

12 Zhang, X., Lu, J., Yuan, S., et al., "A novel method for identification of lithium-ion battery equivalent circuit model parameters considering electrochemical properties," *Journal of Power Sources*, 2017, 345: 21–29.

13 Hu, X., Li, S., and Peng, H., "A comparative study of equivalent circuit models for Li-ion batteries," *Journal of Power Sources*, 2012, 198: 359–367.

14 Kjell, M. H., Malmgren, S., Ciosek, K., et al., "Comparing aging of graphite/LiFePO$_4$ Cells at 22°C and 55°C – electrochemical and photoelectron spectroscopy studies," *Journal of Power Sources*, 2013, 243: 290–298.

15 Danzer, M., Liebau, V., and Maglia, F., "Aging of lithium-ion batteries for electric vehicles," in B. Scrosati, J. Garche, and W. Tillmetz (eds), *Advances in Battery Technologies for Electric Vehicles*, Cambridge, UK: Elsevier, 2015, pp. 359–387.

16 Nelson, P., Bloom, I., Amine, K., et al., "Design modeling of lithium-ion battery performance," *Journal of Power Sources*, 2002, 110(2): 437–444.

17 He, H., Xiong, R., Zhang, X., et al,, "State-of-charge estimation of the lithium-ion battery using an adaptive extended Kalman filter based on an improved thevenin model," *IEEE Transaction on Vehicular Technology*, 2011, 60(4): 1461–1469.

Part II

Li-Ion Batteries

4

Introduction to Li-Ion Batteries

Lithium-ion (Li-ion) batteries are the most promising candidate for all available electrical energy storage options. It is an interesting exercise to trace back the research and development phases of Li-ion from the last decade and its technology potential in the future. This chapter will cover the basic history of technology, component structure, popular recipes of electrodes, separators, and electrolytes.

Two basic classifications are Li-ion and Li-metal batteries based on function, usage, and the electrodes used. Li-metal batteries are primary disposable batteries using Li-metal and Li-compounds as electrodes, whereas Li-ion batteries are rechargeable batteries where active Li-ions move between the anode and cathode [1]. Li-ion batteries use an intercalated Li-compound as the electrode material instead of metallic Li.

With increasing demand for a high-energy density, longer life, and lightweight battery, Li-ion becomes the superior solution among existing electrochemical energy storage technologies.

4.1 Components of Li-Ion Batteries: Electrodes, Electrolytes, Separators, and Cell Packing

Li-ion battery characteristics vary with the combination of electrodes and electrolytes used. The potential market and varied applications have created a large number of opportunities for various electrode combinations. Each pair of electrodes provides diverse capabilities and ensures improvement in existing performances, such as energy and power densities, operating voltage, current and temperature range, self-discharge, life expectancies, and degradation of electrodes [2].

The cathode mostly limits the energy capability while the anode and electrolytes mainly contribute to limiting the power capacity. Like in any electrochemical system, interface regions play a more dominating role in affecting the performance of the system rather than individual components. In a graphite anode, the prime reason for irreversible capacity loss is a solid electrolyte interface (SEI) layer between the electrolyte and electrodes (anode). The SEI layer also broadly affects charge transfer kinetics and storage properties [3–5]. Surface formulation, coating of positive electrodes, and using appropriate additives are a few of the recently proved excellent techniques used to control and optimize the SEI layer and to moderately enhance the performance.

On the other hand, at the surface of positive electrodes, interfacial reactions and the growth of a passivation layer upon cycling have been researched. The research results have established a paramount importance of using different materials as positive electrodes and have found significant differences to the performance degradation of the battery upon aging and cycling [6]. The surface reactions at the positive electrode/electrolyte interface have been clearly demonstrated in subsequent literature [7]. However, experimental conditions of formation, growth, and modification, as well as their subsequent influences on the electrochemical performance, remain unclear.

Battery Management System and its Applications, First Edition. Xiaojun Tan, Andrea Vezzini, Yuqian Fan, Neeta Khare, You Xu, and Liangliang Wei.

Most popular cathode materials used are $LiMn_2O_4$, NCA, $LiNi_{1/3}Mn_{1/3}Co_{1/3}O_2$, $LiNi_{1/2}Mn_{1/2}O_2$, and $LiFePO_4$. Surface treated graphite, mesophase pitch, and Titanate are used as anodes. Most Li-ion batteries use organic solvents as the electrolyte, which is $LiFP_6$ mixed with carbonates, and have a low electrical resistance [8]. Although these organic solvents are still used, they are thermally unstable. Polymers and inorganic compounds in gel and solid phase electrolytes offer a compact thermally stable system and hence are safe. However, they offer high resistance and slow down the kinetics, which increase self-discharge and affect power capacity adversely.

Figure 4.1 describes a systematic approach in order to improve the Li-ion chemistry, particularly new types of electrodes. Mostly, these six steps are common regardless of the type of material, crystal structure, or operating mechanism: (a) dimension reduction, (b) composite formation, (c) doping and functionalization, (d) morphology control, (e) coating and encapsulation, and (f) electrolyte modification.

a) Dimension reduction at the nano particle level of transition metal oxides has the potential to improve kinetic properties, surface reactivity, and higher intercalation of ion/electrons. It also subdues the mechanical stress during intercalation. Higher intercalation and surface reactivity provide a high rate of charging and discharging capabilities, whereas less mechanical stress at the anode during the intercalation ensures a longer life.
b) Composite formation: conductive media increases the kinetics of the system.
c) Doping and functionalization are used in order to increase ion kinetics and thermal stability.
d) Morphological changes control ion transportation, increase reactivity, and provide structural stability.
e) Coating and encapsulation stabilize the solid electrolyte interface (SEI) and passivation layer formation. This protects the electrolyte decomposition and electro surface from degradation.
f) Electrolyte modification and additives help to suppress side reactions and, as a result, it reduces the solubility of active material and decomposition in the electrolyte.

Table 4.1 is a summary of the frequently used electrodes and their properties. The most common combination is Li-metal compound as the cathode, such as Li-cobalt, Li-manganese, Li-nickel manganese cobalt, Li-nickel cobalt oxide, and Li-iron phosphate with carbon-based graphite as the anode [9].

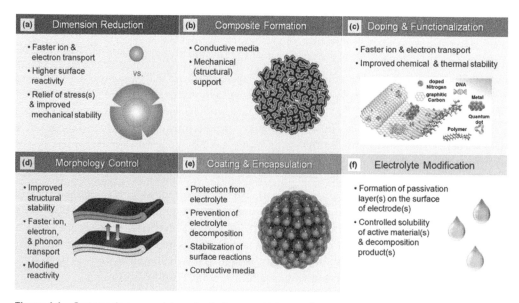

Figure 4.1 Systematic approach in order to improve Li-ion performance.

Table 4.1 Summary of Li-ion electrode properties.

		Li-ion cobalt	Li-ion manganese	Li-ion NCM	Li-ion NCO	Li-ion phosphate	Lithium titanium
Energy density (Wh/kg)		150–200	100–170	160–300	85–130	140–190	50–80
Power density (W/kg)		150–440	400–2000	380–860	200–1600	250–3300	100–700
Pulse power density duration of 2 s (Wh/kg)		n/a	n/a	up to 2300	up to 2300	up to 6000	up to 3600
Cycle life (to 80% of initial capacity)		500–1000	300–700	1000–2000	1000	1000–4000	3000–7000
Temp. range °C	Discharging	–20 to 60	–40 to 60	–40 to 60	–30 to 60	–40 to 60	–40 to 60
	Charging	0 to 45	0 to 65	20 to 55	0 to 40	20 to 55	–30 to 45
Self-discharge per month		5%	5%	–	–	5%	5%
Cell voltage (average)		3.7	3.8	3.7	3.65	3.3	2.4
Max. load current	Constant	1 to 3 C	up to 18 C	2.5 to 5 C	3 to 20 C	3 to 20 C	2 to 10 C
	Pulse (2 s)	n/a	n/a	10 C	15 to 25 C	10 to 50 C	10 to 18 C
Charge rate	Normal	1 C	1 to 4 C	0.5 to 1 C	1 C	0.5 to 2 C	1 C
	Fast	n/a	n/a	n/a	2 to 4 C	3 to 5 C 15 s pulse to 15 C	10 C
Safety		Low safety, thermally unstable	Low safety, thermally unstable	Low safety, thermally unstable	Low safety, thermally unstable	Good safety, thermally stable, not flammable	Good safety, thermally stable

Recently, manufacturers have been using a few graphite alternatives like titanate for extending the life of the Li-ion battery by compromising energy densities. Medium gray cells in the table relate to a moderate to high performance, but dark gray cells represent poor features of the electrode chemistries.

4.2 Li-Ion Electrode Manufacturing

Li-ion electrode manufacturing is quite an involved process. Most generic manufacturing process flow of the anode and cathode is given in Figure 4.2. The Li-ion electrode manufacturing process is similar to that used for nickel cadmium cells and nickel metal hydride cells. The anode and cathode follow similar steps, but their ingredients differ. As a first step, the anode uses a graphite or carbon mix slurry whereas the cathode uses a lithium compound and binder mix slurry. Then precisely cleaned foils of copper for the anode and aluminum for the cathode are added to the coating machine. These foils work as current collectors for respective electrodes. The slurry mix spreads on to the surface of the foil as it passes into the coating machine. The coating and foil runs are precisely controlled to avoid any differences at the coating process. The next step is to control the thickness of the coated material to match the gravimetric or volumetric energy storage capacity of the anode and cathode using compressing techniques. Compressing is followed by drying and slitting processes. Slitting machines cut the foil into narrower strips suitable for different sizes of electrodes and lengths. Any burrs on the edges of the foil strips could give rise to internal short circuits in the cells. The entire anode and cathode manufacturing processes run into different clean rooms to avoid any contamination, which can ruin the cell completely.

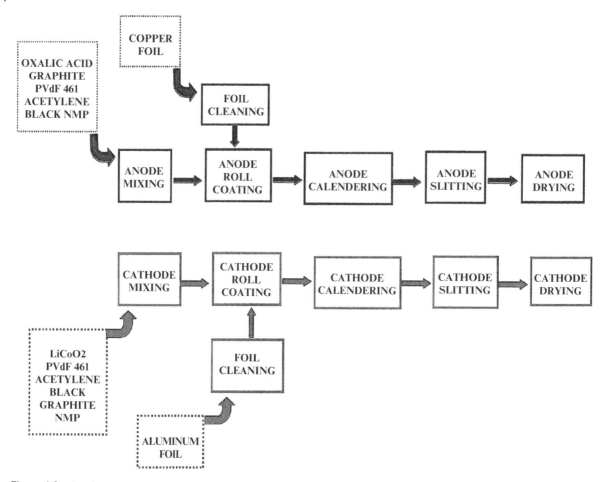

Figure 4.2 Anode and cathode manufacturing process flow.

4.3 Cell Assembly in an Li-Ion Battery

The cell assembly slightly differs for prismatic and cylindrical cells. Plates need to be stacked together for a prismatic cell and to be wound together for cylindrical cells. Cell assembly steps for both prismatic and cylindrical cells are shown in Figure 4.3. Before entering a drying room, the anode, cathode, and separator are stacked/wound together with appropriate connectors and then the electrode stack/jellyroll is inserted for prismatic and cylindrical cells respectively. Cell stacked/wound structures undergo procedures for subassembly of connections, terminals, vents, and safety devices. The final step before the drying room is to be heat sealed or welded. Welding is the only option used for a cylindrical cell. Later in the drying room, vacuum heat dry and vacuum filling of the electrolyte are done and the case is sealed. Labeling, formation cycle, and pre-testing for shipping and packaging are other steps that manufacturers have to follow before they can ship the cell,

Table 4.2 summarizes commercially available Li-ion cell assemblies.

The first stage in the assembly process is to build the electrode subassembly in which the separator is sandwiched between the anode and the cathode. Two basic electrode structures are shown in Figure 4.3: a stacked structure for use in prismatic cells and a spiral wound structure for use in cylindrical cells.

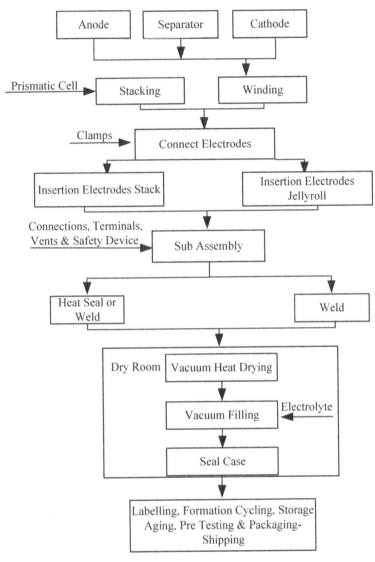

Figure 4.3 Cell assembly steps for both prismatic and cylindrical cells.

4.4 Safety and Cost Prediction

The Li-ion is a well-received candidate for energy storage applications. Over the years technology has an exponential research growth rate. The overwhelming research results and varied chemistry created a wide scope for use. However, safety and cost for high power applications are still among the hot issues to address in Li-ion batteries.

The energy stored in a cell can self-heat it to over 500 °C. Safety incidents in portable devices are rated at ~ 1 in 5 million. Safety problems must be resolved or government safety action is threatened (concern for fire on an aircraft). An explosive thermal runaway can result from self-heating reactions of the anode and/or cathode with

Table 4.2 Type of commercially available Li-ion cells.

Plastic case

Robust. Easy packaging, Flammable, Inexpensive. Stacked or jelly roll electrodes. Retains heat, Poor thermal dissipation, Sizes up to 1000 Ah

Used by: Thunder Sky, International Battery, China Hipower, HuanYu
Credit: Thunder Sky Winston Battery

Cylindrical steel case

Robust. High energy density cells but packs are wasteful of space, Space allows cooling air flow, Expensive, Sizes up to 200 Ah

Used by: Gaia, PHET, LifeBatt, BAK, SAFT, A123
Credit: GAIA

Prismatic metal case (Steel/Aluminum)

Robust, Easy packaging, Good space utilization, Expensive, Jelly roll or stacked electrodes, High energy density, Good heat dissipation, Sizes up to 200 Ah

Used by: BYD, HYB, Lishen, Toshiba, Varta
Credit: Komachine Co

Pouch cell – also known as Lipo cells

Vulnerable, Inexpensive, Design freedom on dimensions, Difficult packaging, High energy density but reduced by support packaging needed, Prone to swell and leak, Less danger of explosion (cell bursts), Good heat dissipation, Made in very high volumes, Economical for small volumes, Sizes up to 240 Ah

Used by: Kokam, BAHUP, ATL, Yoku, EIG, Enerdel, LG, and many others
Credit: HYB BATTERY CO., LTD

Small cylindrical and prismatic cells

Metal cans, Low-cost products in standard shapes, Made in very high volumes, Complex and expensive packaging and BMS electronics due to low unit capacity, Sizes up to 5 Ah

Used by: ATL, BAK, B&K, BYD, Lishen, Panasonic, Sony, Sanyo, Toshiba, Samsung, Valence, and many others
Credit: Panasonic Corporation of North

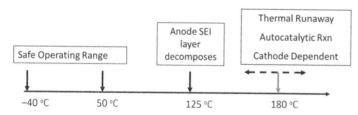

Figure 4.4 Safety operation temperatures.

the electrolyte, initiated by internal or external short circuits, voltage, and temperature excursions outside safe areas (see Figure 4.4).

Another problem limiting use is the cost. For Li-ion batteries to take their place in widespread commercialization of hybrid electric vehicles (HEVs), plug-in hybrid electric vehicles (PHEVs), and full electric vehicles (EVs), the system cost must still be reduced by $3e^4$ to about \$125 per kWh. Three important ways to achieve significant system cost reductions are to:

1) Reduce the electrode processing cost associated with the costly organic solvent and primary solvent drying time.
2) Substantially increase the electrode thicknesses to ~2 the current "power" levels (to $3.5e^{4.5}$ mAh/cm^2) while preserving power density.
3) Reduce the formation time associated with the anode solid electrolyte interface (SEI) layer. These Li-ion cell fabrication steps contribute significantly to the current overall pack cost.

References

1 Hasan, R. and Scott, J. B., "Fractional behaviour of rechargeable batteries," in *2016 Electronics New Zealand Conference*, Electronics New Zealand Inc., 2016.

2 Peterson, S. B., Apt, J., and Whitacre, J. F., "Lithium-ion battery cell degradation resulting from realistic vehicle and vehicle-to-grid utilization," *Journal of Power Sources*, 2010, 195(8): 2385–2392.

3 Liaw, B. Y., Jungst, R. G., Nagasubramanian, G., et al., "Modeling capacity fade in lithium-ion cells," *Journal of Power Sources*, 2005, 140(1): 157–161.

4 Arora, P., White, R. E., and Doyle, M., "Capacity fade mechanisms and side reactions in lithium-ion batteries," *Journal of the Electrochemical Society*, 1998, 145(10): 3647–3667.

5 Ramadass, P., Haran, B., Gomadam, P. M., et al., "Development of first principles capacity fade model for Li-ion cells," *Journal of the Electrochemical Society*, 2004, 151(2): A196–A203.

6 Dubarry, M., Truchot, C., Liaw, B. Y., et al., "Evaluation of commercial lithium-ion cells based on composite positive electrode for plug-in hybrid electric vehicle applications. Part II. Degradation mechanism under 2C cycle aging," *Journal of Power Sources*, 2011, 196: 10336–10343.

7 Methekar, R. N., Northrop, P. W. C., and Chen, K., "Kinetic Monte Carlo simulation of surface heterogeneity in graphite anodes for lithium-ion batteries: Passive layer formation," *Journal of the Electrochemical Society*, 2011, 158(4): A363–A370.

8 Wang, Y., Zhang, C., Chen, Z., et al., "A novel active equalization method for lithium-ion batteries in electric vehicles," *Applied Energy*, 2015, 145: 36–42.

9 Christensen, J. and Newman, J., "A mathematical model of stress generation and fracture in lithium manganese oxide," *Journal of the Electrochemical Society*, 2006, 153(6): A1019–A1030.

5

Schemes of Battery Testing

The battery characteristic tests are an important part of the battery management system development, and have often been neglected in the past. However, the battery characteristic tests are extensive. Several representative test items are selected and described in this chapter. It is noted that the power battery characteristic tests are not completely equivalent to the product tests specified in current automotive industry standards or by ISO. The test items specified in industry standards are designed to verify whether the battery product is qualified and can be sold on the market, while the battery characteristic tests discussed in this chapter are designed to understand the characteristics of the "managed" battery in order to design a practical battery management system.

5.1 Battery Tests for BMS Development

5.1.1 Test Items and Purpose

The power battery characteristic tests discussed in this chapter are related but yet different from the general performance tests. On the one hand, many characteristic test items are the same as the performance test items. On the other hand, there is a difference in the purpose and significance between the characteristic test and the performance test. The performance tests are designed to evaluate the battery performance, while the characteristic tests are designed to test the battery management object after selecting the battery type and model, so as to understand the battery characteristics for the purpose of developing BMS software and hardware [1]. The characteristic tests discussed in this chapter cover many aspects. Possible test items involved are given below:

1. **Actual capacity test**

The actual capacity refers to the charge a battery can actually release. When the battery is delivered, the battery manufacturer provides the rated capacity, which is measured at a specific discharge rate and a specific temperature according to relevant standards [2]. Before the development of BMS, the battery should have been sampled to measure its actual capacity at different temperatures and discharge rates, so that the SoC estimation algorithm in the BMS could be applicable under different working conditions and at different ambient temperatures. Furthermore, this characteristic test is very important for battery balancing management, battery charge, discharge energy control, and other aspects.

2. **Charge and discharge efficiency test**

The charge and discharge efficiency test refers to measuring the ratio of the effectively discharged energy to the effectively charged energy of the battery. Similarly, the charge and discharge efficiencies vary with the temperature and the discharge rate, and so they should be measured at different temperatures and different

Battery Management System and its Applications, First Edition. Xiaojun Tan, Andrea Vezzini, Yuqian Fan, Neeta Khare, You Xu, and Liangliang Wei.

discharge rates. The charge and discharge efficiency test is important for SoC estimation and charge and discharge energy control.

3. Discharge rate characteristic test

The discharge rate characteristic test refers to testing the maximum current discharged from the battery during its working condition. The maximum discharge current is related to the operating ambient temperature and the SoC of the battery. Generally, the higher the ambient temperature, the greater the activity of the material in the battery and the larger the maximum discharge current are. That is, the higher the SoC, the larger the maximum discharge current will be. This characteristic test is important for the battery safety protection function and charge and discharge energy control.

4. Electromotive force curve and equivalent internal resistance test

The electromotive force and internal resistance are the important factors that affects power battery's external characteristics. They are related to the operating ambient temperature and the SoC of the battery [3, 4]. In practice, the two items can be tested simultaneously. In the process of SoC estimation, it is often necessary to use the open circuit voltage of the battery to estimate the SoC. Therefore, it is very important to obtain the functional relationship between the electromotive force and the SoC of the battery (SoC-EMF) curve for an estimation of the SoC.

5.1.2 Standardization of Characteristic Tests

Special power battery test standards have been issued in many countries and can be roughly divided into two categories. One is applicable to the battery manufacturers to check whether their batteries are qualified and sellable in the market. Another is applicable to the electric vehicle manufacturers to test whether the selected power batteries can meet the performance requirements of their electric vehicles. Although these two categories of test standards have different purposes, their test items are not completely different. Two categories of test standards share many of the same test items. Due to the different test purposes, the same test items may be tested using different test methods. The test results may be evaluated using different indicators. For example, although battery capacity and cycle life tests are specified in two categories of test standards, the test standards applicable to the battery manufacturers focus on testing whether the battery capacity and the cycle life reach their nominal values at a specific temperature and a constant discharge rate, while the test standards applicable to the electric vehicle manufacturers focus on testing the performance of the battery under actual operating conditions at various limiting temperatures [5, 6].

Four vehicle power battery test standards are listed in this section. The first test standard is applicable to battery manufacturers and the other three test standards are applicable to electric vehicle manufacturers.

1. Auto industry standard of China GB/T 31486-2015 – electrical performance requirements and test methods for a traction battery of an electric vehicle

This standard describes how to test a single battery and a battery module consisting of multiple batteries respectively. The test items cover battery appearance, polarity, size and mass, discharge capacity, charge retention and capacity recovery, and cycle life and safety performance. The test standard also specifies a simple simulation test and a vibration test to verify whether the battery module matches the performance of the electric vehicle. The GB/T 31486-2015 standard provides the specific test procedures for all test items and also the indicators of the test items that should be reached by the battery and the battery module during test.

2. SAE J1798 – recommended procedure of battery performance evaluation for electrical vehicles

Founded in 1905, the SAE (Society of Automotive Engineers) is the largest automotive engineering academic organization in the world. The standards developed by SAE are authoritative and widely used in the automotive industry and also in other industries, a considerable number of which has been used as national standards in the United States.

SAE J1798 lists a number of battery performance tests, including static capacity tests, charge retention test, charging acceptance test, peak power capacity test, and dynamic capacity test [7]. The standard also stipulates test module selection, test conditions, test temperature, sensor location, sampling frequency, measurement accuracy, and other details. It is a very rigorous test standard.

3. **IEC 61982-3 – secondary battery for electric vehicles on a standard road: Part III: performance and service life tests**

Founded in 1906, the IEC (International Electrotechnical Commission) was the world's first international electrical standardization body responsible for international standardization in electrical and electronic engineering fields.

The IEC 61982-3 standard is applicable to the battery system used in small and low-speed urban electric vehicles, but is not applicable to the battery system used in special-purpose vehicles such as public transport, garbage collecting vehicles, motorcycles, and large commercial vehicles [8].

The standard specifies a verification test according to the performance requirement of the electric vehicle, mainly including three basic tests, capacity test, power test, and service life test, and also provides some optional tests, such as maximum power test, battery resistance test, charge test, and operating voltage range test. This test standard can help electric vehicle manufacturers to determine whether the batteries under test can meet the performance requirements of the electric vehicles to be developed, and provide a basis for electric vehicle manufacturers to compare and screen the desired batteries from multiple brands.

4. **PNGV electric vehicle test manual**

This test standard is a battery test standard developed by the INEEL (Idaho National Engineering and Environmental Laboratory) for the PNGV (Partnership for a New Generation of Vehicles).

The PNGV formulated performance targets of the energy storage systems for power-assisted and two-mode hybrid electric vehicles [9]. The battery tests specified in this standard are designed to test whether the battery under test can meet the vehicle performance goals set in the PNGN.

This standard defines the static capacity test, hybrid pulse power characteristic test, available energy test, self-discharge test, and other items. Unlike other test standards, this standard not only defines the test methods of the above test items, but also provides the test data and results to be recorded and an analysis of some results. Therefore, this standard is not only a standard test manual, but is also a reference book describing battery characteristics and principles.

The introductions of standards written above are the briefing of these four test standards, those who are interested in details can refer to the original text. Through the introduction of these test standards, it is easy to see that although a lot of domestic and oversea battery test standards are available. They are only designed for battery manufacturers or electric vehicle manufacturers. Almost no battery test standards are specifically designed for battery management system developers. Because of this, when some battery operating characteristics tests, the battery equilibrium potential test, voltage rebound characteristics test, electromotive force hysteretic characteristics test, etc., are required, battery management system developers should design practical test methods based on current test standards.

In general, the battery characteristics tests required by BMS developers can be carried out according to the following three requirements:

1. **Test based on actual conditions of the BMS development**

The lithium-ion (Li-ion) battery products are constantly updated and more powerful battery products are launched every year. Their characteristics vary with the manufacturers, therefore, during the BMS development, various characteristic tests need to be performed for the managed object – the power battery.

2. **Use existing standards as much as possible**

As some battery characteristic test items are the same as the test items of the battery industry and the vehicle industry, existing test standards should be complied in order to avoid unnecessary development.

3. **It is important that a unified characteristic test standard should be developed as soon as possible**
There are three parts in the test including:

1) The developers of electric vehicles and their parts should prepare unified internal standards. During power battery selection and BMS development, the battery samples from different manufacturers and the different batches are evaluated to formulate unified test standards, which can not only save the test cost, but also obtain comparable test results.

2) If different research and development units adopt a unified test standard, it is beneficial to the industrial development of a battery management system that is an emerging auto part.

3) The unified test standard is beneficial to the production and manufacture of test equipment.

5.1.3 Some Issues on Characteristic Tests

1. **Optimization of test steps**

It is necessary to optimize the test steps for the battery characteristic test to obtain the results of multiple test items by using one test step process as much as possible. For example, in Section 5.2, the power battery capacity test and the charge and discharge efficiency are conducted through the same test process; in Section 5.4, the electromotive force curve test and the equivalent internal resistance test are also conducted through the same test process.

The benefits of such optimization include:

1) Reduce the number of test samples. The high-power batteries are usually expensive, and reduction of the number of test samples is equivalent to saving the test cost.

2) Save test time. Many characteristic tests are time consuming. If different test items can be combined for testing in the same test step, the time will be saved while the development period of the power battery management system will be shortened.

3) Improve the equipment utilization rate. The test equipment is exclusively required by many test items and improvement of the equipment utilization rate can reduce the number of items required for the laboratory.

2. **Setting of the temperature and charge and discharge rate**

Many characteristic test items are required to determine the temperature and charge and discharge rate. Ideally, the test should be performed at each temperature point and charge and discharge rate as many times as possible, but this would require a very long test time. Therefore, in practice, representative temperature points and the charging and discharging rates can be selected for testing. Two suggestions are given below.

First, a small number of samples are selected for large scale and high intensity testing in order to accurately understand the temperature and charge and discharge rate change rule of a certain battery type. For example, the test is performed at 5°C intervals from −40 to 60°C with a total of 21 tests and at every 0.1 C charge and discharge rate once from 0.1 to 1.0 C with a total of 10 tests. Two hundred and ten tests are performed by combining the above two conditions. Although a higher cost and more time are required for such tests, the characteristic change rule of the battery under different working conditions can be clearly understood to provide a sufficient basis for subsequent BMS development.

Second, some representative working conditions are selected to carry out a daily test, as not all working conditions are necessary. For example, during the battery cycle characteristic test, if a deep charging and deep discharging cycle test is carried out at 0.1 C, the cycle process will be slow and 500 cycles may not be completed in one year, so a 1 C charge and discharge rate is selected for the cycle test. When the temperature characteristic test is performed by sampling the batteries from different batches to verify whether there is a larger performance difference between different batches of the batteries, a representative temperature value may be selected for testing. According to some issued standards, the representative temperature values may be −20°C, 20°C, and 55°C. However, in any case, once selected for the tests, the selected working conditions are fixed as the internal specification of the laboratory needs to obtain comparable test results over different periods.

3. How to fully charge the battery

When carrying out a power battery test, it is often necessary to fully charge the battery in order to test its maximum discharge performance. "Fully charged" can be defined as "not to be continuously charged otherwise damaging the battery" and abstractly as "a full reaction of all substances participating in the chemical reaction during charging." However, in practice, the above two definitions are impractical. Therefore, the "fully charged" is defined by electrical indicators: when the charge current approaches zero at a specified threshold voltage, the battery is deemed to have been fully charged. Since it is difficult to quantize "approaches zero," a smaller current value is specified for this purpose in many test standards; if the charge current is less than the specified value, the battery is deemed to have been fully charged. For example, some issued standards specify that if the charge current is less than 0.033 C, the battery is deemed to have been fully charged.

In order to save the charging time and ensure that the battery can be fully charged, the "two-stage method" can be generally used for charging the battery. In the first stage, the battery is charged at a constant current, the maximum charging current value stipulated by the manufacturer, until the voltage difference between the battery poles reaches the cut-off voltage. In the second stage, the battery is charged at a constant voltage, the upper limit of charging voltage specified by the manufacturer. At this time, the charging current will be automatically adjusted according to the change in the internal resistance of the battery. The second phase ends when the charge current value is lower than the above-mentioned "smaller current value."

4. How to fully discharge the battery

There is often a requirement to fully discharge the battery during the battery tests in order to evaluate the quantity of electric charge stored in the battery. Qualitatively, "fully discharged" means that the chemical reaction has been fully performed without over-discharging; in other words, all Li-ions in the crystal on the positive pole have disengaged from Li^+. Similarly, it is difficult to determine the above chemical state. Therefore, fully discharged is defined by the electrical indicators: the battery is discharged at a very small discharge current until the battery voltage reaches the lower limit of the voltage specified by the manufacturer. In practice, there is a requirement to specify the "very small discharge current;" for example, the battery is discharged when the small discharge current is 0.02 C during the specified tests.

5.1.4 Contents of Other Sections of This Chapter

Summary of other sections of this chapter:

1) Sections 5.2 to 5.4 provide common battery characteristic test items for the BMS development, together with their descriptions, test procedures, and typical test results. It is noted that due to the large number of battery characteristic test items, this chapter only lists some possible ones, and the R&D personnel can select or formulate more test methods and procedures according to specific work requirements.
2) Sections 5.5 and 5.6 provide evaluation methods for battery cycle characteristics. Section 5.5 describes how to obtain the aged battery using the deep charge and discharge cycle, while Section 5.6 describes how to evaluate the performance of the aged battery using the test items described in Sections 5.2 to 5.4 after a certain cycle period. Section 5.6 also provides the time estimated for the test items.
3) The characteristic tests in this chapter are for the $LiFePO_4$ battery.

5.2 Capacity and the Charge and Discharge Rate Test

This section describes two test items. First, test the actual battery capacity. The battery is sampled to test its actual capacity at different temperatures and different discharge rates for the purpose of comparison with the rated capacity stated by the battery manufacturer, thus providing a basis for SoC estimation, battery balancing management, battery charge and discharge energy control, etc. Second, test the charge and discharge rate at different temperatures and different discharge rates.

5.2.1 Test Methods

This section provides the test methods used by the laboratory to test the capacity and the charge and discharge rate during R&D. In practical work, the R&D teams can modify some parameters used in the tests, such as the temperature, the discharge rate, etc., as needed. The specific operation procedures are given below.

1) Respectively control the test temperature to $T = \{0°C, 20°C, 40°C\}$ and the charge and discharge rate to $r = \{0.2\,C, 0.5\,C, 1.0\,C\}$, and test the battery nine times at different temperatures and different discharge rates according to permutation and combination.
2) During the tests, maintain the constant ambient temperature around the tested battery by a suitable thermostat. During the tests, maintain the constant discharge current by an electronic load with a constant-current discharge function.
3) Perform each test according to the following steps:

---

a) Fully charge for the first time
Fully charge the battery by the charging method defined in Section 5.1.3 (3). The charge cut-off voltage is 3.6 V (or voltage recommended by the manufacturer).
b) Suspend for the first time
Maintain the battery in the state of no charge and no discharge until the difference between the temperature of the negative pole and the designated test temperature T is not more than 2°C.
c) Discharge
Discharge the battery at the discharge rate r until the battery voltage is 2.2 V (or voltage recommended by the manufacturer).
d) Suspend for the second time
Maintain the battery in the state of no charge and no discharge until the difference between the temperature of the negative pole and the designated test temperature T is not more than 2°C.
e) Fully charge for the second time
Fully charge the battery by the charging method defined in Section 5.1.3 (3). The charge cut-off voltage is 3.6 V (or voltage recommended by the manufacturer).

---

4) Record the voltage, current, and temperature measured every second.
5) Calculate the total discharge capacity using the following equation:

$$Q_d = \frac{1}{3600} \int_0^{t_d} I_d(\tau)d\tau \,(\text{Ah}) \qquad (5.1)$$

where $I_d(\tau)$ is the current measured in real time during the discharge and t_d is discharge time. Due to discharge at a constant current during the test, the total discharge capacity may also be calculated by the following equation:

$$Q_d = \frac{I_d t_d}{3600}(\text{Ah}) \qquad (5.2)$$

6) Calculate the total discharged energy using the following equation:

$$W_d = \int_0^{t_d} I_d(\tau)U_d(\tau)d\tau \,(\text{J}) \qquad (5.3)$$

where $U_d(\tau)$ and $I_d(\tau)$ are the voltage and current measured in real time during the discharge respectively and t_d is the discharge time. Due to discharge at a constant current during the test, the total discharged energy may also be calculated using the following equation:

$$W_d = I_d \int_0^{t_d} U_d(\tau) d\tau \, (\text{J})$$ (5.4)

7) Calculate the total charge capacity using the following equation:

$$W_c = \int_0^{t_c} I_c(\tau) U_c(\tau) d\tau \, (\text{J})$$ (5.5)

where $U_c(\tau)$ and $I_c(\tau)$ are the voltage and current measured in real time during the charge respectively and t_c is the time required for full charge for the second time.

8) Calculate the charge and discharge rate using the following equation:

$$\eta = W_d / W_c \times 100\%$$ (5.6)

5.2.2 Test Report Template

1) The test result of the battery at different temperatures and different discharge rates, such as the charge and discharge capacity and the charge and discharge rate, can be tabulated after test, as shown in Tables 5.1 and 5.2 (the test sample is a 100 Ah battery produced by a manufacturer).
2) As a byproduct of this test, the operating voltage curve of the battery at different discharge rates may be drawn, as shown in Figure 5.1.

Table 5.1 Discharge capacity of a battery at different temperatures and different discharge rates (Ah).

		Charge and discharge rate (C)		
		0.2	0.5	1
Temperature (°C)	0	100.49	97.61046536	101.9931171
	20	107.6312459	107.8673678	107.890404
	40	108.633324	108.3568886	107.804018

Table 5.2 Energy, charge and discharge rate of a battery at different temperatures and different discharge rates.

		Charge and discharge rate (C)								
		0.2			0.5			1		
		Wd (J)	Wc (J)	H (%)	Wd (J)	Wc (J)	H (%)	Wd (J)	Wc (J)	H (%)
Temperature (°C)	0	1947600	2088803	93.24	1864800	2156835	86.46	1900800	2280230	83.36
	20	2152800	2267537	94.94	2113200	2307994	91.56	2066400	2330438	88.67
	40	2181600	2274158	95.93	2131200	2303004	92.54	2084400	2318318	89.91

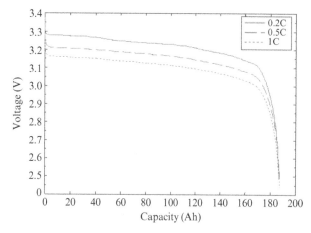

Figure 5.1 Discharge voltage and capacity relation curve of a single battery at different discharge rates.

5.3 Discharge Rate Characteristic Test

This section describes the test of the maximum discharge current and SoC relationship of the power battery at different temperatures. Generally, the higher the ambient temperature, the greater the activity of the material in the battery and the greater the current discharged in same SoC. At the same time, at the same temperature, the higher the SoC of the battery, the higher the electromotive force of the battery, the smaller the internal resistance, and the greater the maximum discharge current.

5.3.1 Test Method

1) Respectively set the test temperature to $T = \{0°C, 20°C, 40°C\}$.
2) During the tests, maintain the constant ambient temperature around the tested battery using a suitable thermostat. During the tests, the discharge current regularly drops with the decrease of SoC by the electronic load with a discharge current regulation function.
3) Perform each test according to the following steps:

> a) Set the initial discharge rate to r
>   The maximum discharge rate r used in this test is set according to the actual operating current peak of the electric vehicle or is recommended by the battery manufacturer.
> b) Fully charge the tested battery
>   Fully charge the battery using the charging method defined in Section 5.1.3 (3). The charge cut-off voltage is 3.6 V (or the voltage recommended by the manufacturer).
> c) Suspend
>   Maintain the battery in the state of no charge and no discharge until the difference between the temperature of the negative pole and the designated test temperature T is not more than 2°C.
> d) Discharge at a high discharge rate
>   Discharge the battery at a continuous current at the discharge rate r for 30 s. Enter step (f) if the battery voltage is less than 2.2 V (or the minimum voltage threshold value recommended by the manufacturer) within 30 s; otherwise enter step (e).

e) Suspend
   Maintain the battery in the state of no charge and no discharge until the difference between the temperature of the negative pole and the designated test temperature T is not more than 2°C, and return to step (d).
f) Suspend and reduce the discharge rate
   Maintain the battery in the state of no charge and no discharge until the difference between the temperature of the negative pole and the designated test temperature T is not more than 2°C, and set $r = r - 1$.
g) Determine whether another cycle is required
   If $r > 0$, return to step (d); otherwise, start step (h).
h) Fully discharge
   Fully discharge the battery by the method defined in Section 5.1.3 (4).
i) Record the voltage and current measured every second.

5.3.2 Test Report Template

1) The discharge rate characteristic of the battery at different temperatures can be obtained, as shown in Table 5.3.
2) The results of Table 5.3 are shown in Figure 5.2.

Table 5.3   Discharge rate characteristic of a 100 Ah battery at different temperatures.

| Charge accumulation (Ah) | Temperature (°C) | | | | | |
| | 0 | | 20 | | 40 | |
	Discharge rate (C)	Maximum current (A)	Discharge rate (C)	Maximum current (A)	Discharge rate (C)	Maximum current (A)
97.5	3	300	3	300	3	300
95	3	300	3	300	3	300
...	...	...	...	...	...	...
27.5	3	300	3	300	3	300
25	2	200	2	200	2	200
22.5	2	200	2	200	2	200
20	2	200	2	200	2	200
17.5	2	200	2	200	2	200
15	2	200	2	200	2	200
12.5	2	200	2	200	2	200
10	1	100	2	200	2	200
7.5	–	–	1	100	2	200
5	–	–	–	–	1	100
2.5	–	–	–	–	1	100

Note: the initial value of r used in the test is 3 (C).

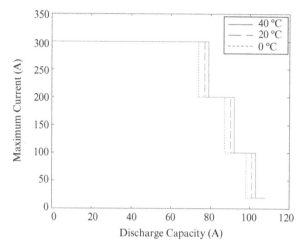

Figure 5.2 Discharge rate characteristic curve of a 100 Ah battery at different temperatures.

5.4 Charge and Discharge Equilibrium Potential Curves and Equivalent Internal Resistance Tests

This section describes the electromotive force curve test and the equivalent internal resistance test, namely measurement of the electromotive force value and the equivalent internal resistance spectrum of the battery at different temperatures and different SoC values. The two test items are combined to save test time and cost.

Since the open-circuit voltage of the power battery is hysteretic and the equivalent internal resistance is different during charging and discharging, the tests are respectively performed in the process of charging and discharging.

5.4.1 Test Method for Discharge Electromotive Force Curve and Equivalent Internal Resistance

1) Respectively set the test temperature to $T = \{0°C, 20°C, 40°C\}$.
2) During the tests, maintain the constant ambient temperature around the tested battery using a suitable thermostat. During the tests, maintain the constant discharge current using an electronic load with a constant-current discharge function.
3) Perform each test according to the following steps:

> a) Fully charge
>   Fully charge the battery using the charging method defined in Section 5.1.3 (3). The charge cut-off voltage is 3.6 V (or voltage recommended by the manufacturer).
> b) Suspend
>   Maintain the battery in the state of no charge and no discharge until the difference between the temperature of the negative pole and the designated test temperature T is not more than 2°C.
> c) Discharge at a large current
>   Discharge the battery at a continuous current at a discharge rate of 0.5 C for 300 s. Enter step (e) if the battery voltage is less than 2.2 V (or a minimum voltage threshold value recommended by the manufacturer) within 300 s; otherwise enter step (d).

> d) Suspend
> Maintain the battery in the state of no charge and no discharge for 3600 s and return to step (c).
> e) Discharge at a trickle current
> Discharge the battery at a continuous current at a discharge rate of 0.01 C for 300 s. Enter step (g) if the battery voltage is less than 2.2 V (or a minimum voltage threshold value recommended by the manufacturer) within 300 s; otherwise enter step (f).
> f) Suspend
> Maintain the battery in the state of no charge and no discharge for 600 s and return to step (e).
> g) End the test.

4) Record the voltage and current measured every second.
5) Record the discharge voltage at the last second of step (c) (e.g. 300th, 4200th, 8100th, ... second) as the discharge operating voltage U_{dn} ($n = 1, 2, 3, ...$), the discharge voltage of the last second of step (d) (e.g. 3900th, 7800th, 11700th, ... second) as the discharge open-circuit voltage U_{ocvdn} ($n = 1, 2, 3, ...$). Calculate the equivalent discharge resistance of the battery using the following equation:

$$R_{idn} = \left(U_{ocvdn} - U_{dn}\right)/I_d \ (\Omega)$$ (5.7)

6) Calculate the discharge capacity at the operating voltage U_{dn} ($n = 1, 2, 3, ...$) using the following equation:

$$Q_{dn} = \frac{1}{3600}\int_0^{t_{dn}} I_{dn}(\tau)d\tau$$ (5.8)

where $I_{dn}(\tau)$ is the current measured during the discharge and t_{dn} is the discharge duration. Due to discharge at a constant current during the test, the discharge capacity may also be calculated using the following equation:

$$Q_{dn} = \frac{I_{dn}t_{dn}}{3600}(Ah)$$ (5.9)

7) Draw the discharge equilibrium potential curve.
Use the discharge capacity as the horizontal axis and the voltage as the vertical axis to draw the discharge equilibrium potential curve. Use the discharge capacity as the horizontal axis and the internal resistance as the vertical axis to draw the discharge equivalent internal resistance spectrum curve.

5.4.2 Test Method for Charge Electromotive Force Curve and Equivalent Internal Resistance

1) Respectively set the test temperature to $T = \{0°C, 20°C, 40°C\}$.
2) During the tests, maintain the constant ambient temperature around the tested battery using a suitable thermostat. During the tests, maintain the constant charge current using a charger with a constant-current charge function.
3) Perform each test according to the following steps:

> a) Fully discharge
> Discharge the battery by the method defined in Section 5.1.3 (4). The discharge cut-off voltage is 2.2 V (or voltage recommended by the manufacturer).

b) Suspend

Maintain the battery in the state of no charge and no discharge until the difference between the temperature of the negative pole and the designated test temperature T is not more than 2°C.

c) Charge at large current

Charge the battery at a continuous current at a discharge rate of 0.5 C for 300 s. Enter step (e) if the battery voltage is more than 3.6 V (or a minimum voltage threshold value recommended by the manufacturer) within 300 s; otherwise enter step (d).

d) Suspend

Maintain the battery in a state of no charge and no discharge for 3600 s and return to step (c).

e) Fully charge at a low current

Charge the battery at a continuous current at a discharge rate of 0.01 C for 300 s. Enter step (g) if the battery voltage is more than 3.6 V (or a minimum voltage threshold value recommended by the manufacturer) within 300 s; otherwise enter step (f).

f) Suspend

Maintain the battery in a state of no charge and no discharge for 600 s and return to step (e).

g) End the test.

4) Record the voltage and current measured every second.

5) Record the charge voltage at the last second of step (c) (e.g. 300th, 4200th, 8100th, ... second) as the charging operating voltage U_{cn} ($n = 1, 2, 3, ...$), the charge voltage of the last second of step (d) (e.g. 3900th, 7800th, 11700th, ... second) as the charge open-circuit voltage U_{ocvcn} ($n = 1, 2, 3, ...$). Calculate the equivalent charge resistance of the battery using the following equation:

$$R_{icn} = (U_{ocvcn} - U_{cn}) / I_c \ (\Omega) \tag{5.10}$$

6) Calculate the charge capacity at the operating voltage U_{cn} ($n = 1,2,3, ...$) using the following equation:

$$Q_{cn} = \frac{1}{3600} \int_0^{t_{cn}} I_{cn}(\tau) d\tau \tag{5.11}$$

where $I_{cn}(\tau)$ is the current in real time measured during the charge and t_{cn} is the charge duration. Due to the charge at a constant current during the test, the charge capacity may also be calculated using the following equation:

$$Q_{cn} = \frac{I_{cn} t_{cn}}{3600} \ (Ah) \tag{5.12}$$

7) Draw a charge equilibrium potential curve.

Use the charge capacity as the horizontal axis and the voltage as the vertical axis to draw the charge equilibrium potential curve. Use the charge capacity as the horizontal axis and the internal resistance as the vertical axis to draw the charge equivalent internal resistance spectrum curve.

5.4.3 Discussion of the Test Method

It is worth discussing some details of this test. The first is about whether different discharge rates have an effect on the measurement of the electromotive force. In many experiments, the electromotive force is often measured by "titration;" in other words, the battery is discharged at a very small discharge rate. However, the measurement period is longer, so the discharge at a small discharge rate is not applicable to testing different batches.

Furthermore, for a theory of electrochemistry, without damage to the battery, the electromotive force of the battery is only related to the SoC and temperature of the battery, not to the discharge rate. Therefore, it is possible to discharge the battery at any discharge rate within the safe discharge rate range. The discharge rate of 0.5 C is recommended herein for heat dissipation. In the process of discharge, the battery may heat up, thus affecting the test temperature conditions. According to the actual work experience, during discharge at the discharge rate of 0.5 C, the heat generated in the battery test can be quickly removed, so that the battery temperature is close to the set test ambient temperature.

The second is about standing time. This test is designed to obtain the electromotive force of the battery. By definition, the electromotive force is measured when the charge carried in all polarized capacitors is released. In the operation, the electromotive force may be defined as follows: if the open-circuit voltage of the battery does not change within one hour, the charge in the polarized capacitors is considered to have been released. At this time, the open-circuit voltage is equal to the electromotive force of the battery and the next test can be started. According to an actual work experience, as the standing time is more than 2 hours during such an operation, a 1 hour standing time is recommended. However, a 1 hour standing time is not really enough, as the measured open-circuit voltage is not equal to the electromotive force. However, after careful analysis of the electrochemical microscopic mechanism of the battery, it can be found that, in fact, the voltage rebound of the battery can be equivalent to a three-order resistance capacitance network in the standing process. Therefore, the electromotive force of the battery can be obtained by fitting in mathematical algorithms according to the voltage data within 1 hour in order to save test time.

5.4.4 Test Report Template

Figure 5.3(a) to (d) shows the charge and discharge equilibrium potential curve and equivalent internal resistance curve of a 100 Ah power battery produced by a manufacturer at different temperatures.

5.5 Battery Cycle Test

5.5.1 Features of Battery Cycle Test

The cycle test is designed to test the performance and characteristics of the battery in the life cycle by many charge and discharge cycles. The cycle test is a relatively special test in all battery tests for the following causes:

1. Longer test time

Only one or two deep charge and discharge cycles are required for the above-mentioned test items, and thousands of deep charge and discharge cycles are required for the power battery cycle test. Typically, it may take several months or more than one year to test one cycle test item. Therefore, high-stability test equipment and a high-reliability test program are required for this test.

2. Irreversibility of each test sample

Ideally, the characteristics of the power battery in charge and discharge cycles can be obtained at different temperatures. After 500 cycles in one battery sample at 20°C, for example, it may be necessary to replace the the battery sample to obtain the characteristics of the battery at 0°C or 40°C.

3. Selection of the discharge rate

It is often hoped that the BMS developer will obtain cyclic characteristics of the battery at different discharge rates. However, it is impractical to select a smaller discharge rate. For example, if the discharge rate of 0.1 C is selected for the charge and discharge cycle test, it takes at least 10 hours to complete the charge and discharge

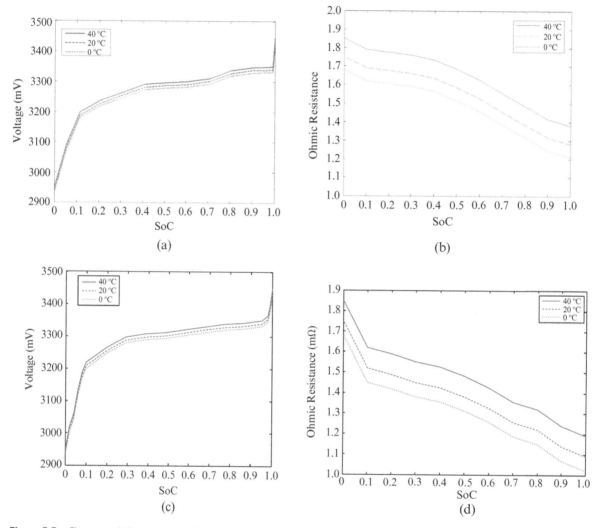

Figure 5.3 Charge and discharge equilibrium potential curves and equivalent internal resistance curve of a power battery at different temperatures: (a) discharge equilibrium potential curve; (b) discharge equivalent internal resistance curve; (c) charge equilibrium potential curve; (d) charge equivalent internal resistance curve.

process, and almost 1 day to complete one charge and discharge cycle, so less than 500 cycles are completed in one year (365 days). Such a cycle test is insignificant. Therefore, in the actual operation process, the cycle test is often performed using two methods: a fixed rate method and a working condition method, which are described below.

5.5.2 Fixed Rate Cycle Test Method

This section describes the cycle test of the power battery at a fixed rate. The charge and discharge cycle tests are performed for the battery sample at a constant temperature and fixed rate in order to observe the general characteristics of the battery in the cycle and to draw the capacity change characteristic, internal resistance change characteristic, and charge efficiency change characteristic curves. These test results provide a basis for the SoC estimation, battery balancing management, battery charge and discharge energy control, etc.

1) Set the temperature of each cycle test to $T = 20°C$.
2) During the tests, maintain the constant ambient temperature around the tested battery using a suitable thermostat. During the tests, maintain the constant charge and discharge current using a switching power supply with a charging function and an electronic load with a constant-current discharge function or a special cycle test instrument with a constant-current charge and discharge function.
3) Perform the cycle test according to the following steps:

> a) Set the number of test cycles.
> b) Charge
>   Fully charge the battery using the charging method defined in Section 5.1.3 (3). Maintain a 1 C charge rate during the charge at a constant current and a 3.6 V charge cut-off voltage (or a voltage recommended by the manufacturer).
> c) Suspend
>   Maintain the battery in a state of no charge and no discharge until the difference between the temperature of the negative pole and the designated test temperature T is not more than 2°C.
> d) Discharge
>   Discharge the battery at a continuous current at a discharge rate of 1 C and stop the discharge when the battery voltage is 2.2 V (or a value recommended by the manufacturer).
> e) Suspend
>   Maintain the battery in a state of no charge and no discharge until the difference between the temperature of the negative pole and the designated test temperature T is not more than 2°C. The cycle number $n = n - 1$.
> f) Determine whether the test is continuously performed
>   If $n > 0$, return to step (d); otherwise end the test.

4) Record the voltage, current, and temperature measured every second.
5) Calculate the total charge energy of the cycle n ($n = 1, 2, 3, ...$) by the following equation:

$$W_{cn} = \int_0^{t_{cn}} I_{cn}(\tau) U_{cn}(\tau) d\tau \, (\text{J}) \tag{5.13}$$

where $U_{cn}(\tau)$ and $I_{cn}(\tau)$ are respectively the real time voltage and current measured during the charge and t_{cn} is the time required for a full charge for n times.

6) Calculate the total discharge capacity of the cycle n ($n = 1, 2, 3, ...$) by the following equation:

$$Q_{dn} = \frac{1}{3600} \int_0^{t_{dn}} I_{dn}(\tau) d\tau \, (\text{Ah}) \tag{5.14}$$

where $I_{dn}(\tau)$ is the real time current measured during the discharge and t_{dn} is the discharge time. Due to discharge at a constant current during the test, the discharge capacity may also be calculated using the following equation:

$$Q_{dn} = \frac{I_{dn} t_{dn}}{3600} \, (\text{Ah}) \tag{5.15}$$

7) Calculate the total discharged energy of the cycle n ($n = 1, 2, 3, ...$) using the following equation:

$$W_{dn} = \int_0^{t_{dn}} I_{dn}(\tau) U_{dn}(\tau) d\tau \, (\text{J})$$ (5.16)

where $U_{dn}(\tau)$ and $I_{dn}(\tau)$ are respectively the real time voltage and current measured during the discharge and t_{dn} is the discharge time. Due to discharge at a constant current during the test, the total discharged energy may also be calculated using the following equation:

$$W_{dn} = I_{dn} \int_0^{t_{dn}} U_{dn}(\tau) d\tau \, (\text{J})$$ (5.17)

8) Calculate the charge efficiency of the cycle n ($n = 1, 2, 3, ...$) using the following equation:

$$\eta_n = W_{dn} / W_{cn} \times 100\%$$ (5.18)

5.5.3 Cycle Test Schemes Based on Standard Working Conditions

This section describes the standard working condition cycle test of the power battery. The charge and discharge cycle tests are performed for the battery sample at a constant temperature and under standard working conditions in order to observe the operation characteristics of the battery and to draw the capacity change characteristic, internal resistance change characteristic, and charge efficiency change and capacity change characteristic curves. These test results will provide a basis for the SoC estimation, battery balancing management, battery charge and discharge energy control, etc.

1. Differences and similarities between the standard working condition test and the fixed rate test

As described in Sections 5.2 to 5.4, the power battery has the discharge rate characteristic and the internal resistance characteristic. During the actual operation of the electric vehicle, the battery performance is limited and is affected by these characteristics. These characteristics should be tested using the standard working condition cycle to verify whether the battery meets the actual operation requirement of the electric vehicle. In addition, if the cycle test is conducted at fixed charge and discharge rates, it is impossible to know the effect of current fluctuations on the battery life during actual operation of the electric vehicle, because, in practice, after a full charge, the battery is discharged at a variable discharge rate rather than the constant discharge rate that occurs during operation of the electric vehicle (a short-time charge process is sometimes caused by the braking energy recovery).

This is like the oil consumption test performed for an orthodox car at a constant speed and under working conditions, the cycle test is also performed for the battery at a fixed discharge rate. The fixed rate cycle test corresponds to the oil consumption test at a constant speed, while the working condition cycle test corresponds to the oil consumption test under specific working conditions. The results obtained from the two test methods are different and have different reference significance.

In summary, the standard working condition cycle test described in this section has similarities and differences from the cycle test at a fixed charge rate, described in Section 5.5.2. Their similarities are given below.

First, they have basically the same charging process. During the standard working condition cycle test, although on the premise that working conditions change, the current is changed upon the discharge, while the battery is charged basically at "constant current – constant voltage" in two stages. This charging process is completely the same as that used in the fixed rate cycle test and is consistent with the actual charging process of the electric vehicle. The battery is charged in two stages to ensure that the battery is fully charged and both the charge and the discharge depth reach their ultimate value in each cycle while exhibiting the best performance of the battery in the cycle.

Second, they both limit the battery operating temperature during tests. As mentioned above, the temperature has a major effect on the operating characteristic of the power battery, so the fixed rate and working condition cycle tests must be performed at a certain temperature. The test results vary with the temperature.

The above described the similarities while their differences are given as below:

First, during the standard working condition cycle test of the battery, a short-time charging process may be added to simulate the braking energy recovery during operation of the electric vehicle. At present, the motors used in most of the electric vehicles have a braking energy recovery function. In other words, in the process of change from a high speed to a medium speed, some kinetic energy will be converted into electric energy, which is equivalent to addition of a short-time charging process upon the discharge in the whole cycle. During the fixed rate cycle test, it is impossible to add the charging process to the discharge.

Second, during the fixed rate cycle test, the (charge) discharge per second is fixed and unchanged; during the standard working condition cycle test, the (charge) discharge per second is determined by the power spectrum and may be the same or different from the current of the next second. Therefore, two cycle test methods have different equipment requirements. For the fixed rate cycle test, the equipment is only required to supply a constant charge and a discharge current, and the current direction is changed only by the control end. For the working condition cycle test, the equipment is required to support the charge and discharge at variable currents. Generally, a response time is required for the equipment to respond to the control instruction, the current change range cannot be too large, and the switching frequency cannot be too high. Generally, it is feasible to choose 1 second as the execution cycle of the control instruction.

2. Working condition of the car and battery

The driving cycle of the car refers to the "time-speed" course of a certain car type according to the traffic feature of a region, which is established by data analysis after investigating the actual driving cycle of the car. At present, a typical driving cycle includes a USA FTP 75 driving cycle, a European ECE+EUDC driving cycle, and a Japan 10.15 driving cycle. The GB/T 18386-2005 driving cycle is used for some tests in China.

As an example, a typical GB/T 18386-2005 driving cycle is shown in Figure 5.4. Each GB/T 18386-2005 driving cycle consists of four basic urban cycles and one basic suburb cycle. As shown in the figure, the typical GB/T 18386-2005 driving cycle specifies the relationship between time and speed. The relationship between time and displacement and the relationship between time and acceleration can be calculated. Therefore, the driving cycle can be discretized to generate a table, as shown in Table 5.4.

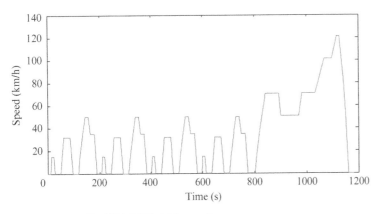

Figure 5.4 GB/T 18386-2005 driving cycle.

Table 5.4 Discretized GB/T 18386-2005 driving cycle.

Time (s)	Acceleration (m/s^2)	Speed (m/s)	Unit displacement (m)
1	0	0	0
2	0	0	0
...	...	...	...
13	1.04	1.04	1.5808
14	1.04	2.08	4.2016
15	1.04	3.12	7.8624
16	1.04	4.16	12.563
...	...	...	...
1156	−1.39	7.35	11195
1157	−1.39	5.96	11202
1158	−1.39	4.57	11208
1159	−1.39	3.18	11212
1160	−1.39	1.79	11215
1161	−1.39	0.4	11216
...	...	...	...

However, all of the above cycles are mechanical working conditions, while an electrical working condition is required for the actual battery test. In other words, the output power per second, rather than the speed per second of the car, is required for the battery test. In short, the driving cycle corresponds to the "road spectrum" while the "power spectrum" is required for the battery test. It is noted that the "power spectrum" strictly corresponds not to the isolated voltage and current information per second, nor to the consumed power of the battery, but to the "output power" of the battery. Since each battery inevitably has an internal resistance, for vehicles the "road spectrum" corresponds to the external output power of the battery, excluding the power consumed by the internal resistance of the battery.

The electrical working condition is obtained by two methods.

First, the electrical working condition is generated by the standard working condition of the car according to the actual parameters of the electric vehicle. For example, if the total weight, wind resistance coefficient, tire friction coefficient, mechanical transmission loss, motor efficiency, etc. have been determined for the electric vehicle, the output power per second of the battery pack can be calculated according to the standard working condition. If such power is divided by the number of batteries, the approximate output power spectrum required for a single battery test can be obtained. Table 5.5 shows the power spectrum of a single power battery generated according to the GB/T 18386-2005 driving cycle of a car type, where the operating current is calculated according to the actual voltage at a certain time.

It is emphasized again that the output power per second is controlled during the above test. Since the voltage is uncertain during the actual operation of the battery, it is necessary to calculate the working current per second based on the real-time measured voltage.

Second, this method is directly used for a known electrical standard working condition for the battery cycle test. The IEC 61982-3-2001 standard provides the electrical working condition for the battery cycle test, as shown in Table 5.6.

Table 5.5 Power spectrum of a single power battery generated according to the GB/T 18386-2005 driving cycle of a car type.

Time (s)	Power (W)	Voltage (V)	Current (A)
1	0	3.345	0
2	0	3.345	0
...	...	...	...
13	14.819	3.345	4.430194
14	29.639	3.33	8.900601
15	44.462	3.328	13.35998
16	59.29	3.327	17.82086
...	...	...	...
1156	−110.85	3.367	−32.9225
1157	−89.911	3.34	−26.9195
1158	−68.957	3.338	−20.6582
1159	−47.991	3.328	−14.4204
1160	−27.017	3.325	−8.12541
1161	−6.0375	3.322	−1.81743
...	...	...	...

Note: the voltage and current provided in the table are the possible data recorded during the tests and are not universal.

Table 5.6 Electrical working condition for the battery cycle test of the IEC 61982-3-2001 standard.

Step	Duration (s)	Power of battery pack (%)	Power of battery pack (kW)	Power of single battery (kW)
1	16	0	0	0
2	28	−12.5	−6	−0.06
3	12	−25	−12	−0.12
4	8	+12.5	6	0.06
5	16	0	0	0
6	24	−12.5	−6	−0.06
7	12	−25	−12	−0.12
8	8	+12.5	6	0.06
9	16	0	0	0
10	24	−12.5	−6	−0.06
11	12	−25	−12	−0.12
12	8	+12.5	6	0.06
13	16	0	0	0
14	36	−12.5	−6	−0.06
15	8	−100	−48	−0.48
16	24	−62.5	−30	−0.3
17	8	+25	12	0.12
18	32	−25	−12	−0.12
19	8	+50	24	0.24
20	44	0	0	0

3. Specific test operations

It should be noted that the standard working condition cycle test is almost identical to the fixed rate cycle test in terms of test procedures and data processing methods. For example, the temperature condition of the cycle test (such as $T = 20°C$) should be selected, and the charge, energy, charge efficiency as well as discharge efficiency of each cycle should be recorded.

However, the standard working condition cycle test has the following two differences from the fixed rate cycle test:

First, in step (d), the fully charged battery is discharged at a constant discharge current until the battery voltage reaches the lower limit during the fixed rate cycle test and the discharge current changed according to a test requirement during the standard working condition cycle test.

For clarity, the steps of the standard working condition cycle test of LiFePO$_4$ batteries are detailed as follows:

---

a) Set the number of the test cycles.
   Set the number of the test cycle to *n*.
b) Fully charge for the first time.
   Fully charge the battery using the charging method defined in Section 5.1.3 (3). The large cut-off voltage is 3.6 V (or the voltage recommended by the manufacturer).
c) Suspend
   Maintain the battery in the state of no charge and no discharge until the difference between the temperature of the negative pole and the designated test temperature *T* is not more than 2°C.
d) Discharge
   Discharge the battery at the operating current meeting the standard working condition. The discharge power per second is selected from the "list of predefined standards working conditions." If the braking energy recovery test is performed for the battery, the battery is charged when the discharge current shown in the table is negative. The battery is repeatedly discharged from the first to last discharge power shown in the "list of working conditions" until the battery voltage is 2.2 V (or the value recommended by the manufacturer).
e) Suspend
   Maintain the battery in the state of no charge and no discharge until the difference between the temperature of the negative pole and the designated test temperature T is not more than 2°C. The cycle number *n* = *n* – 1.
   Determine whether the test is continuously performed.
f) If *n* > 0, return to step (d); otherwise end the test.

Second, during data processing, in addition to the calculation based on the equation described in Section 5.5.2, two calculation processes should be added: (1) according to the duration of step (d), calculate the number of standard working conditions to be completed after the full charge of the battery. If the duration of a standard working condition is T_d and the duration of step (d) is t_{dn} in the cycle *n*, the number of completed cycles under the standard working condition is calculated by the following equation:

$$C_n = t_{dn} / T_d \tag{5.19}$$

Finally, calculation of the equivalent SoR of the cycle *n* may be made using either of the following two equations (where L is the equivalent SoR under the standard working condition in km):

$$R_n = LC_n \tag{5.20}$$

$$R_n = Lt_{dn} / T_d \tag{5.21}$$

Table 5.7 Energy charged and discharged and efficiency of the battery cycle test.

Cycle number	Capacity (Ah)	Discharge W_d (J)	Charge W_c (J)	Efficiency η (%)
1	105.93	1225400	1222800	99.05
2	104.6	1211200	1221700	98.91
3	104.33	1203400	1217200	98.71
...	...	...	...	...

5.5.4 Test Report Template

1. Cycle test data table

Table 5.7 shows the data obtained from the fixed rate cycle test, including the cycle number, discharge capacity, discharged energy, charged energy, and energy charged and discharged efficiency.

It should be noted that the data of the test report template given in Table 5.7 are obtained from the fixed rate cycle test of a certain brand of battery. In fact, the format of the test report for the standard working condition cycle test is exactly the same as that in Table 5.7, except that the test result data obtained by the same power battery are different due to the different cycle conditions.

2. Curve obtained from the fixed rate cycle test

Figure 5.5 shows the capacity curve obtained from 600 cycles of a 100 Ah battery at 0.3 C. The horizontal axis represents the cycle number and the vertical axis is the maximum discharge capacity of the battery in each cycle. As shown in the figure, the battery capacity declines monotonously with the increase in the cycle number. In addition, it can be seen from the figure that the capacity loss rate of the battery in the first 200 cycles is greater than that after 200 cycles.

Figure 5.6 corresponds to Figure 5.5 and shows the charge and discharge efficiency of the power battery in each cycle. As shown in the figure, battery capacity also decreases monotonously mainly for a continuous increase of the ohmic resistance in the battery due to battery aging.

Figure 5.5 Capacity curve obtained from the battery cycle test.

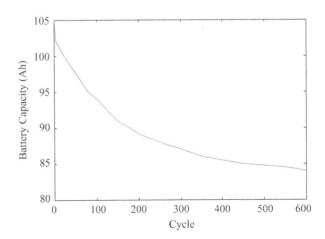

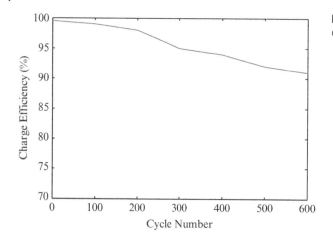

Figure 5.6 Charge and discharge efficiency curve obtained from the battery cycle test.

3. **Curve obtained from the standard working condition cycle test**

Figure 5.7 shows the result of the GB/T 18386-2005 working condition cycle test of the battery. Figure 5.7(a) shows the number of working conditions completed for the battery in each cycle and Figure 5.7(b) shows the equivalent SoR of the electric vehicle converted according to the working conditions.

5.6 Phased Evaluation of the Cycle Process

Section 5.5 described how to perform an aging cycle for the battery at a certain temperature and current. Each cycle requires testing of the charge and discharge capacity, energy, and efficiency of the battery. However, it is impossible to obtain all performance characteristics of the battery using these tests. Therefore, after each phase of the cycle test, the test is suspended to evaluate the current state of the battery and identify the performance and characteristics after this stage of the cycle.

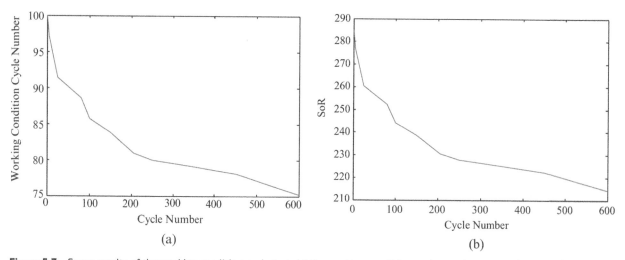

Figure 5.7 Some results of the working condition cycle test: (a) the working condition cycle number curve of each cycle; (b) the SoR curve of each cycle.

5.6.1 Evaluation Method

This section discusses some detailed problems on the phased evaluation in the battery cycle.

1. How to divide the cycle phase

The phased evaluation is designed to understand the characteristic changes of the battery after the aging cycle. How often is the phased evaluation made? When the phased evaluation period is determined, consideration is given to the following two aspects. On the one hand, in the case of a low frequency and long evaluation interval (for example, once every 500 cycles), it is possible to cause data loss and fail to fully understand the characteristics of the battery at each stage of the aging cycle test [10]. On the other hand, in case of a short evaluation interval and high frequency (such as once every 20 cycles), it is possible to affect the effect of the aging cycle test. Therefore, the phased evaluation period is determined reasonably according to the battery use cycle number provided by the different battery manufacturers. For example, if 1000 use cycles are provided by the manufacturer, the phased evaluation should be made once every 100 cycles during the aging cycle test. Similarly, if 500 use cycles are provided by the manufacturer, the phased evaluation should be made once every 50 cycles during the aging cycle test.

2. Test items of phased evaluation

The phased evaluation is designed to understand the general characteristic of the battery after the aging cycle. In order to minimize the impact of the phased evaluation on the aging cycle test, the test items should be selected as carefully as possible, and there should not be too many selected test items. Three test items described in Sections 5.2 to 5.4 cover all the basic characteristics of the battery, such as the battery capacity, the charge and discharge efficiency, the discharge rate characteristic, the equilibrium potential curve, and the equivalent internal resistance. Therefore, the phased evaluation may be conducted as described in the three sections.

3. Selection of test parameters for a phased evaluation

However, in the test items described in Sections 5.2 to 5.4, the test parameters, such as the temperature and the charge and discharge rate, are not strictly limited. If the tests are conducted respectively at different temperatures and different charge and discharge rates, excessive tests may be required, which may affect the effect of the aging cycle test. Furthermore, in order to ensure the comparability of the phased evaluation, it is necessary to limit the test parameters for each evaluation, and a similar evaluation should be conducted at the same temperature and the same charge and discharge rate.

For example, the following provision may be established: for the battery capacity and charge–discharge rate test specified in Section 5.2, take $T = 20°C$ and the charge–discharge rate as $r = 1.0$ C; for the discharge rate characteristic test specified in Section 5.3, take $T = 20°C$; and for the test specified in Section 5.4, the prescribed charge–discharge equilibrium potential curve and equivalent internal resistance test, take $T = 20°C$ and the charge–discharge rate $r = 1.0$ C.

5.6.2 Estimation of the Test Time

The cycle test is time consuming and is typically conducted through a programmable automatic control system. The estimation of the test time is necessary to help the tester predict when the results of the cycle tests will be obtained. As mentioned above, in the testing process, the cycle phase should be properly planned. Take the fixed rate cycle as an example; if $n = 100$ is selected as a cycle period, the operations provided in Table 5.8 should be conducted in each cycle.

In this table, although the phased evaluation item 3 includes two test items (charge test and discharge test), two test items only constitute one cycle. The operations described in the table are applicable to the fixed rate cycle. The time required for each step of the tests is estimated as follows:

Table 5.8 Cycle test operation description.

Cycle number	Operation type	Operation
1	Phased evaluation item 1	Battery capacity and charge and discharge efficiency test
2	Phased evaluation item 2	Discharge rate characteristic test
3	Phased evaluation item 3	Charge electromotive force and equivalent internal resistance tests and discharge electromotive force and equivalent internal resistance tests
4∼100	Phased evaluation of 97 cycles	Fixed rate cycle test
101	Phased evaluation item 1	Battery capacity and charge and discharge efficiency test
102	Phased evaluation item 2	Discharge rate characteristic test
103	Phased evaluation item 3	Charge electromotive force and equivalent internal resistance tests and discharge electromotive force and equivalent internal resistance tests
104∼200	Phased evaluation of 97 cycles	Fixed rate cycle test
201	Phased evaluation item 1	Battery capacity and charge and discharge efficiency test
202	Phased evaluation item 2	Discharge rate characteristic test
203	Phased evaluation item 3	Charge electromotive force and equivalent internal resistance tests and discharge electromotive force and equivalent internal resistance tests
204∼300	Phased evaluation of 97 cycles	Fixed rate cycle test
...	...	...

Table 5.9 Time estimation of the battery capacity and charge and discharge efficiency test.

Operation	Estimated time (h)	Note
Fully charge for the first time	2	1 h for constant current,1 h for constant voltage, similarly hereinafter
Suspend for the first time	0.5	0.5 h temperature recovery time according to experience, similarly hereinafter
Discharge	1	
Suspend for the second time	0.5	
Fully charged for the second time	2	

Subtotal: 6 h

1) Time estimation of the battery capacity and charge and the discharge efficiency test (according to Section 5.2.1, the charge and discharge rate r is 1.0 C) (see Table 5.9).
2) Discharge rate characteristic test (according to Section 5.3.1, the maximum discharge rate is 3 C) (see Table 5.10).
3) Time estimation of the charge electromotive force and equivalent internal resistance tests (according to Section 5.4.2, the charge and discharge rate r is 1.0 C) (see Table 5.11).
4) Time estimation of the discharge electromotive force and equivalent internal resistance tests (according to Section 5.4.1, the charge and discharge rate r is 1.0 C) (see Table 5.12).
5) Time estimation of the fixed rate cycle test (according to Section 5.5.2, the charge and discharge rate r is 1.0 C) (see Table 5.13).
6) Estimation of the total time required for 100 fixed rate cycle tests (including a phased evaluation) (see Table 5.14).

Table 5.10 Time estimation of the discharge rate characteristic test.

Operation	Estimated time (h)	Note
Fully charge for the first time	2	
Suspend for the first time	0.5	
Discharge at 3 C	0.66	0.33×2 (stop for 30 s every 30 s discharge)
Suspend for the second time	0.5	
Discharge at 3 C	1	0.5×2
Suspend for the third time	0.5	
Discharge at 1 C	2	1.0×2
Suspend for the fourth time	0.5	
Fully discharge	1	

Subtotal: 8.66 h

Table 5.11 Time estimation of the charge electromotive force and equivalent internal resistance tests.

Operation	Estimated time (h)	Note
Fully charge for the first time	2	
Suspend for the first time	0.5	
Charge at large current	22	$2 + 20 \times 1$ $2 + 20 \times 1$ (0.5 C constant current, charge time 2 h, 1 h suspension per charge, and 20 charge phases (empirical value))
Charge at small current	5	0.25×20 0.25×20 (0.01 C constant current, 300 s charge and 600 s suspension per stage with 0.25 h time, 20 charge stages (empirical value))

Subtotal: 29.5 h

Table 5.12 Time estimation of the discharge electromotive force and the equivalent internal resistance tests.

Operation	Estimated time (h)	Note
Fully charge for the first time	2	
Suspend for the first time	0.5	
Discharge at large current	22	$2 + 20 \times 1$ $2 + 20 \times 1$ (0.5 C constant current, discharge time 2 h, 1 h suspension per charge, and 20 charge phases (empirical value))
Discharge at small current	0.25×20	0.25×20 0.25×20 (0.01 C constant current, 300 s discharge and 600 s suspension per stage with 0.25 h time, and 20 charge stages (empirical value))

Subtotal: 29.5 h

Table 5.13 Time estimation of the fixed rate cycle test.

Operation	Estimated time (h)	Note
Fully charge for the first time	2	
Suspend for the first time	0.5	0.5 h temperature recovery time according to experience, similarly hereinafter
Fully discharge for the first time	2	1 h for constant current,1 h for trickle current, similarly hereinafter
Suspend for the first time	0.5	

Subtotal: 5 h

Table 5.14 Time estimation of the 100 fixed rate cycle tests.

Operation sequence	Operation	Estimated time (h)	Note
1	Battery capacity and charge and discharge efficiency test	6	
2	Discharge rate characteristic test	8.66	
3	Charge and discharge electromotive force and equivalent internal resistance tests and discharge electromotive force and equivalent internal resistance tests	59	29.5×2
4 ~ 100	Fixed rate cycle test	485	5×97

Total: 558.66 h, approximately 23.3 days

Table 5.15 Energy (J) charged and discharged and efficiency diagnosis.

	W_d (J)	W_c (J)	η (%)
New battery	1105400	1116200	99.03
After 100 cycles	1024800	1048000	97.78
After 200 cycles	965630	999580	96.60
...	...	...	...

Table 5.16 Capacity (Ah) diagnosis.

	Capacity (Ah)
New battery	95.507
After 100 cycles	88.991
After 200 cycles	84.607
...	...

It takes 23.3 days to complete a phase of 100 cycle tests (including phased evaluation). The above estimation is applicable to the fixed rate cycle test. For the standard working condition cycle test, the time should be estimated according to the single-step time, which is not detailed herein due to limited space.

5.6.3 Test Report Template

After the phased evaluation, the performance degradation of the battery, such as efficiency, capacity, etc., can be tabulated as shown in Tables 5.15 and 5.16, which only indicate the results of three diagnoses respectively corresponding to the tests of a new battery and after 100 cycles and 200 cycles.

Figure 5.8 shows the battery capacity loss obtained by the diagnostic test of the battery capacity. The curve in the figure is obtained after fitting.

Figure 5.9 shows the result obtained by the diagnostic test of the battery charge and discharge efficiency. The curve in the figure is obtained after fitting.

Figure 5.10 shows the result obtained by the diagnostic test of the discharge rate characteristic.

Figure 5.11 shows the result obtained by the diagnostic test of the discharge equilibrium potential.

Figure 5.12 shows the result obtained by the diagnostic test of the equivalent internal resistance.

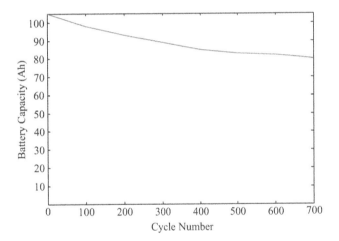

Figure 5.8 Diagnostic test curve of the discharge capacity (discharge rate 1.0 C, temperature $T = 20°C$).

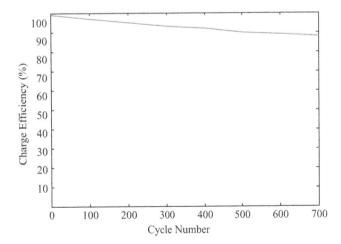

Figure 5.9 Diagnostic test curve of charge efficiency ($T = 20°C$).

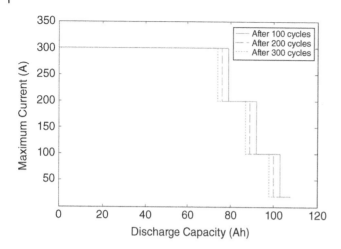

Figure 5.10 Diagnostic test curve of the discharge efficiency (T = 20°C).

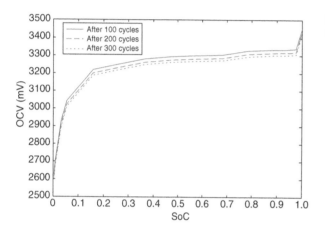

Figure 5.11 Diagnostic test curve of the discharge equilibrium potential (T = 20°C).

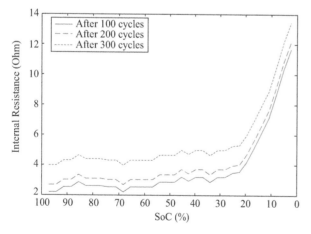

Figure 5.12 Diagnostic test curve of the internal resistance (T = 20°C).

References

1 Idaho National Laboratory, *Battery Test Manual for Electric Vehicles*, USA: Department of Energy, 2015.

2 Gabano, J. D., "Fractional identification algorithms applied to thermal parameter estimation," *IFAC Proceedings Volumes*, 2009, 42(10): 1316–1321.

3 Zhu, M., Hu, W., and Kar, N. C., "The SOH estimation of LiFePO$_4$ battery based on internal resistance with Grey Markov Chain," in *2016 IEEE Transportation Electrification Conference and Expo (ITEC)*, IEEE, 2016, pp. 1–6.

4 Yuan, H.-F. and Dung, L.-R., "Effect of external resistance on SOH measurement of LFP cells," in *2015 IEEE PES Asia-Pacific Power and Energy Engineering Conference (APPEEC)*, IEEE, 2015, pp. 1–5.

5 Bloom, I., Cole, B., Sohn, J., et al., "An accelerated calendar and cycle life study of Li-ion cells," *Journal of Power Sources*, 2001, 101(2): 238–247.

6 Wang, J., Liu, P., Hicks-Garner, J., et al., "Cycle-life model for graphite-LiFePO$_4$ cells," *Journal of Power Sources*, 2011, 196: 3942–3948.

7 Society of Automotive Engineers US, "*Recommended practice for performance rating of electric vehicle battery modules,*" SAE J1798-1997, 1997.

8 IEC 61982-3-2001, "*Secondary batteries for the propulsion of electric road vehicles – Part 3: Performance and life testing (traffic compatible, urban use vehicles).*"

9 US Department of Energy, "*PNGV battery test manual.*" DOE/ID-10597, Rev. 3-2001, Washington, DC: Department of Energy, 2001.

10 Belt, J., Utgikar, V., and Bloom, I., "Calendar and PHEV cycle life aging of high-energy, lithium-ion cells containing blended spinel and layered-oxide cathodes," *Journal of Power Sources*, 2011, 196(23): 10213–10221.

6

Test Results and Analysis

This chapter provides the characteristic test and degradation test results of the $LiFePO_4$ and $Li(NiCoMn)O_2$ power batteries widely used in electric vehicles and their contrasting analysis to provide a reference for BMS research and development.

6.1 Characteristic Test Results and Their Analysis

6.1.1 Actual Test Arrangement

After the sample of a new battery is obtained, the BMS developers are required to make a characteristic test for the battery sample to obtain enough parameters. The comparatively complete characteristic tests are given in Table 6.1. The test items mainly include the battery capacity, the equilibrium potential curve, and the equivalent internal resistance curve. In order to obtain the characteristics of a certain type of battery at different temperatures, a –10 to 40°C test ambient temperature range was selected and a set of data was collected every 10°C. In fact, the temperature range of the characteristic test should basically cover the actual operating temperature range of the battery. The temperature range from –10 to 40°C mentioned in this section is only applicable to the specific application of a certain battery type. In the actual work, the test temperature range can be selected according to different practical applications. The specific steps of the characteristic test have been discussed in detailed in Chapter 5 and will not be repeated in this chapter, which mainly provides the characteristics test results and their analysis. Since $LiFePO_4$ and $Li(NiCoMn)O_2$ batteries are widely used in engineering, they are tested and the parameters of the battery samples are shown in Tables 6.2 and 6.3.

   For the purpose of distinction, the $LiFePO_4$ cell test samples are numbered with A + Arabic numerals and the $Li(NiCoMn)O_2$ ternary cell test samples are numbered with B + Arabic numerals in this book. The tests include the characteristic test and degradation test described in this chapter. Furthermore, in order to avoid unreliable test

Table 6.1 Battery characteristic test scheme.

- Test items: battery capacity, equilibrium potential curve, and equivalent internal resistance curve
- Test ambient temperature: –10°C, 0°C, 10°C, 20°C, 30°C, and 40°C
- Test procedure: see Chapter 5
- Type of test sample: $LiFePO_4$ and $Li(NiCoMn)O_2$
- Parameters of the test sample: see Tables 6.2 and 6.3

Battery Management System and its Applications, First Edition. Xiaojun Tan, Andrea Vezzini, Yuqian Fan, Neeta Khare, You Xu, and Liangliang Wei.
© 2023 China Machine Press. All rights reserved. Published 2023 by John Wiley & Sons Singapore Pte. Ltd.

Table 6.2 Parameters of the LiFePO$_4$ power battery test sample.

Battery type	Nominal capacity	Nominal voltage	Charge cut-off voltage	Discharge cut-off voltage	Sample arrangement
LiFePO$_4$	15 Ah	3.2 V	3.7 V	2.2 V	A1, A2

Table 6.3 Parameter of the Li(NiCoMn)O$_2$ power battery test sample.

Battery type	Nominal capacity	Nominal voltage	Charge cut-off voltage	Discharge cut-off voltage	Sample arrangement
Li(NiCoMn)O$_2$	32 Ah	3.7 V	4.2 V	2.5 V	B1, B2

results caused by the poor consistency of the battery products and test equipment error, two independent cell samples are usually selected for respectively testing a certain test item.

It is worth mentioning that, in order to ensure data accuracy and eliminate the effect of such factors as insufficient activation of the new battery and the test equipment error, the cell samples are generally charged and discharged in multiple cycles during the capacity test. If the capacity deviation is not more than 2% in three consecutive cycles, the average of the last three test results is taken as the test result. In order to ensure data stability during equilibrium potential and equivalent internal resistance tests, the cells are placed for 2 hours after each charge and discharge cycle to provide enough time for the completion of the chemical reaction and achievement of a stable state in the tested cells, in order to avoid the mutual effect between the tests.

6.1.2 Characteristic Test Results of the LiFePO$_4$ Battery

According to the test arrangement described in the previous section, this section analyzes the characteristic test results of the LiFePO$_4$ battery and summarizes the general rules for the capacity, equilibrium potential curve, and equivalent internal resistance curve of the LiFePO$_4$ battery at different temperatures.

Capacity test

The capacity of the LiFePO$_4$ battery at different temperatures is shown in Figure 6.1, where Figure 6.1(a) shows the absolute capacity, namely the result of the charge and discharge capacity test at different temperatures; Figure

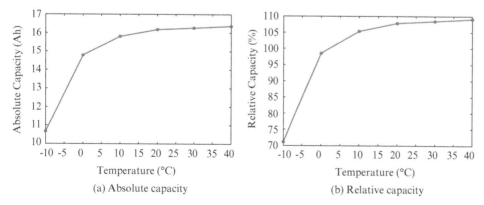

(a) Absolute capacity   (b) Relative capacity

Figure 6.1 Absolute capacity and relative capacity of the LiFePO$_4$ battery at different temperatures.

6.1(b) shows relative capacity, namely the ratio of the capacity test results at different temperatures to the nominal rated capacity stated by the manufacturer. As shown in the figure, the available capacity of the LiFcPO$_4$ battery decreases with a drop in temperature. However, relatively, the decrease in the battery capacity with the temperature drop is not obvious between 20°C and 40°C, but obvious when it is below 20°C.

Specifically, the tested battery with a nominal capacity of 15 Ah has an absolute capacity of 16.36 Ah and a relative capacity of 109.07% at 40°C and an absolute capacity of 10.67 Ah and a relative capacity of 71.13% at –10°C. The above results mean that for the LiFePO$_4$ battery, the available capacity at –10°C (such as use in winter) is decreased by nearly 40% compared with the available capacity at 40°C (such as use in summer).

Furthermore, it can be seen from Figure 6.1 that the capacity shows an accelerating trend of decrease with the temperature drop. With a 10°C temperature interval, the decrease is the most obvious and the relative capacity difference is about 27.41% between –10°C and 0°C. The above results show that the available capacity of the battery decreases significantly at low temperatures. However, if thermal management measures are taken for the battery working at low temperatures and the temperature is increased to more than 10°C, the available capacity of the battery will increase.

Charge and discharge equilibrium potential test

Figure 6.2 shows the charge and discharge equilibrium potential curves of the LiFePO$_4$ battery sample at different temperatures. During the period from SoC = 0 to SoC = 100%, it seems that the equilibrium potential difference at different temperatures is not obvious, so it is necessary to locally amplify different SoC periods, as shown in Figures 6.3 and 6.4. Figure 6.3 shows the comparison of the charge equilibrium potential, which is the local amplification analysis of Figure 6.2(a). Figure 6.4 shows the comparison of the discharge equilibrium potential, which is the local amplification analysis of Figure 6.2(b).

Contrasting analysis of the charge equilibrium potential of the LiFePO$_4$ battery at different temperatures

As shown in Figure 6.3, in order to facilitate the analysis of the difference in the equilibrium potential of the battery at different temperatures, the horizontal axis (SoC value) is divided into three periods, [0, 10%], [10%, 90%], and [90%, 100%], for comparison.

Figure 6.3(a), (c), and (e) respectively represent the absolute value of the equilibrium potential of the LiFePO$_4$ battery in three periods, which can be regarded as a local amplification of Figure 6.2. Figure 6.3(b), (d), and (f) are the relative difference values based on the data at 40°C, which can directly reflect the effect of different temperatures on the equilibrium potential value.

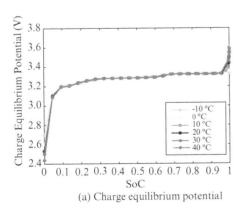

(a) Charge equilibrium potential

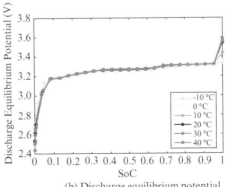

(b) Discharge equilibrium potential

Figure 6.2 Charge and charge equilibrium potential of the LiFePO$_4$ battery at different temperatures.

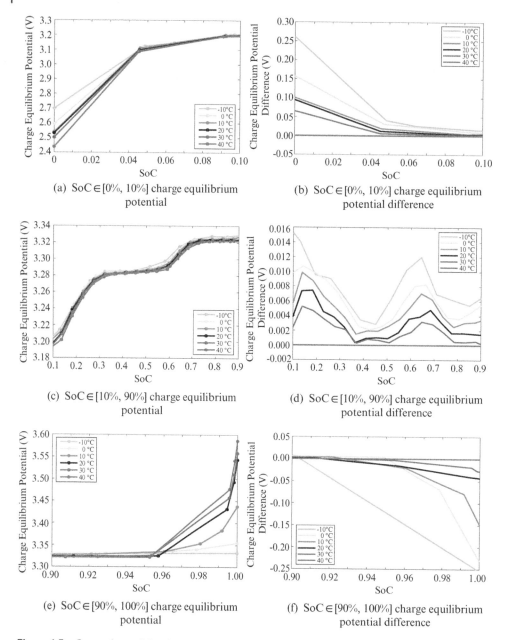

Figure 6.3 Comparison of the charge equilibrium potential of the LiFePO$_4$ power battery at different temperatures and during different SoC periods.

As shown in Figure 6.3, the charge equilibrium potential of the LiFePO$_4$ battery presents the following rules:

1) As shown in Figure 6.3(a) to (d), during the SoC$\in$ [0, 90%] period, the charge equilibrium potential curve at low temperatures is basically higher than that at high temperatures.
2) In contrast with the above rule, as shown in Figure 6.3(e) to (f), during the SoC$\in$ [90%, 100%] period, the charge equilibrium potential curve at low temperatures is basically lower than that at high temperatures.
3) At the beginning and end of the SoC period, the equilibrium potential difference increases with the SoC approaching 0 or 100%.

Contrasting analysis of discharge equilibrium potential of the LiFePO₄ battery at different temperatures

Similar to Figure 6.3, Figure 6.4 shows the discharge equilibrium potential of the battery during different SoC periods. As shown in the figure, the discharge equilibrium potential of the LiFePO₄ battery presents the following rules:

1) As shown in Figure 6.4(a) and (b), during the SoC∈ [0, 10%] period the discharge equilibrium potential curve at low temperatures is basically higher than that at high temperatures.
2) In contrast with the above rule, as shown in Figure 6.4(c) to (f), during the SoC∈ [10%, 100%] period, the discharge equilibrium potential curve at low temperatures is basically lower than that at high temperatures.

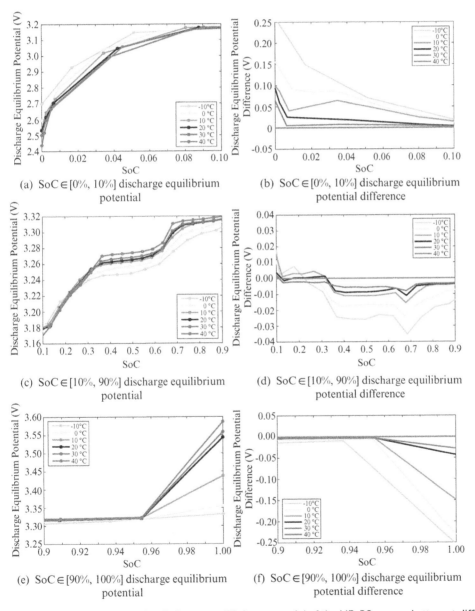

(a) SoC ∈ [0%, 10%] discharge equilibrium
potential

(b) SoC ∈ [0%, 10%] discharge equilibrium
potential difference

(c) SoC ∈ [10%, 90%] discharge equilibrium
potential

(d) SoC ∈ [10%, 90%] discharge equilibrium
potential difference

(e) SoC ∈ [90%, 100%] discharge equilibrium
potential

(f) SoC ∈ [90%, 100%] discharge equilibrium
potential difference

Figure 6.4 Comparison of the discharge equilibrium potential of the LiFePO₄ power battery at different temperatures and during different SoC periods.

3) At the beginning and end of the SoC period, the equilibrium potential difference increases with the SoC approaching 0 or 100%.

According to the above discussion, it can be seen that the charge equilibrium potential and discharge equilibrium potential of the LiFePO$_4$ battery shows basically a consistent rule during SoC∈ [0, 10%] and SoC∈ [90%, 100%] periods, that is, the lower the temperature, the "gentler" the equilibrium potential change is. For example, as shown in Figure 6.4(e), at 40°C, the equilibrium potential is 3.587 V during SoC = 100% and 3.319 V during SoC = 90%, with a difference of 268 mV, while at –10°C, the equilibrium potential difference is only 26 mV.

There is a difference between the charge and discharge equilibrium potential of the LiFePO$_4$ battery mainly during the SoC∈ [10%, 90%] period. To visually show the effect of temperature, the data at –10°C and 40°C are selected here to compare the charge and discharge equilibrium potentials under these two temperature conditions during the [10%, 90%] period, as shown in Figure 6.5.

It can be seen from Figure 6.5 that the charge equilibrium potential at –10°C is higher than that at 40°C. Conversely, the discharge equilibrium potential at 40°C is higher than that at –10°C. This means that for the LiFePO$_4$ battery, the lower the temperature, the more obvious the hysteretic effect will be and the greater the charge and discharge equilibrium potential difference will be.

According to the above analysis, the impact of the temperature on the equilibrium potential of the LiFePO$_4$ battery varies with the SoC periods, showing different rules and differences. In engineering, the BMS developers are more concerned about the impact of the temperature on the accuracy of battery state estimation, for example:

1) If the temperature difference is ignored, and only one equilibrium potential table is checked for all temperatures, the sum of SoC estimation will be huge.
2) If only the test data in 40°C is available due to insufficient test samples in the early phase and the data table of 40°C is only checked for 30°C, the estimation will not be accurate.

For this reason, take the charge process as an example (in fact, the discharge process also has similar results); the rules shown in Figure 6.3 are sorted out to generate Table 6.4. As shown in the table, for the different SoC periods, the left column shows "the equilibrium potential difference under conditions such as those of SoC and

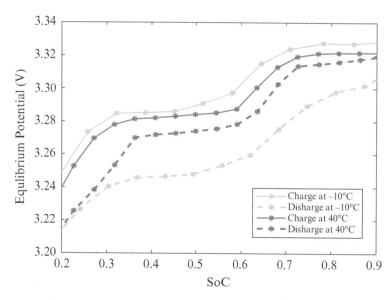

Figure 6.5 Comparison of the charge and discharge equilibrium potential curves of the LiFePO$_4$ battery at –10°C and 40°C.

Table 6.4 SoC estimation error of LiFePO$_4$ power battery resulting from insufficient test data.

Temperature	SoC ∈ [0, 10%]		SoC ∈ [10%, 90%]		SoC ∈ [90%, 100%]	
	Maximum difference (mV)	Reference SoC error (%)	Maximum difference (mV)	Reference SoC error (%)	Maximum difference (mV)	Reference SoC error (%)
−10°C	260.83	1.83	35.35	31.96	−254.26	4.85
0°C	156.92	1.12	24.18	26.18	−231.63	3.96
10°C	10.02	0.71	15.52	15.93	−149.16	2.52
20°C	9.43	0.68	11.02	9.86	−42.88	1.01
30°C	6.48	0.47	7.98	8.54	−27.94	0.67
40°C	0	0	0	0	0	0

different temperatures based on data at 40°C" and the right column shows "the possible SoC estimation error with test data at a certain temperature (40°C)."

It can be seen from Table 6.4 that the temperature has an impact on the equilibrium potential of the LiFePO$_4$ battery mainly during the SoC ∈ [10%, 90%] period, which is commonly referred to as the "platform period" of the LiFePO$_4$ battery in engineering. Since the equilibrium potential of the battery changes slowly with the SoC during this period, the small potential difference may lead to a large SoC estimation error. As shown in the table, if the data at 40°C is used for an estimation at 30°C, the SoC estimation error is 8.54%. If the temperature impact is ignored and the data at 40°C is used for an estimation at –10°C, the SoC estimation error is 31.96%.

Of course, Table 6.4 is based on the test data at 40°C. If it is based on the data at 20°C, the error may be smaller when the data at 20°C is used for an estimation at –10°C. However, during use of the data at 20°C for an estimation at more than 20°C, the maximum SoC estimation error is about 10%. Therefore, it is very necessary to test the battery at different temperatures and obtain the full details of the characteristic parameters of the battery in different operating environments.

3) DC equivalent internal resistance test

The DC equivalent internal resistance curve of the LiFePO$_4$ battery is obtained at different temperatures according to the test arrangement and test steps described above, as shown in Figure 6.6. Figure 6.6(a) shows the charge equivalent internal resistance, while Figure 6.6(b) shows the discharge equivalent internal resistance. According to the equivalent internal resistance test procedure, when approaching full charge or the discharge state, the

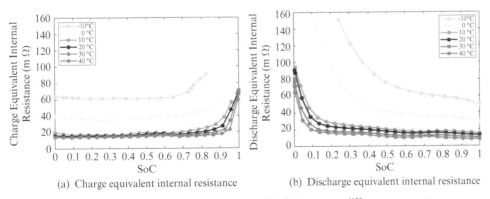

(a) Charge equivalent internal resistance   (b) Discharge equivalent internal resistance

Figure 6.6 DC equivalent internal resistance of the LiFePO$_4$ battery at different temperatures.

cut-off voltage is reached before the sufficient charge and discharge time, the equivalent internal resistance value will be too small to truly reflect of the internal resistance. Therefore, only the effective range of the equivalent internal resistance is considered here.

According to the analysis of the horizontal axis, the charge equivalent internal resistance of the LiFePO$_4$ battery has the opposite growth trend with the discharge equivalent internal resistance:

1) At the same test temperature, the charge equivalent internal resistance increases with an SoC increase, with a slow increase during the SoC$\in$ [0, 90%] period and a rapid increase during the SoC$\in$ [90%, 100%] period.
2) In contrast with the charge equivalent internal resistance, the discharge equivalent internal resistance increases with a SoC decrease, with the largest increase during the SoC$\in$ [0, 10%] period.
3) At the same temperature, the maximum equivalent internal resistance appears when the battery is fully charged or discharged and the maximum discharge internal resistance is larger than the maximum charge internal resistance.

According to an analysis of the vertical axis, the DC equivalent resistance of the LiFePO$_4$ battery increases with the temperature drop, and the increase rate increases with each temperature drop. To intuitively analyze the impact of the temperature on the DC equivalent internal resistance of the LiFePO$_4$ battery, the temperature characteristics of the DC equivalent internal resistance of the battery are analyzed by the charge internal resistance at SoC = {10%, 50%, 80%} and the discharge internal resistance at SoC = {20%, 50%, 90%}, as shown in Figure 6.7. The sample point of the SoC is selected mainly for, as shown in Figure 6.6, the charge internal resistance, which increases with the SoC increase. The 10%, 50%, and 80% correspond to the low, medium, and high levels of SoC respectively, which can reflect the correlation between the DC equivalent internal resistance and the temperature at different SoC levels. If SoC = 90% is selected as the sample for the high SoC level, the effective sample point cannot be obtained in certain low temperature environments (such as –10°C). On the contrary, the discharge equivalent resistance increases with the SoC decrease, with the opposite trend to that of charge test process. Therefore, in order to ensure the integrity of the sample points, the SoC points symmetric with those of the charge process are selected as samples, i.e. 20%, 50%, and 90%.

According to a comparison of Figure 6.7(a) and (b), regardless of a low, medium, or high SoC, the internal resistance variation trend of the LiFePO$_4$ battery is basically the same at different temperatures. In addition, it can be seen that the charge equivalent internal resistance at SoC = 10% is similar to the discharge equivalent internal resistance at SoC = 90%, the charge equivalent resistance at SoC = 50% is similar to the discharge equivalent resistance at SoC = 50%, and the charge equivalent resistance at SoC = 80% is similar to the discharge equivalent resistance at SoC = 20%.

For the BMS engineers, if the temperature and internal resistance relationship model can be built and the corresponding functional model can be obtained, then the DC equivalent internal resistance of the battery at different temperatures can be estimated by using the obtained model and local test data.

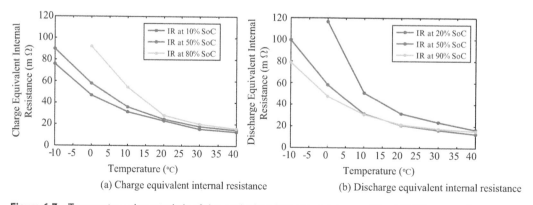

(a) Charge equivalent internal resistance   (b) Discharge equivalent internal resistance

Figure 6.7 Temperature characteristic of the equivalent internal resistance of the LiFePO$_4$ power battery.

According to the test results in Figures 6.6 and 6.7, function fitting is performed for the test data. Through further observation, it can be found that with the temperature change, the DC equivalent internal resistance of the battery varies roughly in the form of an exponential function or power function. The exponential function $f1(x) = ae^{bx} + ce^{dx}$ and the power function $f2(x) = p(x + q)^s + t$ are selected here in order to fit the experimental data. A more appropriate function is determined by comparing the fitting error. Where the function input x is the temperature, the output f is the equivalent internal resistance, and the variables a, b, c, d, p, q, s, and t are the undetermined parameters. The fitting results are shown in Tables 6.5 and 6.6.

Tables 6.5 and 6.6 respectively show the charge resistance and the discharge resistance, the fitting results in two different function forms, and specific parameter values. In order to visually compare the fitting effect of the two functions, the correlation coefficient and the maximum relative error are shown in Tables 6.7 and 6.8.

From the data in Tables 6.7 and 6.8, it can be seen that both functions can represent the relationship between the DC equivalent internal resistance and temperature, of which the fitting effect of the second-order exponential function is better. This is determined by the morphological characteristics of the two functions. When s is

Table 6.5 Charge internal resistance curve fitting result of two functions.

	$f1(x) = ae^{bx} + ce^{dx}$				$f2(x) = p(x + q)^s + t$			
SoC	a	b	c	d	p	q	s	t
80%	3.646	−0.160	42.585	−0.032	5.593e + 04	28.754	−1.897	−3.820
50%	38.598	−0.064	17.995	−0.012	1.1164e + 04	38.996	−1.386	−13.370
10%	88.217	−00.062	2.729	0.028	858.752	26.282	−0.784	−19.961

Table 6.6 Discharge internal resistance curve fitting result of two functions.

	$f1'(x) = ae^{bx} + ce^{dx}$				$f2'(x) = p(x + q)^s + t$			
SoC	a	b	c	d	p	q	s	t
90%	25.550	−0.077	20.380	−9.704e−3	9.963e + 05	43.980	−2.513	2.647
50%	43.230	−0.074	12.030	−2.532e−3	4.603e + 05	34.240	−2.386	1.551e−09
20%	52.170	−0.214	64.380	−33.690e−3	1780 e + 04	1.770	−1.173	1.137e−07

Table 6.7 Charge internal resistance curve fitting result comparison of two functions.

	$f1(x) = ae^{bx} + ce^{dx}$		$f2(x) = p(x + q)^s + t$	
SoC	Correlation coefficient R-square	Max. relative error (%)	Correlation coefficient R-square	Max. relative error (%)
80%	0.9939	8.4177	0.9933	11.4086
50%	0.9984	4.0904	0.9979	12.0845
10%	0.9991	8.1193	0.9990	6.3311

Table 6.8 Discharge internal resistance curve fitting result comparison of two functions.

	$f1'(x) = ae^{bx} + ce^{dx}$		$f2'(x) = p(x+q)^s + t$	
SoC	Correlation coefficient R-square	Max. relative error (%)	Correlation coefficient R-square	Max. relative error (%)
90%	0.9990	3.7105	0.9986	10.8704
50%	0.9981	6.5908	0.9988	8.0170
20%	0.9984	6.1718	0.9980	8.1439

negative, there is a vertical asymptote $x = -q$, which makes the power function form more suitable for the curve with a vertical upward trend. From the test data, it can be seen that the equivalent internal resistance of the LiFePO$_4$ battery increases rapidly with the temperature drop, but does not show an obvious vertical rise, so better results can be obtained by using the exponential function.

Through the above analysis, it can be seen that the internal resistance value at different temperatures can be estimated using a specific exponential function, so as to estimate the power characteristics and heating of the battery, which is of great significance for the estimation of SoF, SoP, and thermal management in practical applications.

6.1.3 Characteristic Test Results of the Li(NiCoMn)O$_2$ Ternary Battery

The characteristic test results of the LiFePO$_4$ battery are discussed in the previous section. This section provides the characteristic test results of the Li(NiCoMn)O$_2$ ternary battery, and analyzes the rules of charge and discharge capacity, equilibrium potential curve, and DC equivalent internal resistance curve.

Capacity test

Figure 6.8 shows the capacity of the Li(NiCoMn)O$_2$ sample battery at different temperatures, where Figure 6.8(a) shows the absolute capacity, which is obtained according to the actual charge and discharge of the battery. Figure 6.8(b) is the relative capacity, which is obtained by actual charge and discharge and rated capacity of the battery. The relative capacity curve can visually show the capacity loss of the battery with temperature.

It can be seen from Figure 6.8 that, similar to the LiFePO$_4$ battery, the capacity of the Li(NiCoMn)O$_2$ battery decreases with the temperature drop on the whole, the battery has a good capacity retention rate at 20–40°C, and degradation is accelerated with the temperature drop from 10°C. Specifically, the absolute capacity and relative

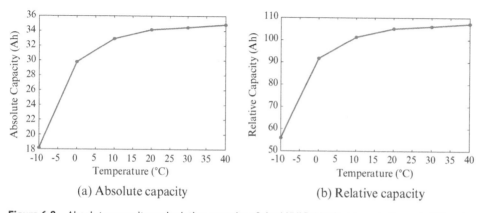

(a) Absolute capacity

(b) Relative capacity

Figure 6.8 Absolute capacity and relative capacity of the Li(NiCoMn)O$_2$ power battery at different temperatures.

capacity of the tested sample are 34.85 Ah and 107.23%, respectively, at 40°C, while the absolute capacity and relative capacity of the tested sample are 18.17 Ah and 55.90%, respectively, at –10°C.

For the tested samples, the capacity of the Li(NiCoMn)O_2 battery decreases with temperature roughly the same as the LiFePO$_4$ battery samples, with a significant decrease rate in the low temperature range. For example, the difference is about 35.74% in the relative capacity at –10°C and 0°C.

Charge and discharge equilibrium potential test

Figure 6.9 shows the charge and discharge equilibrium potential of an Li(NiCoMn)O_2 battery at different temperatures, where Figure 6.9(a) shows the charge equilibrium potential and Figure 6.9(b) shows the discharge equilibrium potential. Because the voltage range is large in the whole SoC period, it is impossible to effectively reflect the difference of the equilibrium potential at different temperatures. Therefore, the relative difference of the equilibrium potential at different temperatures is compared based on the equilibrium potential at 40°C, as shown in Figure 6.10. In this test, unlike the test results of the LiFePO$_4$ battery described in the previous section, the equilibrium potential of the Li(NiCoMn)O_2 battery is relatively consistent in different periods, so the whole SoC period is analyzed here.

As shown in Figure 6.9, the temperature has a certain impact on the equilibrium potential curve of the Li(NiCoMn)O_2 battery. As a whole, the equilibrium potential curve rotates clockwise around a central point as the temperature drops. This phenomenon is also shown in Figure 6.10. Relative to the data at 40°C, the equilibrium potential at other temperatures has a positive error during the lower SoC and a negative error during the higher SoC as the temperature drops.

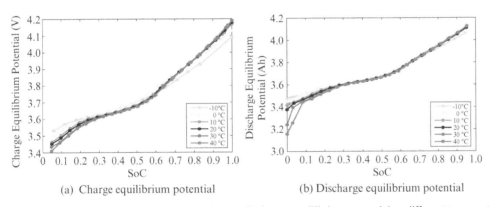

(a) Charge equilibrium potential (b) Discharge equilibrium potential

Figure 6.9 Li(NiCoMn)O_2 power battery charge–discharge equilibrium potential at different temperatures.

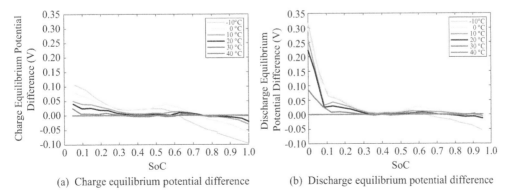

(a) Charge equilibrium potential difference (b) Discharge equilibrium potential difference

Figure 6.10 Comparison of the equilibrium potential difference of the Li(NiCoMn)O_2 battery at different temperatures.

For the charge and discharge equilibrium potential, the extreme error value at the same temperature appears roughly during 0 SoC and 100% SoC. For the charge equilibrium potential, the positive and negative extreme error values are roughly the same; for example, the difference in the equilibrium potential below −10°C and at 40°C is 106.32 mV during 0 SoC and −97.08 mV during 100% SoC. For the discharge equilibrium potential, the positive error during 0 SoC is significantly larger than the negative error during 100% SoC; for example, the difference in the equilibrium potential below −10°C and at 40°C is 317.15 mV during 0 SoC and −52.78 mV during 100% SoC.

To analyze the impact of the equilibrium potential difference on battery management accuracy further, according to the analysis process described in the previous section, extreme error values of equilibrium potential at different temperatures and the SoC error caused by them are recorded, as shown in Table 6.9.

It can be seen that, according to the comparison between Tables 6.4 and 6.9, the temperature has a small effect on the SoC estimation error of the $Li(NiCoMn)O_2$ battery; the extreme error values appear during 0 SoC and 100% SoC. For the test samples described in this section, the $Li(NiCoMn)O_2$ battery is less sensitive to temperature than the $LiFePO_4$ battery, so it is relatively easy to estimate the SoC of the $Li(NiCoMn)O_2$ battery.

DC equivalent internal resistance test

Figure 6.11 shows the DC equivalent internal resistance of the $Li(NiCoMn)O_2$ battery sample at different temperatures, where Figure 6.11(a) shows the charge equivalent internal resistance and Figure 6.11(b) shows the discharge equivalent internal resistance.

As shown in Figure 6.11, according to the horizontal axis, the $Li(NiCoMn)O_2$ battery (Figure 6.11b) has a similar discharge resistance rule to the $LiFePO_4$ battery, that is, the discharge resistance gradually increases with the decrease of SoC, and suddenly increases during $SoC \in [0, 10\%]$. However, for the change of charge internal resistance (Figure 6.11(a)), the $Li(NiCoMn)O_2$ battery does not show a significant increase during the higher SoC range. For example, the charge internal resistance at the last test point at 40°C, at which the SoC was 97.88% and the battery is approximately fully charged, is approximately the same as that during other SoC periods, and this feature is significantly different from the $LiFePO_4$ battery.

At the vertical axis, the resistance of the $Li(NiCoMn)O_2$ battery increases with the temperature drop. For example, the charge internal resistance is increased respectively by 0.991 mΩ, 1.633 mΩ, 3.467 mΩ, 6.438 mΩ, and 13.891 mΩ for every decrease of 10°C from 40°C. With the temperature drop, the increase of the equivalent internal resistance is accelerated, causing a higher polarization voltage difference during charging and discharging, and reaching the charge and discharge cut-off voltages in advance. Therefore, it can explain why the available capacity decreases at low temperatures in the capacity test. From the perspective of capacity and internal resistance, the operating temperature of the battery is controlled between 20°C and 40°C as far as possible in order to maintain the best performance of the battery.

Table 6.9 SoC estimation error of the $Li(NiCoMn)O_2$ battery caused by insufficient test data.

Temperature	SoC∈ [0, 30%]		SoC∈ [30%, 70%]		SoC∈ [70%, 100%]	
	Maximum difference (mV)	Reference SoC error (%)	Maximum difference (mV)	Reference SoC error (%)	Maximum difference (mV)	Reference SoC error (%)
−10°C	371.15	12.54	22.31	8.94	97.08	7.56
0°C	298.21	8.06	37.59	6.76	59.77	5.52
10°C	259.78	4.85	13.31	3.83	30.98	2.85
20°C	222.23	3.97	9.90	2.87	21.64	2.21
30°C	85.82	2.45	5.05	1.66	1.42	1.63
40°C	0	0	0	0	0	0

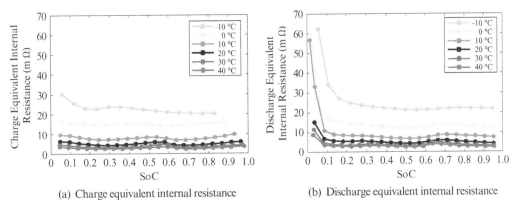

Figure 6.11 Charge and discharge equivalent internal resistance of the Li(NiCoMn)O$_2$ battery at different temperatures: (a) charge equivalent internal resistance; (b) discharge equivalent internal resistance.

6.1.4 Characteristics Comparison of the Two Battery Types

According to the characteristic test results provided in Sections 6.1.2 and 6.1.3, it can be seen that the LiFePO$_4$ power battery has the same characteristics with and different characteristics from the Li(NiCoMn)$_2$ power battery. The similarities and differences of two battery types are now analyzed.

Similarities

1) Impact of temperature on capacity
According to the capacity test results of the two battery types, it can be seen that the actual capacity of the LiFePO$_4$ battery and the Li(NiCoMn)$_2$ battery decreases as the temperature drops, with a small change at 20–40°C in the capacity of the LiFePO battery. From the perspective of capacity, the optimal operating temperature of the two battery types is 20–40°C.

2) Variation tendency of the equilibrium potential
For the LiFePO$_4$ battery and the Li(NiCoMn)$_2$ battery, the equilibrium potential curves are generally monotonically increasing, that is, the equilibrium potential of the battery increases with the SoC increase. In addition, at the same temperature and during the same SoC, the charge equilibrium potential is higher than the discharge equilibrium potential and there is a certain difference between the two values.

3) Impact of temperature on the equilibrium potential
During a low SoC (SoC: 0–10%), the lower the temperature, the higher the equilibrium potential will be. During a high SoC (SoC: 90–100%), the lower the temperature, the lower the equilibrium potential will be.

4) Impact of temperature on the DC equivalent internal resistance
During the same SoC, the lower the ambient temperature, the greater the DC equivalent resistance of the battery will be. The increase rate of the internal resistance increases with the temperature drop.

Differences

1) Variation tendency of the equilibrium potential curve
The equilibrium potential curve of the LiFePO$_4$ power battery has a different variation tendency during different SoC periods. During a medium SoC (SoC: 10–90%), the equilibrium potential of the battery increases slowly and there is a so-called "platform period." During low (0–10%) and high (90–100%) SoC, the equilibrium potential of the battery changes rapidly.

On the contrary, the equilibrium potential curve of the $Li(NiCoMn)_2$ power battery has basically the same variation tendency during different SoC periods and the relationship between the equilibrium potential and the SoC is roughly linear [1, 2].

2) Variation tendency of the DC equivalent internal resistance

The DC equivalent internal resistance of the $LiFePO_4$ power battery shows a monotonical variation tendency with the SoC, where the charge DC equivalent internal resistance increases with the SoC increase and the discharge DC equivalent internal resistance increases with the SoC decrease.

The $Li(NiCoMn)_2$ power battery has a similar discharge DC equivalent internal resistance with the $LiFePO_4$ battery, that is, it increases with the SoC decrease but the variation tendency of its charge DC equivalent internal resistance with SoC is not obvious, and the charge DC equivalent internal resistance is basically the same during different SoC periods.

3) Hysteresis characteristic

Both the $LiFePO_4$ battery and the $Li(NiCoMn)_2$ battery have hysteretic voltage phenomenon, that is, the charge and discharge equilibrium potential curves do not completely coincide with some differences between them. However, the $LiFePO_4$ power battery has different hysteretic properties from the $Li(NiCoMn)_2$ power battery.

The hysteretic difference in the charge and discharge equilibrium potentials of the $LiFePO_4$ power battery and the $Li(NiCoMn)_2$ power battery at different temperatures are shown in Figures 6.12 and 6.13.

As shown in Figure 6.12, the hysteretic difference of the $LiFePO_4$ battery has a layered change with the temperature, that is, the lower the temperature, the greater the difference between the charge and discharge equilibrium potentials. As shown in Figure 6.13, the charge and discharge equilibrium potential difference of the $Li(NiCoMn)O_2$ power battery has no significant change, especially above 10°C, and the hysteretic difference between the charge and discharge balance potential of the battery is basically the same.

6.2 Degradation Test and Analysis

In this section, four factors that may affect battery degradation, including the charge ratio (C_{rate}), starting SoC (SoC_{start}), depth of discharge (DOD), and operating temperature (T), are selected to analyze the impact of these factors on the calendar life and cycle of life of the battery [3]. The degradation rule is discussed in two aspects,

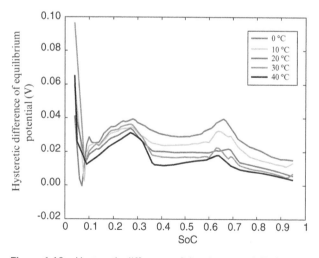

Figure 6.12 Hysteretic difference of the charge and discharge equilibrium potential of the $LiFePO_4$ battery.

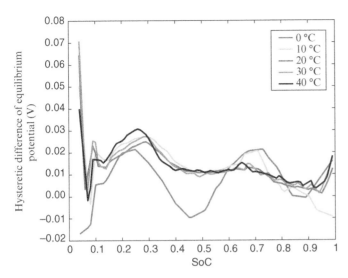

Figure 6.13 Hysteretic difference of the charge and discharge equilibrium potential of the Li(NiCoMn)$_2$ battery.

such as the capacity and the internal resistance. Section 5.5 introduced detailed calendar and cyclic degradation test procedures.

It should be emphasized that it is not recommended to represent the degradation of the battery only by the capacity loss, so it is not recommended to represent the state of the battery after the degradation using Equation (6.1). SoH emphasizes the health state of the battery rather than the capacity loss of the battery:

$$\text{SoH} = \frac{C_n}{C_0} \tag{6.1}$$

In brief, when the battery capacity after the degradation is discussed in this section, the capacity state after the degradation is represented by C_n/C_0, where C_0 and C_n respectively represent the initial capacity and current evaluation capacity after n cycles of the battery [4].

6.2.1 Capacity Change Rule During Battery Degradation

1. Capacity loss of the LiFePO$_4$ battery

Table 6.10 describes the test arrangement of the LiFePO$_4$ battery cycle degradation test, where T, C-rate, SoC$_{start}$, and DOD are respectively changed in the control variable method to investigate thoroughly the impact of various factors on the battery capacity during battery degradation.

Figure 6.14 shows the capacity loss of the LiFePO$_4$ power battery test samples in the cycle test. The capacity loss can be expressed in various forms. The cycle number or the cumulative discharged amperage may be selected for the horizontal axis; the actual capacity (C_n) or the ratio of the degraded capacity after the cycle to the initial capacity (C_n/C_0) may be selected for the vertical axis. The cumulative discharge ampere-hour q, rather than the cycle number, is selected for the horizontal axis based on the following two main advantages. First, because SoC$_{start}$ and DOD may vary with the test samples, the corresponding cycle number is different within the same test time and the cumulative discharged ampere-hour is more comparable. Second, the accumulative mileage of the car mainly depends on the accumulative discharged amperage, which has more reference significance in practical applications. The ratio of the degraded capacity after a cycle to the initial capacity (C_n/C_0), rather than the actual capacity (C_n), is selected for the vertical axis to show the degradation state of the power battery so that the different samples are comparable.

Table 6.10 Cycle test arrangement of the LiFePO$_4$ battery.

Battery number	T(°C)	C-rate	SoC$_{start}$	DOD
A101	20	1 C	100%	80%
A102	20	1 C	100%	30%
A103	20	0.5 C	100%	80%
A104	20	0.5 C	100%	30%
A105	20	0.5 C	100%	80%
A106	20	0.5 C	100%	30%
A107	20	1 C	80%	60%
A108	20	1 C	60%	60%
A109	20	0.5 C	65%	30%
A110	20	1 C	65%	30%
A111	40	1 C	66%	33%
A112	40	0.5 C	33%	33%
A113	40	1 C	66%	66%
A114	40	1 C	100%	66%
A115	40	1 C	80%	60%
A116	40	1 C	65%	30%
A117	40	1 C	100%	30%
A118	40	1 C	30%	30%

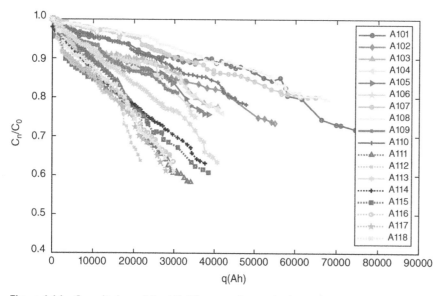

Figure 6.14 Capacity loss of the LiFePO$_4$ power battery in the cycle test.

1) Impact of different C-rates on capacity loss under the same other conditions

   Table 6.11 provides the test arrangement for investigation of the impact of the C-rate on the capacity loss. Figure 6.15 shows the impact of the C-rate on the capacity loss. The capacity loss trend is consistent for the samples and the charge and discharge rates of 0.5 C and 1 C have basically the same impact on the capacity loss of the power battery.

2) Impact of different SoC_{start} values on the capacity loss under the same conditions

   Table 6.12 provides the test arrangement for investigation of the impact of SoC_{start} on the capacity loss. Figure 6.16 shows the impact of SoC_{start} on the capacity loss. The degradation rule of the samples is basically the same and the impact of different SoC_{start} on capacity loss is basically the same.

3) Impact of different DODs on the capacity loss under the same other conditions

   The capacity losses of three samples in Table 6.13 have the same other test conditions except the DOD. Figure 6.17 shows the impact of the DOD on the capacity loss. The samples show basically consistent degradation curves, indicating that the degradation rate of samples is basically the same.

4) Impact of different T on the capacity loss under the same other conditions

   As shown in Table 6.14, four samples are divided into two groups to respectively test the impact of T on the capacity loss of the test sample under the same C-rate, SoC_{start}, and DOD. It can be seen from Figure 6.18 that the capacity loss rate of the test sample at 40°C is significantly higher than that at 20°C. In this section, linear fitting is carried out for the test data of four samples and the fitting slope k is used to represent the capacity loss rate. It can be seen from Table 6.15 that, in this test, the absolute value of k at 40°C is about 2.5 times that at 20°C.

Table 6.11 Test arrangement for investigation of the impact of the different C-rates on the capacity loss.

Experimental variable	A102	A104
T	20°C	20°C
C-rate	0.5 C	1 C
SoC_{start}	100%	100%
DOD	30%	30%

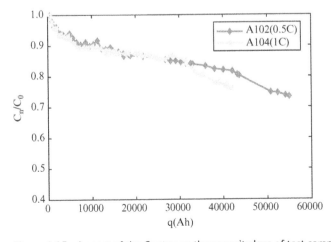

Figure 6.15 Impact of the C-rates on the capacity loss of test samples.

Table 6.12 Test arrangement for investigation of the impact of different SoC_{start} on capacity loss.

Experimental variable	A116	A117	A118
T	40°C	40°C	40°C
C-rate	1 C	1 C	1 C
SoC_{start}	100%	65%	30%
DOD	30%	30%	30%

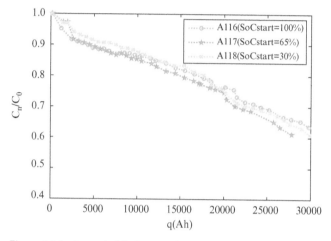

Figure 6.16 Impact of SoC_{start} on the capacity loss of test samples.

Table 6.13 Test arrangement for investigation of the impact of different DODs on the capacity loss.

Experimental variable	A114	A117
T	40°C	40°C
C-rate	1 C	1 C
SoC_{start}	100%	100%
DOD	66%	30%

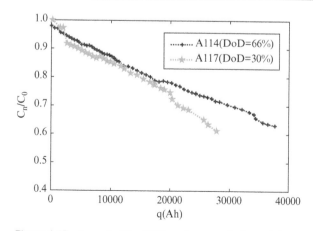

Figure 6.17 Impact of the DOD on the capacity loss of the test sample.

Table 6.14 Test arrangement for investigation of the impact of temperature (T) on the capacity loss.

Factor	A107	A115	A110	A116
T	20°C	40°C	20°C	40°C
C-rate	1 C	1 C	1 C	1 C
SoC_{start}	80%	80%	65%	65%
DOD	60%	60%	30%	30%

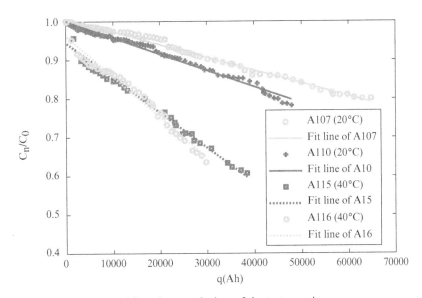

Figure 6.18 Impact of T on the capacity loss of the test sample.

Table 6.15 Linear fitting of the test data.

Fitting slope	A107	A115	A110	A116
$k\,(10^{-6})$	−3.275	−9.121	−4.251	−10.512

2. Capacity loss of the Li(NiCoMn)O$_2$ power battery

Limited by testing time and equipment utilization, the capacity loss rule and features of the LiFePO$_4$ power battery determined in the cycle test are referenced during design of the test condition for the Li(NiCoMn)O$_2$ power battery. The targeted selection of test conditions that may affect the capacity loss of the power battery is conducted to reflect the degradation law of the Li(NiCoMn)O$_2$ power battery as accurately and completely as possible.

As shown in Table 6.16, the impact of the C-rate, SoC_{start}, and DOD is compared to the capacity loss of the Li(NiCoMn)O$_2$ power battery by six test samples. Samples B101 and B102 are used to compare the impact of the C-rate on the capacity loss, samples B102, B103, and B104 are used to compare the impact of SoC_{start} on the capacity loss, and samples B101, B105, and B106 are used to compare the impact of DOD on the capacity loss.

Figure 6.19 provides the impact of the C-rate on the capacity loss of the test sample, where the charge–discharge rate of 0.5 C and 1 C have basically the same impact on the capacity loss of the power battery. Figure 6.20 shows the impact of the DOD on the capacity loss of the test sample, where the sample B101 in 33% DOD has a significantly larger capacity loss rate than the other two samples, because the cycle number of this sample is 3 times that

Table 6.16 Comparison of the test arrangement for impact of the C-rate, SoC_{start}, and DOD on the capacity loss.

Influence factors	B101	B102	B103	B104	B105	B106
T	40°C	40°C	40°C	40°C	40°C	40°C
C-rate	1 C	0.5 C	0.5 C	0.5 C	1 C	1 C
SoC_{start}	100%	100%	67%	33%	100%	100%
DOD	33%	33%	33%	33%	100%	67%

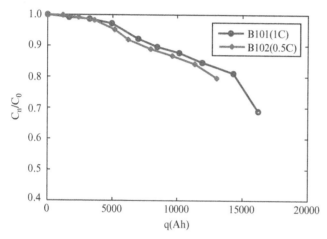

Figure 6.19 Impact of the C-rate on the capacity loss of the test sample.

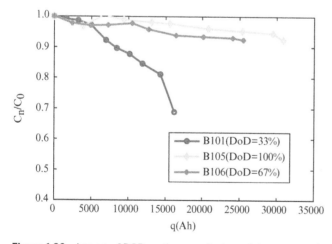

Figure 6.20 Impact of DOD on the capacity loss of the test sample.

of B105 and 2 times that of B106, at the same discharged ampere-hour. For this type of Li(NiCoMn)O$_2$ ternary battery, the degradation degree varies with the cycle number. The higher the cycle number is, the higher the capacity loss is.

Figure 6.21 provides the impact of SoC_{start} on the capacity loss of the test sample, where the sample in 33% SoC_{start} has a significantly larger capacity loss rate than the other two samples. This indicates that, during the

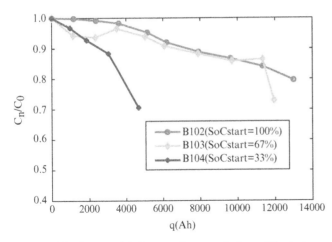

Figure 6.21 Impact of SoC_{start} on the capacity loss of the test sample.

operation of the low battery, the capacity loss of the $Li(NiCoMn)O_2$ power battery is accelerated and the battery cycle life is shortened.

According to the analysis of the capacity loss rule of the battery capacity in this section, it can be seen that the capacity loss of the test samples of two types of power battery is approximately linear in the cycle test. In the cycle test, the impact of different test conditions on the two types of power battery is basically the same and the temperature is the main factor affecting the degradation of the power battery, while other test conditions have a relatively small impact on the degradation of the power battery. Comparatively, the $LiFePO_4$ power battery has a relatively longer service life and discharges a higher cumulative ampere-hour than the $Li(NiCoMn)O_2$ power battery before reaching end of life (EOL). Furthermore, during operation with a small SoC for a long time, the degradation of the $Li(NiCoMn)O_2$ power battery is accelerated, while the $LiFePO_4$ power battery does not show a similar phenomenon.

6.2.2 Internal Resistance Spectrum Change Rule During Battery Degradation

In the whole test, dozens of equivalent internal resistance tests are carried out for the battery samples. In order to show more clearly the change of the battery's equivalent internal resistance, for example, the test is made in A109 for 500 cycles to obtain the test results of the charge and discharge equivalent internal resistance during the phased evaluation, as shown in Figure 6.22. It can be seen that with the increase in cycle number, the equivalent internal resistance spectrum of the battery presents a trend of increasing monotonously. Compared with the discharge equivalent internal resistance, the charge equivalent internal resistance has two advantages in the actual use of the battery: first, the change of the charge equivalent internal resistance with the degradation becomes more obvious; second, it is possible to test the equivalent internal resistance spectrum when charging the battery in practice.

It can be seen from Figure 6.22(a) that, for the result of a certain test, the equivalent internal resistance spectrum shows a trend where the equivalent internal resistance increases with the increase of the cycle number. The charge equivalent internal resistance has a small change when the SoC is below 0.5 and shows an obvious increase trend when the SoC is more than 0.5. Although the degradation of the battery is accurately evaluated by the whole equivalent internal resistance curve, the calculation is complicated, and it is difficult to evaluate the degradation in a practical application.

In order to solve this problem, two relatively simplified parameters are used in this section to represent the change rule of the DC equivalent internal resistance spectrum.

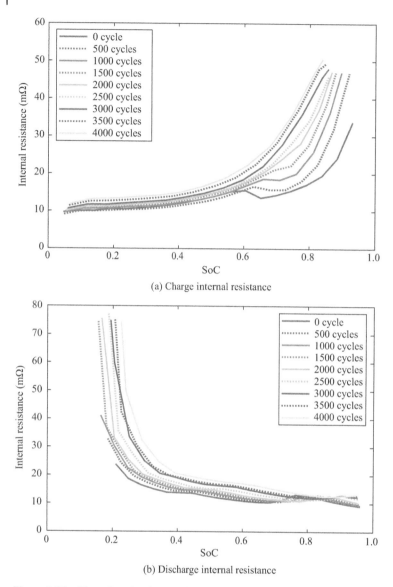

(a) Charge internal resistance

(b) Discharge internal resistance

Figure 6.22 Phased evaluation result of internal resistance of the A109 sample.

1) If a fixed SoC is selected and set to z, the DC equivalent internal resistance at this SoC may be used as the characterization parameter of the whole DC equivalent internal resistance spectrum, as expressed as below:

$$\Upsilon = \gamma(\mathrm{SoC} = z) \qquad (6.2)$$

2) A closed region can be formed with two SoC contour lines, the horizontal axis and the DC equivalent internal resistance curve. The area of this closed region is used as the characterization parameter of the whole DC equivalent internal resistance spectrum, as expressed below:

$$\Psi = \int_{z_1}^{z_2} \gamma(\mathrm{SoC}) d(\mathrm{SoC}) \qquad (6.3)$$

where γ and ψ are the characterization parameters of the DC equivalent internal resistance spectrum, γ (SoC) indicates that the internal resistance is the function on SoC, and z_1 and z_2 represent the contour lines of two SoC values respectively.

In order to compare the effect of representation of the DC equivalent internal resistance spectrum by two simplified parameters, four specific characterization parameters are set in this section according to the above two definitions:

$$\varUpsilon_1 = \gamma(\text{SoC} = 0.5) \tag{6.4}$$

$$\varUpsilon_2 = \gamma(\text{SoC} = 0.8) \tag{6.5}$$

$$\varPsi_1 = \int_{0.5}^{0.8} \gamma(\text{SoC}) d(\text{SoC}) \tag{6.6}$$

$$\varPsi_2 = \int_{0.7}^{0.8} \gamma(\text{SoC}) d(\text{SoC}) \tag{6.7}$$

In Equations (6.6) and (6.7), SoC $\in$ [0.5, 0.8] and SoC $\in$ [0.7, 0.8] are selected, for when SoC is less than 0.5, the DC equivalent internal resistance has a small increase with the degradation of the battery, and if this part is included in ψ, the characteristics of the equivalent internal resistance will be weakened and the sensitivity will be reduced. According to the test experience, in the battery life cycle, when the DC equivalent internal resistance is measured, the SoC cannot be less than 0.8 generally when the external voltage reaches the operating voltage limit of the battery during measurement of the equivalent internal resistance. The schematic diagram for four characterization parameters of the equivalent internal resistance spectrum is shown in Figure 6.23.

According to the characteristics of the charge equivalent internal resistance spectrum, in this section, the function $y = ae^{bx} + c$ is used to fit the characterization parameters to judge the characterization effect of the parameters. The fitting effect can generally be described by the determination coefficient R^2 and the root mean squared error (RMSE) of the fitting result. The function fitting effect of the four parameters is shown in Table 6.17.

It can be found from Table 6.17 that, due to a measurement error, the SoC estimation error and other reasons, if the charge equivalent internal resistance value during a fixed SoC is used to represent the whole internal

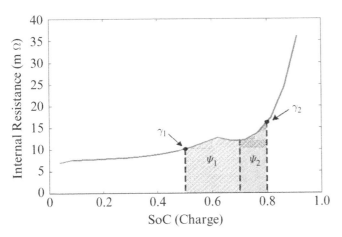

Figure 6.23 Schematic diagram for four characterization parameters of the equivalent internal resistance spectrum.

Table 6.17 Function fitting effect of four parameters.

Battery sample	Υ_1 R^2	RMSE	Υ_2 R^2	RMSE	Ψ_1 R^2	RMSE	Ψ_2 R^2	RMSE
A107	0.911	1.442	0.810	3.942	0.977	0.042	0.771	0.232
A109	0.701	1.095	0.866	2.523	0.946	0.066	0.710	0.403
A110	0.895	1.120	0.882	2.892	0.968	0.053	0.886	0.162
A116	0.814	0.430	0.894	2.726	0.947	0.054	0.540	0.378
A117	0.920	0.323	0.912	3.146	0.953	0.072	0.758	0.327
A118	0.957	0.860	0.952	2.655		0.079	0.874	0.202

resistance curve, it is impossible to reflect the characteristics of the internal resistance curve. However, the impact of all errors can be reduced by characterization of the internal resistance spectrum in the cumulative charge equivalent internal resistance. In general, when the characterization is performed by the cumulative DC charge equivalent internal resistance during SoC $\in$ [0.5, 0.8], the best characterization effect will be achieved. Therefore, ψ_1 is selected as a characterization parameter of the internal resistance spectrum. In order to facilitate the comparison between different test samples, normalization processing will be carried out for ψ_1:

$$\theta = \frac{\Psi_{1,actual}}{\Psi_{1,\,new}} \tag{6.8}$$

where ψ_1, actual and ψ_1, new are respectively the characterization parameters of the internal resistance spectrum calculated from Equation (6.6) under current test conditions and before the cycle test. The characterization parameter θ mentioned hereinafter is also defined by Equation (6.8).

1. Internal resistance change of the LiFePO$_4$ power battery

The LiFePO$_4$ power battery samples, as shown in Table 6.18, were selected to compare the impact of the internal resistance by the cycle test under different conditions. Those six samples are selected because they may be used as the comparison test in four groups. A109 and A110 are used to compare the impact of the C-rate on the equivalent internal resistance of the battery under the same T, SoC$_{start}$, and DOD. A110 and A107 are used to compare the impact of DOD on the equivalent internal resistance of the battery under the same T, SoC$_{start}$, and C-rate. A110 and A116 are used to compare the impact of T on the equivalent internal resistance of the battery under the same C-rate, SoC$_{start}$, and DOD. A116, A117, and A20 are used to compare the impact of SoC$_{start}$ on the equivalent internal resistance of the battery under the same C-rate, T, and DOD.

Table 6.18 Test arrangement for the impact of the cycle test condition on the change of internal resistance.

	A109	A110	A107	A116	A117	A118
T	20°C	20°C	20°C	40°C	40°C	40°C
C-rate	0.5 C	1 C	1 C	1 C	1 C	1 C
SoC$_{start}$	65%	65%	65%	65%	100%	30%
DOD	30%	30%	60%	30%	30%	30%

The variation of the equivalent internal resistance of the battery sample at different degradation degrees can be shown by the characterization parameter θ of the equivalent internal resistance spectrum described in the previous section. The test results of six battery samples are shown in Figure 6.24.

The test results shown in Figure 6.24 can be expressed in a variety of possible ways. According to the description provided in the previous section, the cumulative discharge ampere-hour q is still used as the horizontal axis in this section. It can be seen from the figure that the DC charge equivalent internal resistance of the battery changes in a small range and with the increase in the degradation degree of the battery, the increase rate also rises in the form of a power function and an exponential function. Therefore, θ and q are fitted by the power function and the exponential function respectively, and the fitting result shows a relatively high correlation of θ and q with the quadratic and cubic power functions and the exponential functions. The specific fitting results are shown in Table 6.19.

Since the DC charge equivalent resistance changes in the first quadrant of the function image, the exponential function with a good fitting effect is selected herein to fit the characterization parameter θ of the internal resistance spectrum. The impact of temperature T, C-rate, SoC_{start}, and DOD on the internal resistance of the sample is compared in the above analysis method.

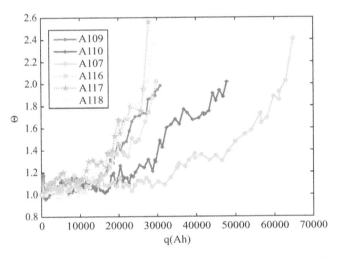

Figure 6.24 Change of characterization parameter θ of the equivalent internal resistance spectrum with cumulative discharge of the sample.

Table 6.19 Fitting result of θ and q correlation obtained by different functions.

Battery sample	$y = ax + b$		$y = ax^2 + bx + c$		$y = ax^3 + bx^2 + cx + d$		$y = ae^{bx} + c$	
	R^2	RMSE	R^2	RMSE	R^2	RMSE	R^2	RMSE
A109	0.777	0.135	0.950	0.064	0.950	0.065	0.946	0.066
A110	0.842	0.118	0.964	0.057	0.964	0.057	0.968	0.053
A107	0.944	0.065	0.968	0.049	0.976	0.043	0.977	0.042
A116	0.852	0.087	0.943	0.056	0.949	0.054	0.947	0.054
A117	0.884	0.112	0.951	0.074	0.951	0.075	0.953	0.072
A118	0.853	0.159	0.978	0.062	0.979	0.061	0.964	0.079

1) Impact of different temperatures T on the change of internal resistance under the same conditions

   A110 and A116 are the test control group using the test temperature as the comparison condition. During the test, the ambient temperature of the battery is strictly controlled by a thermostat. The dissipation rate of the heat generated by the charge and the discharge during the battery cycle is improved by strengthening the air flow in the thermostat to reduce the errors caused by different temperatures. Figure 6.25 shows the fit test results of the samples A110 and A116. It can be seen that, when the cumulative discharge capacity q is the same, θ_{A116} at 40°C is greater than θ_{A110} at 20°C. According to the test of this group, the temperature rise will accelerate the aging of the battery.

2) The impact of different C-rates on a change of internal resistance under the same other conditions

   Figure 6.26 shows the impact of different C-rates on θ. When the cumulative discharge capacity q is the same, the θ_{A109} of the sample A109 at a C-rate of 0.5 C is more than θ_{A110} of the sample A110 at the C-rate of 1 C. Since the C-rate of the sample A110 is twice that of the sample A109, when the cumulative discharge capacity is the same, the test time of the sample A109 is twice that of the sample A110. Considering the impact of working time and calendar life of the battery on its degradation, it is possible that θ_{A109} is larger than θ_{A110} due to the different test time. When the two samples are tested at the same time, that is, q_{A110} is twice as large as q_{A109}, θ_{A110} will be greater than θ_{A109}. This is because, in the same time, the higher the C-rate, the more heat will be generated inside the battery due to internal resistance, and the higher the temperature. Since it takes some

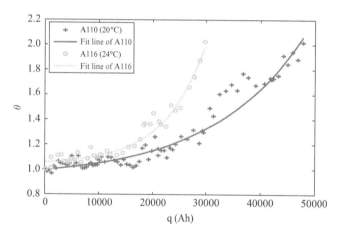

Figure 6.25 Impact of test temperature on θ.

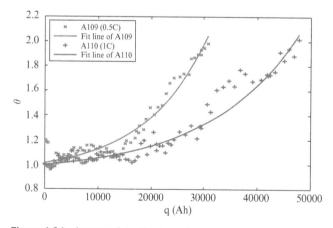

Figure 6.26 Impact of the C-rate on θ.

time to dissipate the heat from the battery, as described in the previous section, the higher temperature will accelerate the degradation of the battery, so the degradation rate will increase accordingly. With the increase of internal resistance, the heat of the battery gradually rises, causing more and more significant θ increases.

3) The impact of different SoC_{start} on the change of internal resistance under the same other conditions

Figure 6.27 shows the change of θ with Q of samples A116, A117, and A118. The SoC_{start} of three samples are respectively 100%, 65%, and 30% and their DOD is 30%.

As shown in the figure, when the cumulative discharge capacity is the same, θ_{A117} and θ_{A118} are slightly greater than θ_{A116}. The degradation rate at 100% and 30% is slightly higher than that at 65% mainly because: when the cycle starts from 100%, the charge internal resistance is large during the SoC period; when the cycle starts from 30%, the discharge internal resistance is large during the SoC period. When it passes through the high resistance zone during the test, the temperature rise caused by the heat inside the battery will be more obvious, thus accelerating the degradation rate of the battery, and during the test, the high resistance causes the more obvious temperature rise by the heat inside the battery, thus also accelerating the degradation rate of the battery. For the same reason described in part 2 of this list, this phenomenon is not obvious at the beginning of the test, but becomes significant as θ increases.

4) The impact of different DODs on a change of internal resistance under the same other conditions

A110 and A107 are used as the test control group to compare the impact of DOD. The test results are shown in Figure 6.28, where θ_{A110} is very similar to θ_{A107} when q is less than 15000 Ah, but after that the increase rate of θ_{A110} rises for the impact of the cycle number on the battery degradation. Since DOD is 30% for the sample A110 and 60% for the sample A107, for the same q, the cycle number of A110 is twice that of A107 and the difference becomes more obvious with the increase in the accumulative discharge capacity of the battery.

2. Changes of internal resistance of the Li(NiCoMn)O$_2$ battery

Compared with the LiFePO$_4$ battery, the Li(NiCoMn)O$_2$ battery has a different internal resistance spectrum curve, as shown in Figure 6.29. The internal resistance of the Li(NiCoMn)O$_2$ battery does not change significantly during the whole charging process. When the battery is nearly full, the charge internal resistance does not increase as significantly as that of the LiFePO$_4$ battery. Similar to the LiFePO$_4$ battery, the internal resistance of the Li(NiCoMn)O$_2$ battery also increases temporarily during charging, and when the battery is further charged, the internal resistance decreases again, which is shown in the "bump" form in the figure.

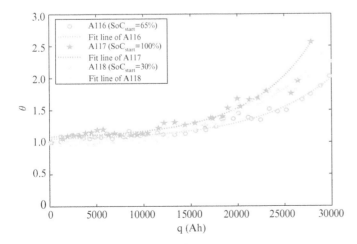

Figure 6.27 Impact of SoC_{start} on θ.

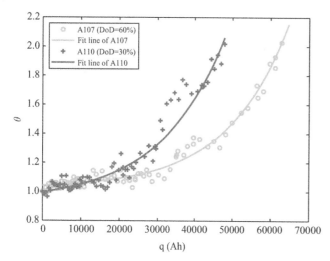

Figure 6.28 Impact of DOD on θ.

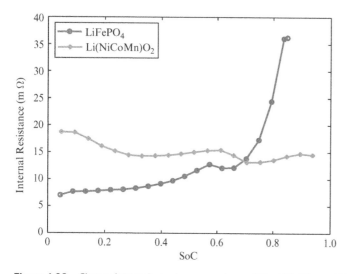

Figure 6.29 Charge internal resistance spectrum of the LiFePO$_4$ and Li(NiCoMn)O$_2$ batteries.

This section discusses the change of the internal resistance during the cycle by the characterization parameter θ of the internal resistance spectrum. In order to verify whether θ is applicable to the Li(NiCoMn)O$_2$ battery, the change in the characterization parameter θ of the internal resistance spectrum is fit for the different test samples by $y = ae^{bx} + c$ in this section. According to the results shown in Table 6.20, θ can also better describe the change of the Li(NiCoMn)O$_2$ battery during the cycle.

The impact of the test conditions on the internal resistance may be compared using the characterization parameter θ of the internal resistance spectrum. Figures 6.30, 6.31, and 6.32 respectively show the impact of the C-rate, DOD, and SoC$_{start}$ on θ.

As shown in the figures, the impact of the C-rate and DOD on θ in the Li(NiCoMn)O$_2$ battery is similar to that in the LiFePO$_4$ battery. At the same discharge Ah, the sample B102 has a higher θ than the sample B101 at the C-rate of 0.5 C due to a longer test time. Since a higher cycle number is required for the sample B101 with 33%

Table 6.20 The θ fitting results of the test samples.

	B101	B102	B103	B104	B105	B106
R^2	0.9461	0.9305	0.9407	0.9983	0.9291	0.817
RMSE	0.0414	0.0736	0.0391	0.0159	0.0242	0.0294

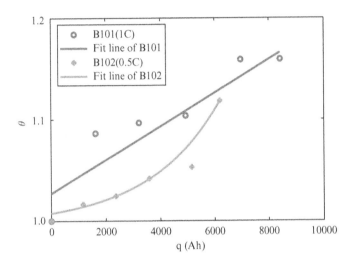

Figure 6.30 Impact of the C-rate on θ.

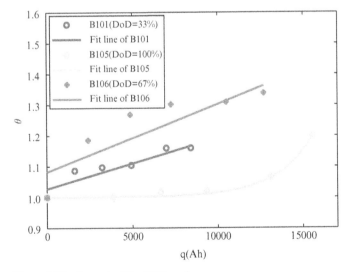

Figure 6.31 Impact of the DOD on θ.

DOD, the increase amplitude of corresponding θ is higher and the sample is close to the EOL. The samples B105 and B106 have a gentler θ change and can be continuously used for the test. The sample B104 with 33% SoC_{start} has a significantly higher θ increase rate than the other two samples, suggesting that the characteristic of the Li(NiCoMn)O$_2$ battery degradation is accelerated for a battery with a low SoC_{start}.

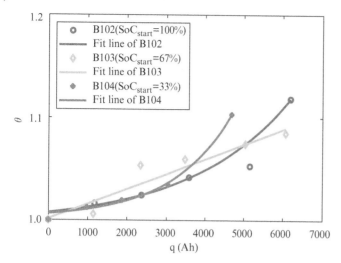

Figure 6.32 Impact of the SoC$_{start}$ on θ.

From the above analysis, it can be seen that the characterization parameter θ of the internal resistance spectrum can be used to characterize the change in the equivalent internal resistance spectrum of two battery types during the cycle test. The characterization parameter θ of the internal resistance spectrum has an exponential relationship with discharge Ah, the temperature is a main cause for the battery degradation, and other test conditions impose an impact on the battery degradation, mainly for their impact on the internal or external temperature during operation of the battery.

In the analysis of capacity loss in Section 6.2.1, only the test temperature T has a significant impact on the capacity loss rate. In the analysis of an internal resistance change in Section 6.2.2, all test conditions affect the change in the characterization parameters of the internal resistance spectrum, since, compared with the battery capacity, the internal resistance is more sensitive to the test conditions and can better reflect the impact of battery degradation on battery performance during the test. Therefore, the relationship between the test conditions and battery degradation cannot be judged completely based on the changes in the battery capacity. In practical applications, the battery degradation state can be more accurately identified by comprehensive analysis of the battery capacity and the internal resistance in order to make a reasonable operation plan for the battery.

6.2.3 Impact of Storage Conditions on Battery Degradation

In this section, the calendar life is tested on the battery to investigate the effect of different storage conditions on the battery degradation [5]. The test scheme is shown in Table 6.21. A total of nine batteries made of the materials A and B described in the previous chapter are selected for testing, and the test conditions include different ambient temperatures and different SoC$_{start}$. Specific steps of the calendar life test are as follows. First, the battery capacity is tested to obtain the current initial capacity of the battery. Then, according to the initial capacity, the SoC of the battery is adjusted to a certain set value and the battery is stored for a long time at a specific temperature. During the test, the battery is evaluated regularly to obtain the rule of battery capacity and internal resistance degradation with storage time.

Figures 6.33 and 6.34 show the capacity and internal resistance degradation rule of the battery under different storage conditions. It can be seen from Figure 6.33 that the higher the storage temperature, the higher the deterioration rate of the battery will be. In the actual tests, the phased evaluation is conducted on the battery samples about every four weeks. In order to reduce the data overlapping, 12 weeks are selected as the presentation period

Table 6.21 Calendar life test arrangement of a power battery.

Battery number	T (°C)	SoC
A119	40	90%
A120	40	90%
A121	40	50%
A122	40	50%
A123	40	20%
B107	40	50%
B108	40	90%
B109	20	50%
B110	30	50%

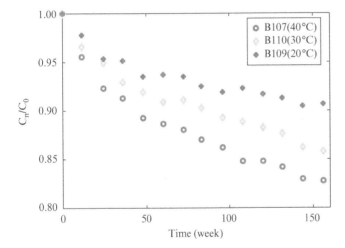

Figure 6.33 Relationship between the storage temperature and battery capacity.

in order to provide more concise and clearer plots. It can be seen from Figure 6.34, during storage of the battery, that the low SoC may increase the internal resistance of the battery. After about 156 weeks of storage, the battery samples with SoC 20% upon storage have a significantly higher internal resistance than the battery samples with SoC 50% and 90%.

The battery is always in a degradation process since its delivery. It can be seen from the above test results that the capacity loss of the battery is slow during storage. In order to more intuitively compare the degradation rate of the battery under the storage conditions and cyclic operating conditions, two samples were selected for comparison at 40°C, as shown in Figure 6.35.

It can be seen from Figure 6.35, at the same ambient temperature, that there is a difference in the capacity loss between the storage test sample A121 and the cycle test sample A116. At the same test time, the capacity loss of the cycle test sample A116 is about three times that of the storage test sample A121, which indicates that, for the tested samples, during storage without charge and discharge operations, the degradation rate of the storage test battery is lower compared with the cycle test sample and the storage time has a limited impact on the battery performance.

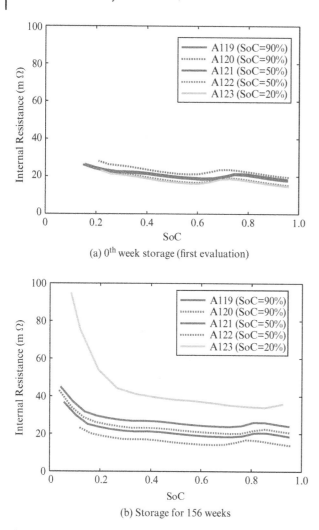

(a) 0th week storage (first evaluation)

(b) Storage for 156 weeks

Figure 6.34 Relationship between the SoC and internal resistance during battery storage.

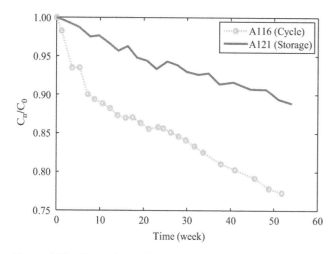

Figure 6.35 Comparison of battery capacity loss under storage and cycle conditions.

References

1 Kalman, R. E., "A new approach to linear filtering and prediction problems," *Journal of Basic Engineering*, 1960, 82: 35–45.

2 Hu, Y. and Yurkovich, S., "Battery cell state-of-charge estimation using linear parameter varying system techniques," *Journal of Power Sources*, 2012, 198: 338–350.

3 Käbitz, S., Gerschler, J. B., Ecker, M., et al., "Cycle and calendar life study of a graphite LiNi1/3Mn1/3Co1/3O2 Li-ion high energy system. Part A: Full cell characterization," *Journal of Power Sources*, 2013, 239: 572–583.

4 Asakura, K., Shimomura, M., and Shodai, T., "Study of life evaluation methods for Li-ion batteries for backup applications," *Journal of Power Sources*, 2003, 119–121: 902–905.

5 Ramasamy, R. P., White, R. E., and Popov, B. N., "Calendar life performance of pouch lithium-ion cells," *Journal of Power Sourcs*, 2005, 141: 298–306.

7

Battery Modeling

This chapter describes the modeling and simulation of the external characteristics of the battery. The external characteristics are actually the relationship between current and voltage in the working process of the lithium-ion (Li-ion) power battery, that is, the volt-ampere characteristics of the battery. The battery model is of great significance in the development of the battery management system. On the one hand, a good battery model can provide a basis for the estimation algorithm of the SoC and, on the other hand, a good power battery model can simulate the characteristics and behavior of the battery to help the development of energy management strategies [1].

7.1 Battery Modeling for BMS

7.1.1 Purpose of Battery Modeling

Generally speaking, battery modeling intends to determine the mathematical relationship between the environmental factors and the characteristic quantities of the battery. The objects to be considered include terminal voltage, operating current, SoC, temperature, internal resistance, electromotive force, SoH, etc. It is of great significance for the development of the battery management system to find their relationships.

On the one hand, various performances of the battery in work can be estimated by the established battery model in order to simulate various battery management strategies (such as energy control strategy, battery balance strategy, etc.), and the effectiveness of the strategies can be verified through software. Compared with the experimental test, it is not required to use the hardware circuit board and the actual battery sample during the simulation verification, which not only saves hardware cost and development time, but also shortens the verification cycle. In the simulation of a balancing control strategy based on energy transfer attached hereinafter, it takes only a few minutes to simulate a particular initial state by software and the battery model, while it can take hours to verify the effectiveness of the strategy using an actual circuit board. The simulation verification has more obvious advantages when there are many samples to be tested.

On the other hand, the accurate battery model is of great significance for the estimation algorithm of the SoC. SoC estimation, a difficult point in the battery management system, is of great significance to the development of other functions of the battery management system. In practice, the internal SoC is often estimated by external physical quantities such as voltage, current, and temperature. If a more accurate external characteristic model can be established, it is very beneficial to find the numerical relationship between the SoC and various physical quantities that can be directly measured, so as to evaluate the internal SoC through the monitored external performance.

Battery Management System and its Applications, First Edition. Xiaojun Tan, Andrea Vezzini, Yuqian Fan, Neeta Khare, You Xu, and Liangliang Wei.

7.1.2 Battery Modeling Requirement of BMS

Three battery modeling requirements are proposed for the BMS:

1. The established model should be an external characteristic model

The established model is not required to describe the internal chemical reaction process of the battery, because the model described herein is established for the development of the BMS rather than analysis of battery design and production issues. The external characteristic model is sometimes called the performance model, which mainly considers the physical quantities such as the terminal voltage, main circuit current, and ambient temperature, which are the external characteristics of the battery.

2. The established model may be an equivalent model

Many examples of circuit models mentioned below belong to the equivalent model. The resistance and capacitance in the circuit are all equivalent elements, and their values only reflect the functional relationship between the voltage and current expressed by the chemical reaction inside the battery.

3. The established model must have stronger practicability

First, the numerical relation reflected by the model should have high accuracy, so as to avoid the "imitational but untrue" embarrassment. Second, the model should be feasible. The feasibility is shown in two aspects: first, the parameters in the model are obtainable and calibrated; second, the model must not be too complex, otherwise it is not easy to realize the model. For example, although the artificial neural network model mentioned below is helpful in identifying the complicated relationship between the SoC and voltage, current, and temperature, due to the high complexity of its algorithm, the artificial neural network model is acceptable for the strategy simulation (usually by a high-speed desktop), but not for the embedded systems with a weaker computing power [2]. Therefore, it cannot be used in real-time BMS.

7.2 Common Battery Models and Their Deficiencies

7.2.1 Non-circuit Models

Electrochemistry model

The battery electrochemistry model refers to a series of electrochemistry equations established to express the battery characteristics according to the electrochemistry process of the battery. By analyzing the electrochemical process of the battery, some scholars established the electrochemical equations corresponding to all areas in the battery system, including the diffusion equation, potential equation, and electrochemical process equation corresponding to positive and negative electrode areas and the diaphragm area [3].

Table 7.1 provides the potential equation corresponding to the positive and negative electrode areas. In the table, $\phi_{s,3}$ is the solid phase potential and $\phi_{e,3}$ is the liquid phase potential, $\sigma_{s,3}$ is the effective conductivity of the solid phase, $D_{s,3}^{eff}$ is the diffusion coefficient, $J(x,t)$ is the current density passing through x, and κ is the solution electrical conductivity.

Although the battery characteristics can be described by the electrochemical model according to the electrochemistry, it is more difficult to establish the electrochemical model due to its more complicated structure. As shown in Table 7.1, only some electrochemical equations of the battery electrochemical model (the positive electrode part) are given and more complicated equations are required for establishment of the model. Moreover, all parameters used in the equations need to be measured using the electrochemical method and the test steps are tedious. In summary, the electrochemical model is not suitable for the development of the battery management system and is generally only applicable to the development and improvement of the battery.

Table 7.1 Potential equation corresponding to the positive and negative electrode areas.

Solid phase	Potential equation	$\dfrac{\partial}{\partial x}\left(D_{s,3}^{eff}\dfrac{\partial}{\partial x}\phi_{s,3}^{eff}(x,t)\right)=j_{s,3}(x,t)$
	Boundary conditions	$\dfrac{\partial}{\partial x}\phi_{s,3}(x,t)=\dfrac{J}{\sigma_{s,3}^{eff}}\quad x=0;\quad \dfrac{\partial}{\partial x}\phi_{s,3}(x,t)=0\quad x=L_c$
Liquid phase	Potential equation	$j_{s,3}(x,t)=\dfrac{\partial}{\partial x}\kappa_{e,3}^{D,eff}\left[\dfrac{\partial \ln C_{e,3}(x,t)}{\partial x}\right]+\dfrac{\partial}{\partial x}\kappa_{e,3}^{D,eff}(x,t)\dfrac{\partial\phi_{e,3}(x)}{\partial x}$
	Boundary conditions	$\dfrac{\partial}{\partial x}\phi_{e,3}(x,t)=0\quad x=0;\quad \dfrac{\partial}{\partial x}\phi_{e,3}(x,t)=\dfrac{J}{\kappa_{s,3}^{eff}}\quad x=L_c$

1) Artificial neural network model

Because of characteristics such as high non-linearity, fault tolerance, and a self-learning habit, the neural network can be used in battery modeling [4, 5]. Generally, the neural network model structure is divided into an input layer (input experimental data), an output layer (output prediction data), and a hidden layer (weighted processing of experimental data). A typical artificial neural network with three inputs and two outputs is shown in Figure 7.1.

If x and y provided in the above figure are replaced by the internal and external characteristic parameters of the battery, the battery model based on an artificial neural network can be obtained. A possible battery model is shown in Figure 7.2.

The artificial neural network model features full use of the characteristics of the neural network, such as non-linearity and self-learning, and combines the experimental data to establish the relationship between various parameters in the battery system. However, it has the disadvantage that a large number of experimental data is needed to predict the battery performance, and greatly depends on the battery's historical data.

7.2.2 Equivalent Circuit Models

As mentioned above, the battery models discussed in this chapter belong to the external characteristic models. In essence, the battery model is used to solve the volt–ampere characteristics for the battery. That is, it can describe the relationship between the voltage and current in the operating process of the battery. On the one hand, this can be expressed by a numerical relationship, but, on the other hand, the battery can be regarded as a two-port network and the circuit network can be used to reflect its volt–ampere relationship. In this sense, an equivalent circuit can be built to describe the volt–ampere relationship of the battery during its operation. Because this circuit follows Thevenin's law, it is sometimes called the Thevenin model. Several representative equivalent circuit models are described below.

Equivalent circuit model based on electron motion theory

According to the internal resistance and electric double layer theory of the battery, an equivalent circuit model may be built for the battery [6], as shown in Figure 7.3. Here R_f is the polarization resistance of the battery, R_Ω is the ohmic resistance of the battery, C_d is an electrical double-layer capacitor, two RC circuits consisting of R_f and C_d

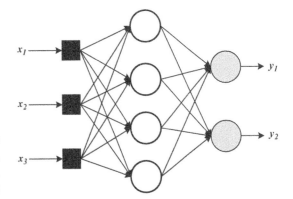

Figure 7.1 Typical artificial neural network with three inputs and two outputs.

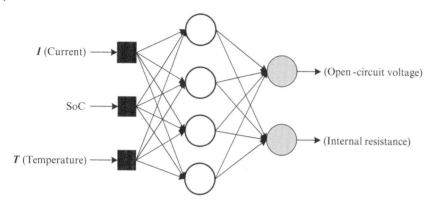

Figure 7.2 Battery model based on an artificial neural network.

are two electrode systems, and R_Ω is the ohmic drop of the battery. This equivalent circuit model describes the characteristics of the battery well, but neglects the potential difference between the two electrodes (the two electrodes deviate from the equilibrium potential by different degrees). In addition, for the Li-ion batteries, this equivalent circuit does not show the rebound voltage characteristics of the battery very well.

PNGV equivalent circuit model

The PNGV model is a standard battery performance model described in the Freedom CAR Battery Test Manual, as shown in Figure 7.4, where R_f is the polarization resistance of the battery, R_Ω is the ohmic resistance of the battery, C_d is the electrical double-layer capacitor, and C_o is the accumulative change of the open-circuit voltage with the time of load current.

The circuit model shown in Figure 7.4 does not reflect the voltage rebound characteristic of the battery, although it includes the polarization of the electrode and the ohmic resistance of the battery. Moreover, the PNGV model has poor practicability due to disadvantages such as a change of each parameter with the SoC and temperature of the battery, complexity in identification of model parameters, and unsuitability for real-time SoC estimation.

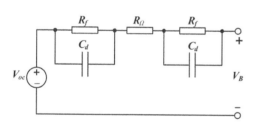

Figure 7.3 Equivalent circuit model based on electron motion theory.

Circuit model with hysteretic characteristics

Figure 7.5 shows the equivalent circuit model considering the hysteretic circuit of the battery. As shown in the figure, the dotted line represents the equilibrium potential of the battery, R_Ω is the ohmic resistance of the battery, R_f is the polarization resistance of the battery, C_d is the electrical double-layer equivalent capacitance, and Z_w is the Weber impedance, which represents the dispersal behavior of the charge particles in the battery and is the capacitive impedance, which similarly represents the parallel RC network with the rebound characteristics of the battery.

In this model, the equilibrium potential of the battery is considered on the basis of the resistance-capacitance equivalent circuit model. That is, the hysteretic voltage of the battery is represented by V_h as part of the equilibrium potential of the battery, so the electromotive force of the battery can be more

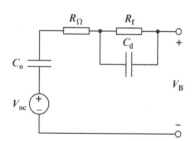

Figure 7.4 PNGV equivalent circuit model.

accurately described. However, the internal resistance spectrum of the battery is required to solve the Weber impedance value and requires a specific electrochemical measurement method during its measurement, for which it is difficult to obtain the required instruments. In addition, although the influence of the hysteretic voltage of the battery is taken into account, this model does not directly reflect the SoC value of the battery, which can only be obtained after an analysis and calculation of the equivalent circuit.

In summary, the existing battery models have the following deficiencies. First, the relationship between the voltage source and SoC is not clear enough. Second, some models are too simple to describe the dynamic characteristics of batteries, such as the voltage rebound characteristic [7]. Third, some models fail to reflect the hysteretic effect of some batteries. The following suggestions are given for battery modeling.

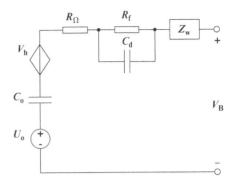

Figure 7.5 Circuit model with hysteretic characteristics.

First, before establishing the battery model, the characteristics of the battery should be fully understood through a characteristic test.

Second, the battery models must be specific. Some models can describe the characteristics of lead-acid batteries well, but they may not be applicable to other battery types, such as Li-ion batteries.

Third, the battery model does not necessarily need to be complex. Although the battery can be more accurately described by the complex models, the complex models are relatively costly and not suitable for the embedded systems that require real-time operation in the electric vehicle [8].

Fourth, it is necessary to study the method for determining the parameters of the model. The form of the model should not be overemphasized, but a set of parameter estimation methods suitable for the model should be given accordingly.

7.3 External Characteristics of the Li-Ion Power Battery and Their Analysis

Section 7.4 provides a new power battery model based on the basic characteristics of the Li-ion battery. In this section, the observed external characteristics of the battery are described and analyzed to provide a basis for the three-order RC model described in the next section. In general, the external characteristics of the power battery include the electromotive force characteristic and the over-potential characteristic, which are described as follows.

7.3.1 Electromotive Force Characteristic of the Li-Ion Battery

EMF (electromotive force) refers to the equilibrium potential of the battery, namely the potential difference between the positive and negative electrodes of the battery system in the balance state, which is an objective physical quantity in the battery system. This section analyzes the EMF characteristics of the Li-ion battery and studies its electrochemical mechanism to provide a basis for obtaining the relationship between SoC and EMF in the battery model.

Since the EMF is the potential of the battery in the balanced state, to obtain the EMF value directly, the voltage at both ends of the battery must be measured when the battery is in a balanced state. Therefore, the following method is generally used: intermittent charging/discharging, that is to say, the battery is charged–discharged for a certain time and then stands for a long enough time to measure the open-circuit voltage of the battery. Then the measured open-circuit voltage value is considered as the EMF value of the battery.

According to the characteristics of the Li-ion battery, after standing for about 8 hours, the battery reaches basically a stable state. However, there is a difference between charging and discharging curves in the equilibrium potential curves obtained using this method. Figure 7.6 shows the equilibrium potential curve upon charging and discharging of a Li-ion battery obtained from the experiments.

Figure 7.6 shows that there is a hysteretic voltage in the Li-ion battery (called the hysteretic voltage because the shape of the battery charging and discharging curves are similar to the magnetic hysteresis curve of a magnetic conductor) and the hysteretic voltage exists in the Li-ion battery, nickel metal hydride batteries, and other types of Li-ion battery. The hysteretic voltage is caused by the electrochemical characteristics of the battery. The reasons for hysteretic voltage are explained in detail later by taking the $LiFePO_4$ battery as an example.

For the $LiFePO_4$ battery, the diffusion coefficient of Li^+ and the movement of Li^+ migration and diffusion will be different when the battery is charged and discharged. Since the positive electrode material of the $LiFePO_4$ battery is crystal $LiFePO_4$ with an olivine structure, its charging and discharging process is the process of Li^+ disengagement from and engagement to the $LiFePO_4$ crystal. The Li^+ disengagement and engagement lead to the volume and structure change of an $LiFePO_4/FePO_4$ crystal in the positive electrode material, namely the phase transition [9]. In order to adapt to this phase transition, during Li^+ disengagement and engagement, the crystal will generate a certain amount of pressure and tension, which are stored in the crystal in the form of "elastic-plastic" energy, while the crystal will behave as an elastic and plastic object.

During the discharging of the $LiFePO_4$ battery, that is, Li^+ engages with $Li_{1-y}FePO_4$ ($FePO_4$ phase), the elastic-plastic energy decreases the battery potential; during charging, that is, Li^+ disengages from Li_xFePO_4 ($LiFePO_4$ phase), the elastic-plastic energy increases the battery potential; the decreased potential $\Delta E_{discharge}$ and the increased potential ΔE_{charge} may be expressed by the following equations:

$$\Delta E_{discharge} = \mathrm{EMF} - E_{discharge} = \frac{Q^t_{discharge} + Q^s_{discharge}}{nF} \tag{7.1}$$

$$\Delta E_{charge} = E_{charge} - \mathrm{EMF} = \frac{Q^t_{charge} + Q^s_{charge}}{nF} \tag{7.2}$$

where $E_{discharge}$ and E_{charge} are respectively the equilibrium potential of the battery upon discharging and charging, EMF is the electromotive force of the battery, $Q^t_{discharge}$, $Q^s_{discharge}$, Q^t_{charge} and Q^s_{charge} are respectively the elastic-plastic energy upon discharging and charging, n is the stoichiometric number upon electrochemical reaction, and F is the Faraday constant.

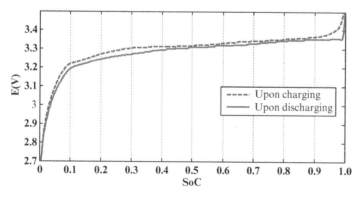

Figure 7.6 Equilibrium potential curve upon charging and discharging of the Li-ion battery. Note: The $LiFePO_4$ battery is exampled here but other Li-ion batteries have similar hysteretic characteristics.

During the charging and discharging process of the LiFePO$_4$ battery, the elastic–plastic energy increases with the increase in charging and discharging depth. According to the above equations, the higher the elastic–plastic energy, the greater the deviation from the EMF will be. Since the "elastic–plastic" energy is different at the beginning of charging and discharging, the phase transition of LiFePO$_4$/FePO$_4$ is different, and different phase transitions generate different "elastic–plastic" energy, which leads to the difference in the equilibrium potential of charging (discharging). In addition, if the battery in the discharging balance state is charged, at the same SoC value, the charging equilibrium potential will be lower than that of the battery in the charging state.

Therefore, according to the experimental charge–discharge equilibrium potential curve and the electrochemical mechanism of the generating hysteretic voltage V_h, the EMF of the battery can be obtained by weighting the charge–discharge equilibrium potential:

Discharging: $V_h = \lambda(\mathrm{OCV}_{charge} - \mathrm{OCV}_{discharge})$ (7.3)

Charging: $V_h = (1 - \lambda)(\mathrm{OCV}_{charge} - \mathrm{OCV}_{discharge})$ (7.4)

$$\mathrm{EMF} = \lambda\mathrm{OCV}_{charge} + (1 - \lambda)\mathrm{OCV}_{discharge}$$ (7.5)

According to the calculation method provided by the above equations and the experimental data, the relation curve of SoC-EMF is calculated as shown in Figure 7.7.

Therefore, the model can be built respectively according to EMF and V_h. Figure 7.8 shows the two parts of the battery equivalent voltage source, EMF and V_h, and the test results can be restored by their combination, which will be discussed in Section 7.4.

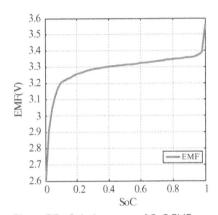

Figure 7.7 Relation curve of SoC-EMF.

7.3.2 Over-potential Characteristics of the Li-Ion Battery

When the electrode reaction occurs in the battery, the electrode potential will deviate from the equilibrium potential. The deviated equilibrium potential is called over-potential. The external characteristics of the battery when it deviates from the balance state are the equivalent impedance and the rebound voltage. In this section, the over-potential characteristics of a battery are discussed from these two aspects.

Equivalent impedance characteristic of battery

When the battery is charged and discharged, the released energy is less than the energy charged into the battery. Table 7.2 shows the ratio of the energy released by the battery to the energy charged into the battery at different charge–discharge rates.

It can be seen from the table that energy loss may be caused in the charging and discharging cycle of the battery, and the value of equivalent impedance calculated with the energy loss is basically the same. Therefore, it can be considered that there is an equivalent resistance in the battery, which will consume a certain amount of energy when the battery is charged and discharged.

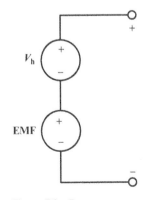

Figure 7.8 Battery equivalent voltage source.

Table 7.2 Charging and discharging efficiency at different charge-discharge rates.

Charge–discharge rate	Charging (J)	Discharging (J)	Efficiency	Equivalent resistance value
5 A	1284672.060	1280946.511	0.997	0.00207
20 A	1255924.267	1240099.621	0.987	0.00220
40 A	1282355.126	1249783.306	0.975	0.00226
60 A	1269754.883	1219726.541	0.961	0.00232

In the operating process of the battery, due to its composition and polarization, a certain voltage drop will occur when the current passes through the battery, which is externally manifested as the equivalent internal resistance of the battery [10]. Figure 7.9 shows the structure diagram of the internal resistance of the battery, which is mainly composed of ohmic resistance and polarization resistance.

The ohmic resistance refers to the resistance caused by the structural characteristics of the battery system and the battery materials, including the contact resistance between the electrode, the electrolyte solution, the diaphragm and other parts, the resistance caused by the surface oxide film of the electrode, and the diaphragm resistance [11]. The contact resistance between all parts may be considered as a constant. In addition, during the operating process of the battery, insoluble substances may be produced on the electrode surface; that is, there may be oxide film, precipitation film and other substances on the electrode surface, resulting in a certain resistance, which can be regarded as a constant for a certain electrode. The diaphragm conducts electricity only when it is immersed in the electrolyte. When the battery is working, the ions will encounter some resistance while passing through the diaphragm. This resistance is called the diaphragm resistance. The diaphragm resistance is mainly related to the resistivity of the diaphragm material and the electrolyte. For a certain diaphragm, its material properties are basically unchanged, while for a certain electrolyte, its resistivity is mainly related to temperature. Therefore, the resistance of the diaphragm and electrolyte can also be approximated as a constant.

For a rechargeable battery, a reversible cell, the electrode, is also reversible. This electrode is called a reversible electrode. When the battery is not working, no current passes through the electrode, the oxidation reaction rate is equal to the reduction reaction rate on the electrode, the net reaction rate is zero, and the battery system is in dynamic balance [12, 13]. When current flows through the electrode, the battery system is no longer in a balanced state, the net reaction rate of the electrode is no longer zero, and the potential of the electrode will deviate from the original potential upon balance. This phenomenon is called electrode polarization. The resistance resulting from the electrode polarization is called the polarization resistance. The causes of electrode polarization are relatively complex. The electrode polarization occurs at the electrode/solution interface, including a series of steps such as adsorption, disengagement, charge transfer, and electrochemical reaction. These steps are complete in series in the electrode reaction. When the rate of a certain step is the slowest, that is, the resistance is the

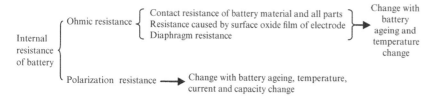

Figure 7.9 Composition structure of the internal resistance of the battery.

maximum, the whole electrode polarization process is controlled by this step. According to the different control steps, the electrode polarization can be divided into electrochemical polarization, concentration polarization, and resistance polarization.

Based on the electrochemical mechanism of the internal resistance of the battery and the external characteristic curve measured by the experiment, the polarization resistance of a certain battery system is mainly affected by the temperature and the ageing degree of the battery, and its value can be regarded as basically unchanged within a certain temperature range and cycle life.

Rebound voltage characteristics of the battery

The external characteristic of battery rebound voltage can be obtained by measuring the open-circuit voltage after the battery changes from a working state to a standing state. Figure 7.10 shows the open-circuit voltage variation curve of the battery after standing at the same discharge rate, different SoC values, and 20°C.

As shown in the above figure, the curve shape is basically the same and different voltages correspond to different SoCs, mainly due to different SoC-EMF relations.

The electrolyte is one of the important components of the battery. The theoretical decomposition voltage of water is 1.23 V. The highest voltage is only about 2 V in a battery made up of aqueous electrolyte (for example, lead–acid batteries). However, as the voltage of the Li-ion battery is 3 ~ 4 V, the aqueous electrolyte is no longer suitable for the Li-ion battery and therefore a non-aqueous electrolyte must be used instead.

The conductivity of the electrolyte is mainly related to the dielectric constant and viscosity of the solvent. The higher the dielectric constant, the greater the number of free lithium ions is and the higher the conductivity of the electrolyte is. The lower the viscosity, the faster the ions move and the higher the conductivity of the electrolyte is. The water has a much higher dielectric constant than other common non-aqueous solvents. As a result, the conductivity of the electrolyte in the Li-ion battery is only a few hundredths of that of the aqueous electrolyte (such as lead–acid or alkaline batteries). Due to the low conductivity, it is impossible for the Li-ion battery to replenish the Li-ion equivalent of the current from the electrolyte during discharging at a high current, causing a voltage drop. When the battery stops discharging and no current flows through the battery, the lithium ions that are not replenished in time enable the battery system to achieve a balance state by two stages, such as diffusion and phase transition, and the external performance of the battery is that the open-circuit voltage rises sharply first

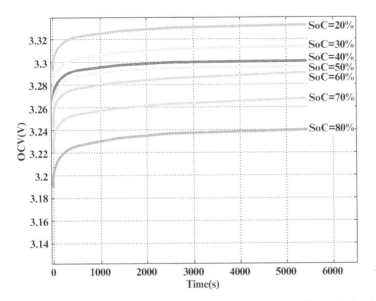

Figure 7.10 Voltage curve at the same discharge rate and different SoC values after standing.

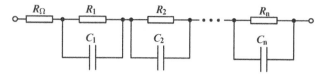

Figure 7.11 The n-order RC network.

and then slowly, until it rises to the balance voltage, that is, the electromotive force of the battery. This phenomenon is called the battery's voltage rebound characteristic.

To sum up, the over-potential of the Li-ion battery shows both resistance and capacitance characteristics, so the over-potential characteristics of the battery can be realized by the resistor–capacitor network. Figure 7.11 shows the structural diagram of the RC network simulating the over-potential characteristics of the battery.

7.4 A Power Battery Model Based on a Three-Order RC Network

In this section, a new power battery model with a high approximation degree, simple structure, and convenient parameter acquisition is established. The high approximation degree is defined relative to the external characteristics of the battery. The external characteristic is actually the relationship between the current and voltage of the Li-ion battery in the process of operation, that is, the volt–ampere characteristics of the battery. The established battery model is suitable for use in the environment with a small temperature change [14, 15].

7.4.1 Establishment of a New Power Battery Model

The traditional equivalent circuit model of the battery is mainly composed of an equivalent voltage source and equivalent impedance. Figure 7.12 shows the circuit of a simple equivalent circuit model, which consists of an ideal voltage source and internal resistance r in series connection.

After analyzing the external characteristics of the battery and its corresponding electrochemical mechanism using a large number of experimental data, the equivalent circuit model of the Li-ion battery can be established, as shown in Figure 7.13. This model has a similar overall structure to the battery model shown in Figure 7.5, and is also composed of the equivalent voltage source and the equivalent impedance. This model not only accurately reflects the rebound voltage characteristics of the Li-ion battery, but also adds the hysteretic voltage characteristics of the Li-ion battery, making the model more perfect and more suitable for the Li-ion battery. The following provides an introduction to the structure of the new model and detailed discussion of the equivalent voltage source and equivalent impedance.

By this model, not only the operating voltage U and the open-circuit voltage (OCV) of the battery can be obtained in real time, but the SoC value of the battery can also be directly estimated in real time. Since the circuit components for the model are common and simple, it is more convenient to implement the model. The circuit components in the simulation software can be directly used to implement this battery model or the mathematical relationship of the circuit model can be obtained after a circuit analysis. The power battery model is then established according to the obtained mathematical relations [16].

The new power battery model mainly consists of two submodules, the equivalent impedance module and the power module of the battery, namely the RC series-parallel network and the OCV power supply as shown in Figure 7.13.

Equivalent voltage source sub-module

The controlled source is a kind of power supply that can be controlled by the current of any other branch in the circuit or the voltage of any single or multiple components. The controlled source can be used to express the hysteretic voltage characteristics of the Li-ion battery and the non-linear relationship between the SoC and EMF [17–19]. Figure 7.14 shows the equivalent voltage source of the established Li-ion power battery model.

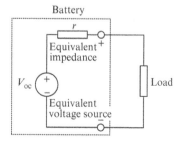

Figure 7.12 Simple equivalent circuit model.

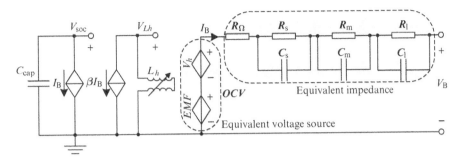

Figure 7.13 New power battery model.

As shown in Figure 7.14, the equivalent voltage source of the Li-ion power battery model consists of branches A, B, and C. The equivalent voltage source OCV represented in branch C is composed of the EMF and V_h. The voltage-controlled voltage source (VCVS) representing the EMF is controlled by the SoC of the battery, and the current-controlled voltage source (CCVS) represents the hysteretic voltage V_h, which is controlled by the SoC and current I_B of the battery.

The capacitance value of the capacitor C_{cap} in branch A represents the rated capacity of the battery. The value of the current flowing through C_{cap} is the value of the current flowing through the battery. That is, for the current controlled current source (CCCS) I_B shown in branch A, the voltage V_{soc} at both ends represents the SoC of the battery. Therefore, the V_{soc} value is between 0 and 1.

The hysteretic voltage of the Li-ion power battery is not only related to the SoC of the battery, but is also affected by the historical current. To determine whether the hysteretic voltage V_h of a battery at a certain time is a charging hysteretic voltage or a discharging hysteretic voltage, the first thing is to find out whether the battery is charging or discharging before this time. Therefore, it is necessary to use the circuit components with a memory effect on the current to indicate that the hysteretic voltage V_h is affected by the historical current. The current flowing through the adjustable inductance L_h, as shown in branch B, is controlled by the current I_B of the battery. The voltage V_{Lh} at both ends of the adjustable inductance indicates that the hysteretic voltage V_h of the battery is the discharging hysteretic voltage or the charging hysteretic voltage; that is, the direction of the hysteretic voltage V_h is decided by V_{Lh}. The hysteretic voltage V_h is controlled by the SoC of the battery, that is, controlled by the voltage V_{soc} at both ends of the capacitor C_{cap} in branch A.

Therefore, the relationship between SoC–EMF and SoC–V_h of the battery can be easily solved by using the controlled source, and the SoC of the battery can be obtained directly and in real time through the established equivalent voltage source.

Equivalent impedance submodule

According to the characteristic shown in the battery rebound voltage curve and the mechanism of the battery rebound characteristic, the equivalent impedance model of the battery can be established. Figure 7.15 shows the open-circuit voltage curve at room temperature after discharging the battery at a discharge rate of 0.5 C.

As shown in Figure 7.15, the rebounding characteristic curve of the battery, i.e. the changeable ΔU of the voltage, can be divided into two parts: rapid voltage rises and slow voltage rises, i.e. ΔU_1 and ΔU_2.

ΔU_1 is mainly affected by the ohmic resistance of the battery. Assuming that the internal resistance of the battery is the series resistance of the ideal power supply, at the moment when the battery circuit is disconnected, the current flowing through the series resistance will suddenly become zero, that is, the open-circuit voltage will have a corresponding jump. Therefore, the

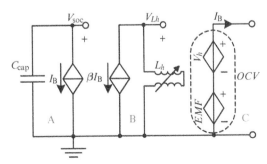

Figure 7.14 Equivalent voltage source of the submodule.

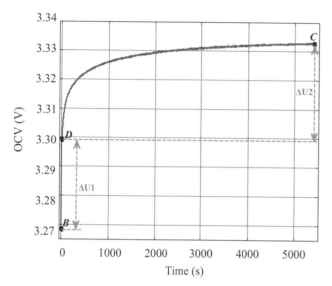

Figure 7.15 Rebound voltage curve.

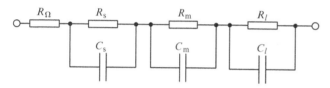

Figure 7.16 Equivalent impedance model.

voltage change of ΔU_1 could be considered as an instantaneous voltage response of the ohmic resistance upon the disconnection.

The voltage change of ΔU_2 is mainly affected by the phase transition of the positive electrode material of the Li-ion battery and the polarization of the battery, i.e. the Li^+ diffusion/migration rate. In ΔU_2, the rise rate of the open-circuit voltage of the battery is quite large within 200s, decreases within 200 to 500s, and becomes low after 500s, after rebounding is started, while the change of the open-circuit voltage of the battery is also small. Therefore, ΔU_2 can equivalently correspond to the voltage response of a three-order RC network.

According to ΔU_1 and ΔU_2, in order to establish the equivalent impedance model with a high degree of approximation with the actual external characteristics of the battery and a relatively simple structure, the equivalent impedance model of the battery can be described by the three-order RC series parallel network shown in Figure 7.16.

The three-order RC network, rather than the two-order or four-order or even higher-order RC network, is used to establish the model mainly because the established equivalent circuit model must have the following characteristics: the external characteristics shown by the established model is highly approximate to the actual external characteristics of the battery, and the structure of the model is simpler. However, these characteristics are contradictory. The higher the approximation degree of the model, the higher the order of the RC network required, and the more complex the structure of the model becomes. Conversely, the simpler the structure of the model, the lower the approximation is. The choice of RC network order is now discussed.

Figure 7.17 shows the OCV curve of the battery output by simulating the over-potential characteristics of the Li-ion battery with RC networks with different orders.

Table 7.3 shows the errors of the OCV of the battery output by the model (compared with the actual measured OCV of the battery) when the RC network model with different orders is applied to the Li-ion battery. As the characteristics simulated by a one-order RC network, as shown in Figure 7.17, are very different from the actual rebound characteristics of the battery, the errors caused by two-order to five-order RC networks are compared here.

According to the comparison of the approximation degree of the external characteristics of the battery fit by the RC network at different orders and their errors, the higher the order of RC network, the higher is the approximation degree of fitting. However, for a three- to five-order RC network, the error change is not large, and considering the complexity of the circuit model, a three-order RC network is more suitable for modeling the Li-ion power battery.

7.4.2 Estimation of Model Parameters

Estimation of equivalent voltage source parameters

In the platform area, EMF is the mean value of the charge–discharge equilibrium potential, and its weight is 0.5. In the non-platform area, SoC has a linear relationship with weight λ, that is, when SoC is $0 \sim 0.1$, λ is $1 \sim 0.5$, and

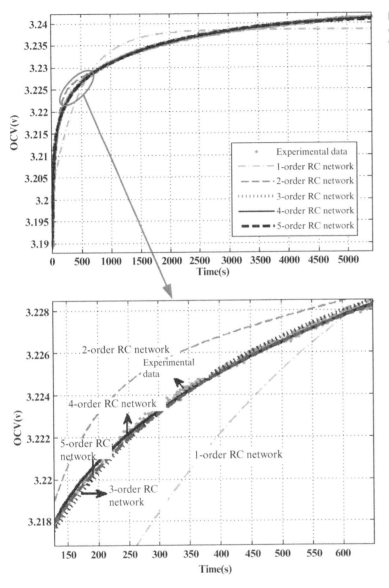

Figure 7.17 Rebound voltage characteristic simulated by RC networks with different orders.

Table 7.3 Errors of external characteristics of battery fit by different orders of RC networks.

Orders of RC network	Standard deviation (V)	Maximum deviation (V)	Average deviation (V)
Two orders	0.0005751	0.00320	0.000225
Three orders	0.0002144	0.00086	0.000163
Four orders	0.0002313	0.00062	0.000115
Five orders	0.0001398	0.00107	0.000121

when SoC is 0.9 ~ 1, λ is 0.5 ~ 1. Table 7.4 shows the calculation method of hysteretic voltage V_h upon discharging and charging in different SoC values.

Since EMF = OCV ± V_h (positive for charging and negative for discharging), after considering the hysteretic voltage V_h, the calculation equations for the EMF value corresponding to different SoC values may be established, as shown in Table 7.5.

Estimation of equivalent impedance parameters

When the battery stops operation, the voltage rebound starts (drops when the charging stops and rises when the discharging stops). The curve obtained after discharging the battery at discharge ratio of 0.5 C for 300 s and standing is shown in Figure 7.18, where the battery is enabled standing before the point A, the battery is discharged during $A \to B$, the battery is enabled standing during $B \to C$, $A \to B$ time is the same as $B \to C$ time, and the standing time is enough long, that is, OCV_C is the equilibrium potential. During $t_{A \to B} = t_\omega = 12 \text{ min} = 720s$, the battery is discharged at a constant current and 0.5 C.

During $A \to B$, there is a zero-state response on the RC network $u(t) = Ri(t)(1 - e^{-t/(RC)})$. According to the equivalent circuit model shown in Figure 7.19, A is considered as the starting point ($t_A = 0$) and the battery voltage at any time t during $A \to B$ is calculated by

Table 7.4 Hysteretic voltage V_h upon discharging and charging at different SoC values.

SoC	V_h upon charging	V_h upon discharging
$0 \to 0.1$	$V_h = 5\, \text{SoC}(OCV_{charge} - OCV_{discharge})$	$V_h = (-5\, \text{SoV} + 1)(OCV_{charge} - OCV_{discharge})$
$0.1 \to 0.9$	$V_h = 0.5\,(OCV_{charge} - OCV_{discharge})$	$V_h = 0.5\,(OCV_{charge} - OCV_{discharge})$
$0.9 \to 1$	$V_h = (5\, \text{SoV} - 4)(OCV_{charge} - OCV_{discharge})$	$V_h = (-5\, \text{SoV} + 5)(OCV_{charge} - OCV_{discharge})$

Table 7.5 Calculation equations for EMF values corresponding to different SoC values.

SoC	EMF
$0 \to 0.1$	$\text{EMF} = (-5\, \text{SoC} + 1)OCV_{charge} + (5\, \text{SoC})OCV_{discharge}$
$0.1 \to 0.9$	$\text{EMF} = 0.5\,(OCV_{charge} + OCV_{discharge})$
$0.9 \to 1$	$\text{EMF} = (-5\, \text{SoC} + 5)OCV_{charge} + (5\, \text{SoC} - 4)OCV_{discharge}$

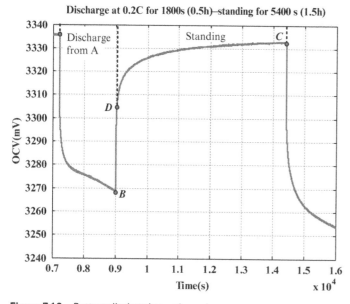

Figure 7.18 Battery discharging–rebound curve.

Figure 7.19 Equivalent circuit of a battery-discharging circuit.

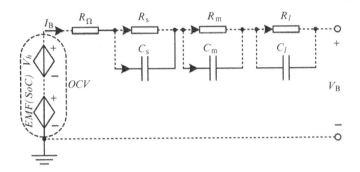

$$u_B(t) = OCV_t - u_s(t) - u_m(t) - u_l(t) - IR_\Omega$$

$$= OCV_t - IR_s\left(1 - e^{-\frac{t}{R_sC_s}}\right) - IR_m\left(1 - e^{-\frac{t}{R_mC_m}}\right) - IR_l\left(1 - e^{-\frac{t}{R_lC_l}}\right) - IR_\Omega \tag{7.6}$$

When the discharging time $t_\omega = 720s$, based on the experimental data and the characteristics of the Li-ion battery, R_s, R_m, and R_l values can be estimated in milliohms, so $t/(R_sC_s)$, $t/(R_mC_m)$, and $t/(R_lC_l)$ values are 10 ~ 100 order of magnitudes, namely $e^{-t_\omega/(R_sC_s)} \approx e^{-t_\omega/(R_mC_m)} \approx e^{-t_\omega/(R_lC_l)} \approx 0$ ($e^{-10} = 4.5400\text{e-}005$), the operating voltage at the point B is expressed by

$$u_B(t_\omega) = OCV_C - IR_s - IR_m - IR_l - IR_\Omega \tag{7.7}$$

Therefore, the voltages of three *RC* networks at the point B are respectively:

$$u_s(t_\omega) = IR_s \tag{7.8}$$

$$u_m(t_\omega) = IR_m \tag{7.9}$$

$$u_l(t_\omega) = IR_l \tag{7.10}$$

It can be seen from Figure 7.18 that the battery is changed from the operating state to the standing state at the point B, that is, at the moment when the battery circuit is disconnected, the battery voltage jumps mainly for the ohmic resistance. According to the established equivalent circuit model, this phenomenon corresponds to the effect of series resistance R_Ω in the equivalent circuit model, that is, during $B \to D$ (the time interval between B and D is very small and they are almost at the same time point), an ohmic drop is caused and the ohmic resistance R_Ω can be calculated using the following equation:

$$R_\Omega = \frac{|U_D - U_B|}{I} \tag{7.11}$$

Estimation of polarization resistance and polarization capacitance parameters

When the circuit of the battery is disconnected, the external current I_B flowing through the battery is zero and the net reaction rate of the electrode will not immediately be zero due to the influence of the electrochemical mechanism of the battery inside it. As shown in Figure 7.18, $B \to C$ area corresponds to the rebound characteristics upon standing. The battery equivalent circuit model upon standing is shown in Figure 7.20.

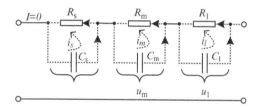

Figure 7.20 Voltage response of equivalent impedance after discharging and standing of the battery.

In Figure 7.20, there is a zero input response in the *RC* network $u(t) = U(0)e^{-t/(RC)}$, as there is an ohmic drop during $B \rightarrow D$, when R_s, C_s, R_m, C_m, R_l, and C_l parameters are calculated by fitting the $U_C(t)$ curve. The data used for fitting should be selected from the data obtained from the next sampling time after the point B to the point C, namely the data obtained during $D \rightarrow C$. If the point D is used as the starting point ($t = 0$), the initial voltage of the *RC* network is the voltage at the point B (or the point D), $u_s(0) = u_s(t_\omega) = IR_s$, $u_m(0) = u_m(t_\omega) = IR_m$, $u_l(0) = u_l(t_\omega) = IR_l$, and the operating voltage of the battery at any time during $D \rightarrow C$ is calculated by the following equation:

$$
\begin{aligned}
u_C(t) &= \mathrm{OCV}_C - u_s(t) - u_m(t) - u_l(t) \\
&= \mathrm{OCV}_C - u_s(0)e^{-t/(R_sC_s)} - u_m(0)e^{-t/(R_mC_m)} - u_l(0)e^{-t/(R_lC_l)} \\
&= \mathrm{OCV}_C - IR_se^{-t/(R_sC_s)} - IR_me^{-t/(R_mC_m)} - IR_le^{-t/(R_lC_l)}
\end{aligned}
\tag{7.12}
$$

Since the standing time is long enough, $e^{-t/(R_sC_s)} \approx e^{-t/(R_mC_m)} \approx e^{-t/(R_lC_l)} \approx 0$ the voltage at the point C is calculated by the following equation:

$$
U_C = u(t_\omega) = \mathrm{OCV}(t_\omega) - u_s(t_\omega) - u_m(t_\omega) - u_l(t_\omega) = \mathrm{OCV}_C
\tag{7.13}
$$

Therefore, the voltage increment during rebound from the point B to the point C is calculated by the following equation:

$$
U_C - U_B = IR_\Omega + IR_s + IR_m + IR_l
\tag{7.14}
$$

This equation indicates that the rebound voltage of the battery depends on the equivalent internal resistance of the battery.

After fitting the rebound voltage and time (OCV-Time) curve according to the function $y = \mathrm{OCV}_\infty - b_1e^{-\tau_1 t} - b_2e^{-\tau_2 t} - b_3e^{-\tau_3 t}$, the parameters b_1, τ_1, b_2, τ_2, b_3, and τ_3 can be obtained. The parameters of the model are calculated by the following equations:

$$
\begin{cases}
R_s = \dfrac{b_1}{I} \\[2mm]
R_m = \dfrac{b_2}{I} \\[2mm]
R_l = \dfrac{b_3}{I} \\[2mm]
C_s = \dfrac{I}{\tau_1 b_1} \\[2mm]
C_m = \dfrac{I}{\tau_2 b_2} \\[2mm]
C_l = \dfrac{I}{\tau_3 b_3}
\end{cases}
\tag{7.15}
$$

7.5 Model Parameterization and Its Online Identification

The high-precision battery model is necessary in order to estimate the battery's SoC. On the one hand, it is necessary to consider whether the model structure conforms to the original electrochemical characteristics of the battery. On the other hand, the model depends on the accuracy of the model parameters. After the model structure is determined, the variation characteristics of the battery with environmental conditions are reflected by the model parameters, so the model parameters should be a function of the external environmental conditions. When the battery behavior is estimated by the model, the model parameter should be upgraded in real time according to the operating conditions. The model parameters can be upgraded according to the test data measured in advance or the real-time voltage and current information of the battery in combination with the optimal estimation theory. According to different application conditions, this section introduces two parameter upgrade methods: offline extension and online identification. In general, the equivalent circuit model is used only to perform approximate simulation by the circuit structure with similar external characteristics, without consideration of the real operation mechanism of the battery, so its accuracy is limited. The accuracy of the model can be improved by optimizing the parameter identification method and adjusting the model structure.

7.5.1 Offline Extension Method of Model Parameters

The offline extension method of a model parameter means that the battery parameters under different conditions are obtained by battery characteristic tests, the zero-response equation of the battery derived according to the battery model, and fitting the rebound voltage value upon standing. A characteristic parameter database is established and the optimal parameter value under the current condition is obtained by the multidimensional query method in the application [20].

According to the DC equivalent resistance test in the battery characteristic test (see Section 5.4 for details), several rebound voltage curves corresponding to different SoC values can be obtained, as shown in Figure 7.21. The model parameters in each SoC value can be obtained by least squares fitting of the voltage response during standing. This section introduces the parameter identification method based on the zero-input response time domain equation by exampling the *n*-order RC network model.

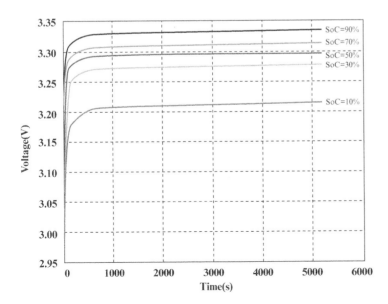

Figure 7.21 Rebound voltage curves in different SoC values.

It can be seen from the analysis of Figure 7.15 that the ohmic resistance of the battery can be obtained by the instantaneous rebound voltage after the current excitation is removed:

$$R_\Omega = \frac{U_D - U_B}{I} \tag{7.16}$$

At the same time, during the standing of the battery, the voltage at two ends of the battery meets

$$u(t) = U_c - IR_1 e^{-t/(R_1 C_1)} - IR_2 e^{-t/(R_2 C_2)} - IR_n e^{-t/(R_n C_n)} \tag{7.17}$$

where $R_1, C_1, R_2, C_2, \ldots R_n$ and C_n are $2n$ specific parameters of the polarization resistance and the polarization capacitance to be determined, I is the discharge current required for the test, and U_c is the voltage after full rebound, which may be considered as a constant; t and $u(t)$ correspond to the sampling data used in the test. These specific parameters can be obtained by least squares non-linear regression.

The model parameters are generally obtained under specific environmental conditions and vary with different temperatures, multipliers, and SoC. Therefore, the model parameters are multidimensional functions designed to function together with different factors. When it is required to extract the model, different conditions can be entered as queries. If the same data with those under current conditions are available, the data may be directly read. However, due to limited testing time and resources, it is generally difficult to test under all environmental conditions. Therefore, how to determine the value under the condition that the sample test cannot cover is a key problem of offline parameter identification. In this section, by means of multidimensional data extension and the cyclic interpolation of known test parameters, the parameters to be measured under arbitrary conditions are obtained in order to determine the model parameters.

First, the values are evenly divided and selected for operation conditions of the battery, such as temperatures of 0°C, 10°C, 20°C, 30°C, 40°C, 50°C, battery capacities of 5 Ah, 10 Ah, 20 Ah, 50 Ah, 100 Ah, and SoCs of 20%, 40%, 60%, 80%, 100%. The above operating conditions of the battery are combined to form 125 operating conditions of the battery. Under these 125 conditions, the actual battery is tested to obtain 125 sets of parameters to be tested. Since the SoC change can be carried out in the same test, the actual test number is 25, and the establishment of a sample database is completed. The larger the data size of the sample database, the more accurately it can describe the real behavior of the battery under different conditions. However, given the testing time and cost and the storage capacity of the existing processing units, the database cannot be too large to cover all conditions.

Since the sample database only contains the parameters to be measured under certain operating conditions, such as temperature, battery capacity, and SoC, if the parameters to be tested under any operating conditions are required, a fitting extension method should be adopted to calculate the parameters. The fitting method can be freely selected according to the actual needs. Considering the computing power of the general processing unit in the project, a piecewise linear interpolation and a cubic spline interpolation are preferred in this section. The piecewise linear interpolation is described as follows.

Assuming that the function value on a known node $a = x_0 < x_1 < \cdots < x_n = b$ is $y_0, y_1, \ldots, y_n$, for any $x \in [a, b]$,

$$\varphi(x) = \frac{x - x_{i+1}}{x_i - x_{i+1}} y_i + \frac{x - x_i}{x_{i+1} - x_i} y_{i+1} \quad (x_i \le x \le x_{i+1}) \tag{7.18}$$

Extrapolation may be performed for $x \notin [a, b]$:

$$\varphi(x) = \frac{x - x_1}{x_0 - x_1} y_0 + \frac{x - x_0}{x_1 - x_0} y_1 \quad (x < x_0) \tag{7.19}$$

$$\varphi(x) = \frac{x - x_n}{x_{n-1} - x_n} y_{n-1} + \frac{x - x_{n-1}}{x_n - x_{n-1}} y_n \quad (x_n < x) \tag{7.20}$$

The cubic spline interpolation

$$\varphi(x) = \frac{(x_i - x)^3}{6h_i} M_{i-1} + \frac{(x - x_{i-1})^3}{6h_i} M_i + (y_{i-1} - \frac{M_{i-1} h_i^2}{6}) \frac{x_i - x}{h_i} + (y_i - \frac{M_i h_i^2}{6}) \frac{x - x_{i-1}}{h_i} \tag{7.21}$$

where

$$h_i = x_i - x_{i-1} \tag{7.22}$$

$$d_i = \mu_i M_{i-1} + 2M_i + \lambda_i M_{i+1} \tag{7.23}$$

$$\mu_i = \frac{h_i}{h_i + h_{i+1}} \tag{7.24}$$

$$\lambda_i = \frac{h_{i+1}}{h_i + h_{i+1}} \tag{7.25}$$

$$d_i = \frac{6}{h_i + h_{i+1}} (\frac{y_{i+1} - y_i}{h_{i+1}} - \frac{y_i - y_{i-1}}{h_i}) = 6f(x_{i-1}, x_i, x_{i+1}) \tag{7.26}$$

It is assumed that a set of basic databases with 125 samples has been obtained through the basic test, as shown in Figure 7.22. In the parameters to be identified in the model, each parameter has a specific relationship with temperature T, battery capacity C, and SoC S. Taking the polarization resistance R_s as an example, according to the existing information, the R_s value is calculated under conditions such as temperature $T = 35°C$, battery capacity $C = 15$ Ah, and SoC $S = 70\%$. Other parameters are obtained by similar methods.

First, the SoC S is fitted to obtain 25 circular reference points in Figure 7.22:

$$R_s(A1) = R_s\left(T = 40°C, C = 100 \text{ AH}, S = 70\%\right)$$
$$= \frac{70 - 80}{60 - 80} R_s\left(T = 40°C, C = 100 \text{ AH}, S = 60\%\right) \tag{7.27}$$
$$+ \frac{70 - 60}{80 - 60} R_s\left(T = 40°C, C = 100 \text{ AH}, S = 80\%\right)$$

$$R_s(A2) = R_s\left(T = 30°C, C = 100 \text{ AH}, S = 70\%\right)$$
$$= \frac{70 - 80}{60 - 80} R_s\left(T = 30°C, C = 100 \text{ AH}, S = 60\%\right) \tag{7.28}$$
$$+ \frac{70 - 60}{80 - 60} R_s\left(T = 30°C, C = 100 \text{ AH}, S = 80\%\right)$$

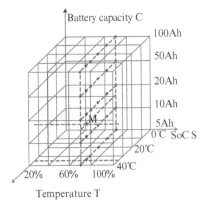

Figure 7.22 Multidimensional extension of model parameters.

Similarly,

$$R_s(A25) = R_s\left(T = 0^\circ C, C = 5\,AH, S = 70\%\right)$$
$$= \frac{70-80}{60-80} R_s\left(T = 0^\circ C, C = 5\,AH, S = 60\%\right) + \frac{70-60}{80-60} R_s\left(T = 0^\circ C, C = 5\,AH, S = 80\%\right) \tag{7.29}$$

where A1, A2, and A25 represent 25 different operating conditions of the battery obtained by combination of SoC $S = 70\%$, temperature T, and battery capacity in turn.

The temperature T is fitted to obtain five rhombic reference points in Figure 7.22:

$$R_s(A26) = R_s\left(T = 35^\circ C, C = 100\,AH, S = 70\%\right)$$
$$= \frac{35-40}{30-40} R_s\left(T = 30^\circ C, C = 100\,AH, S = 70\%\right) + \frac{35-40}{40-30} R_s\left(T = 40^\circ C, C = 100\,AH, S = 70\%\right) \tag{7.30}$$

Similarly,

$$R_s(A30) = R_s\left(T = 35^\circ C, C = 5\,AH, S = 70\%\right)$$
$$= \frac{35-40}{30-40} R_s\left(T = 30^\circ C, C = 5\,AH, S = 70\%\right) + \frac{35-40}{40-30} R_s\left(T = 40^\circ C, C = 5\,AH, S = 70\%\right) \tag{7.31}$$

where A26 and A30 represent five different operating conditions of the battery obtained by combination of SoC $S = 70\%$, temperature $T = 35^\circ C$, and battery capacity C in turn.

The temperature T is fitted to obtain the star-like reference points in Figure 7.22, namely the required value of R_s under the condition of $T = 35^\circ C$, battery capacity $C = 15$ Ah, and remaining power $S = 70\%$:

$$R_s\left(A31\right) = R_s\left(T = 35^\circ C, C = 15\,AH, S = 70\%\right)$$
$$= \frac{15-20}{10-20} R_s\left(T = 35^\circ C, C = 10\,AH, S = 70\%\right) + \frac{15-10}{20-10} R_s\left(T = 35^\circ C, C = 20\,AH, S = 70\%\right) \tag{7.32}$$

where $R_s(A31)$ is the optimal R_s value under the operating condition of $T = 35^\circ C$, battery capacity $C = 15$ Ah, and SoC $S = 70\%$.

In this way, the offline parameter selections can be realized by less piecewise linear fitting of the calculated value. In order to demonstrate the universality of the value extension method, a complete calculation process is introduced herein. In practice, the value may be calculated for fewer points during the linear interpolation. Since the estimation only depends on the information of adjacent points, the value is only calculated for seven adjacent points including the target points, which can greatly reduce the calculation work. The extension calculation of the model parameters depends on the external influence factors. The more influence factors, the longer is the extension step. The relevant steps can be realized through the program and the fitting function can be embedded in the program after the pre-processing. In the actual calculation, without the need to real-time fit, there is only a requirement to perform the data retrieval and some primary operations, which does not require a high computing power of the processor.

In the actual operating conditions, the change of external conditions is relatively slow. Therefore, the calculation frequency of the parameters can be appropriately reduced in the application. For example, the parameters are upgraded and extended every 20 s. In addition to the above conditions, such as temperature, capacity, and SoC, the influence factors of the other parameters, such as the hysteretic effect and battery degradation degree, can be extended through this method, but it is a requirement to accordingly extend and perfect the database and the parameter extension steps.

7.5.2 Online Identification Method of Model Parameters

In the offline extension method introduced in the previous section, the parameter values under any conditions can be obtained by using known parameter data with less calculation work, and the method is applicable to most engineering application scenarios. However, the method has a high requirement for test quantity in the early stage and requires the support of abundant test data. Therefore, in this section, an online identification method of the model parameters is proposed for the condition of a small sample size. Under this condition of no relying on previous data, the model parameters are estimated in real time by using the current and voltage information of a period of time.

According to the state equation of the power battery model, the voltage response of the battery can be expressed by the iterative operation of the equation set, which is a dynamic system using the current as input, model parameters as adjustable variables, and voltage as output. The accurate model parameters should correctly reflect the mapping relation between the input and the output of the system, so that at a specific current output, the output measured through the model operation is consistent with the battery output.

According to the above analysis, the voltage and current data of the past period are selected to obtain the real current as the excitation. The model parameters are adjusted by an heuristic algorithm to minimize the error sum of squares of the estimated voltage and the real voltage, and approximate the real voltage in the sense of the least square, in order to achieve the optimal estimation of parameters. To this end, the objective function is defined as follows:

$$Z = \sum_{k=1}^{N}(y_k - v_k)^2 \tag{7.33}$$

where N is the sample size used for the parameter identification, k is the sampling time, y_k is the real voltage at k time, and v_k is the estimated voltage at k time, which is calculated by the state equation corresponding to the battery model. Each set of model parameters corresponds to a target parameter value, so it can be regarded as an extreme value optimization by using the parameters as independent variables. In order to avoid local optimization and improve the calculation efficiency, a Tree Seed Algorithm (TSA) based on group intelligence is used in this section to identify the model parameters and verify the effectiveness of the algorithm through two dynamic working conditions [21]. The calculation flow of the algorithm is shown in Figure 7.23.

The optimized model parameters are used as the tree seeds in the algorithm, and the tree seeds are firstly initialized by random values in the present range; then

$$T_{i,j} = L_{j,min} + r_{i,j} \times (H_{j,max} - L_{j,min}) \tag{7.34}$$

where $T_{i,j}$ is the initialized value at the No. j dimensional position in the tree i, $L_{j,min}$ is the lower limit set for the value at the No. j dimensional position according to feasible region, $H_{j,max}$ is the upper limit of the value at the No. j dimensional position, and $r_{i,j}$ is a random number in the range of $[0,1]$.

Then the corresponding objective function value is calculated by Equation (7.33), if the function value meets the convergence condition, the current parameter is output; otherwise, the parameters are updated. The tree seeds are upgraded through two strategies:

$$S_{i,j,m} = T_{i,j} + \alpha_{i,j} \times (B_j - T_{r,j}) \tag{7.35}$$

$$S_{i,j,m} = T_{i,j} + \alpha_{i,j} \times \left(T_{i,j} - T_{r,j}\right) \tag{7.36}$$

where $S_{i,j,m}$ is the No. j dimensional position of the seed m in the tree i, B_j is the No. j dimensional position corresponding to the optimal tree in the population at iterations time, and the optimal tree refers to the tree sample minimizing the objective function. $T_{r,j}$ is the No. j dimensional position in the tree r, and r is a random integer that is less than or equal to the number of population NP; $\alpha_{i,j}$ is a random mutagenic factor in the range of $[0,1]$. The strategy selection is controlled by the search factor γ, which is a constant in the range $[0, 1]$. When

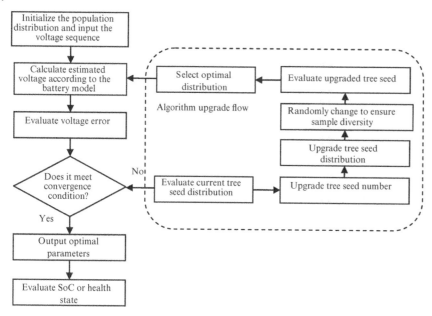

Figure 7.23 Model parameter online identification.

each direction is upgraded, a random number in the range [0, 1] is generated for judgment. If the random number is less than γ, Equation (7.35) is used for an upgrade; otherwise, Equation (7.36) is used for the upgrade.

After the parameters are updated, the objective function value is recalculated. If the function value meets the convergence condition, output the optimal parameter; otherwise, enter the next iteration update. As the samples used for the parameter identification can be obtained online, for example, the data at the first 1000 can be acquired in the data cache in real time to accordingly achieve the online optimization of the model parameters.

In order to verify the accuracy of the different parameter identification methods, Tables 7.6 and 7.7 provide the tests based on two dynamic working conditions, including the dynamic stress test (DST) and the federal urban dynamic schedule (FUDS). These two tests for working conditions are recommended in the USABC manual and can simulate the real operating conditions of city roads, they have strong dynamics, and can be used to verify the accuracy of the model and the parameters.

The current in DST working condition is shown in Figure 7.24(a). For the purpose of this book, the discharge is defined as positive, the charge as negative, and the sampling range is 1000 s. The measured voltage and the voltage estimations calculated by different parameter identification methods are shown in Figure 7.24(b), which provides the comparison of the estimation results obtained by the offline identification, and the online identification based on the genetic algorithm (GA) and the tree seed algorithm (TSA) [22]. In order to more clearly compare the precision of the model with the different parameters, the voltage estimation errors are provided as shown in Figure 7.24(c). The tree seed algorithm error distribution is shown in Figure 7.24(d).

It can be seen from Figure 7.24(b) and (c) that the model parameters obtained by the traditional offline method may have a large estimation error under dynamic working conditions. The error range can be reduced to a certain extent by increasing the sample size of the early phase test and improving the offline database. In such a case, the test cost is increased and a large storage capacity is also required for the processor. On the contrary, the model accuracy can be effectively improved by online parameter identification based on the optimization algorithm. In order to visually compare the model accuracy under different methods, the mean square error and maximum error of the statistical voltage estimation are shown in Table 7.6. It can be seen from the table that after the parameter online identification provided in this section, the RMSE and MAE of the model are 3.28 mV and 12.56 mV respectively, which can basically meet the needs of the battery state estimation. The estimated results under the FUDS test conditions are shown in Figure 7.25 and Table 7.7. The effectiveness of the proposed method is also verified by relevant results.

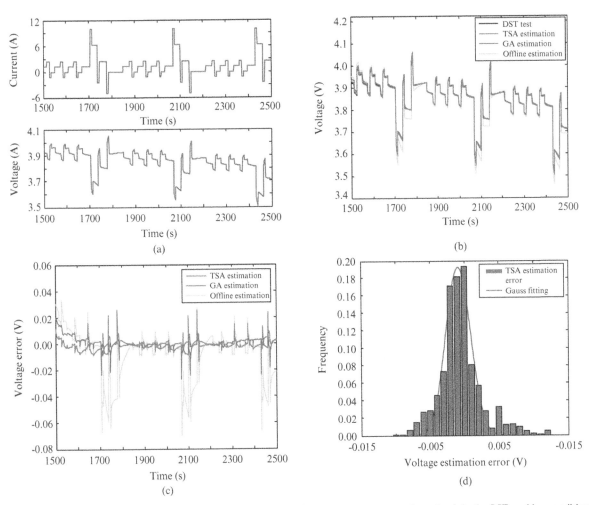

Figure 7.24 Results of the model voltage estimation based on different parameter upgrade methods in the DST working condition.

Table 7.6 Error of the model voltage estimation based on different parameter upgrade methods in the DST working condition.

	Tree seed optimization algorithm	Traditional genetic algorithm	Offline extension method
RMSE (mV)	3.28	6.96	19.69
MAE (mV)	12.56	27.98	70.04

Table 7.7 Error of the model voltage estimation based on different parameter upgrade methods in the FUDS working condition.

	Tree seed optimization algorithm	Traditional genetic algorithm	Offline extension method
RMSE (mV)	3.06	6.32	13.63
MAE (mV)	18.24	30.41	80.45

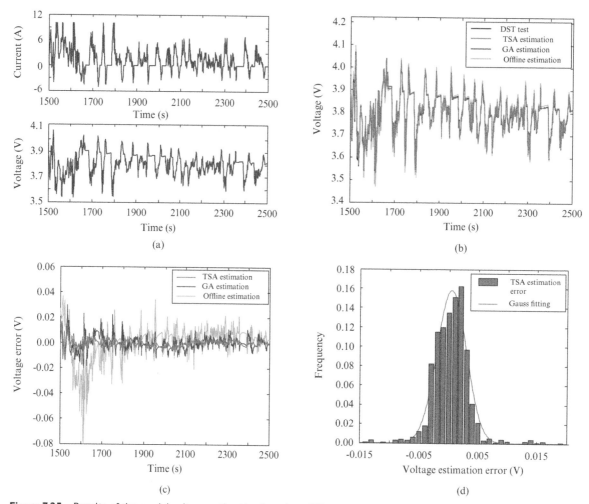

Figure 7.25 Results of the model voltage estimation based on different parameter upgrade methods in the FUDS working condition.

7.6 Battery Cell Simulation Model

Since some parameters of the equivalent circuit model established for the battery vary with such factors as SoC and current of the battery, it is not suitable to establish the battery model only by circuit components. Therefore, the battery model is realized by the standard modules and some components in Matlab/Simulink® [23]. The following provides the specific realization process of the battery model in Simulink and the verification and application of the battery model.

7.6.1 Realization of Battery Cell Simulation Model Based on Matlab/Simulink

Based on the equivalent circuit model of the power battery as shown in Figure 7.13, the corresponding equivalent circuit model is established in Simulink. Figure 7.26 shows the structural diagram of the model simulation.

The constant module and the resistance R_{ohm} in the Simulink model shown in Figure 7.26 are the parameters of the battery model. The ohmic resistance of the battery is represented by the resistance components rather than the mathematical relationship module, mainly for two reasons. First, the ohmic resistance of the battery is basically unchanged if there is no large change in the cycle life of the battery. Second, the addition of circuit components can prevent algebraic rings in the simulation process, avoid the addition of redundant modules (if there is an algebraic ring in the model simulation, it is necessary to add a delay module to eliminate the algebraic rings), and reduce the complexity of the model.

In the model, C_{cap} is the rated capacity of the battery, for example, for the battery with a rated capacity of 100 Ah, $C_{cap} = 100 \times 3600 = 3.6 \times 10^5$ As, the initial SoC is the SoC of the battery before operation, and may be set to any number in the range of 0 ~ 1 as required before simulation. R_s, C_s, R_m, C_m, R_l, C_l, and R_Ω are the parameters corresponding to the equivalent impedance of the equivalent circuit model of the battery.

7.6.2 Model Validation

For the equivalent circuit model of the battery established in the above section, consideration is given to not only the rebound voltage characteristic of LiFePO$_4$ power battery but also the hysteretic voltage characteristic of the battery. Therefore, the accurate SoC value can be obtained from the operating voltage and current of the battery measured in the experiment. In this section, the fifteenth and UDDS working conditions are exampled to verify the accuracy of the SoC estimated by the model.

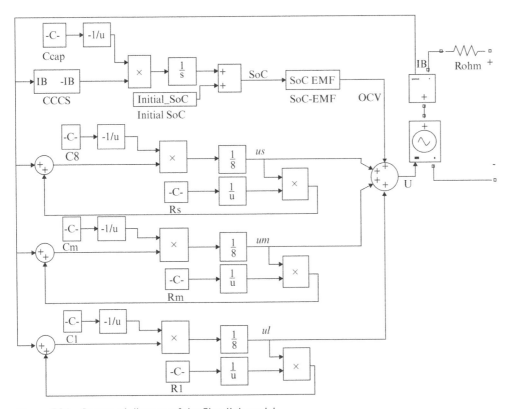

Figure 7.26 Structural diagram of the Simulink model.

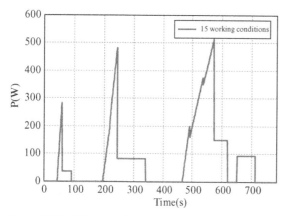

Figure 7.27 Fifteen working conditions.

Battery performance simulation and SoC estimation under fifteen working conditions

The 15 working conditions are the testing working conditions specified in the Chinese standard "Exhaust Contaminant Emission Standards for Lightweight Vehicles" (GB14671.1-93), including four states of cyclic operation: idle, acceleration, constant speed, and deceleration. Therefore, it is of practical significance to verify the effectiveness and accuracy of the model under 15 working conditions. Figure 7.27 shows the power spectrum under the 15 working conditions.

Figure 7.27 shows the operating voltage vs the time change curve of the battery obtained by a model simulation verification under 15 working conditions. Due to the large rated capacity of the used power battery and more cycles of the working condition and a limited picture size, the simulation results of only some of the working condition cycles are provided here. The simulation results are represented by a solid line and the experimental results are represented by a dotted line in the figure.

From the voltage curve shown in Figure 7.28, it can be seen that the external characteristics of the battery obtained by the simulation test of the established battery model under 15 working conditions have a high degree of approximation with the external characteristics of the actual battery. Table 7.8 provides the error between the simulated operating voltage and the real operating voltage of the battery.

Figure 7.29 shows the SoC estimation result of the battery obtained by a model simulation verification under 15 working conditions. The simulation results are represented by solid line and the experimental results are represented by the dotted line in the figure.

It can be seen from Figure 7.29 that accurate SoC estimation of the battery can be achieved using the established model under 15 working conditions. Table 7.9 provides the estimation error of the SoC.

Based on the above simulation results, the established battery model shows the external characteristics of the battery with a high degree of approximation with the real external characteristics of the battery during verification under 15 working conditions; the estimated SoC value is accurate. The feasibility and effectiveness of the model have been verified.

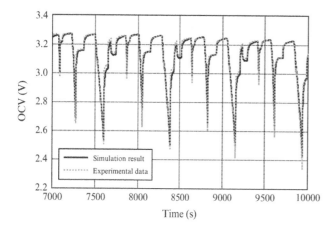

Figure 7.28 Simulated external characteristics under 15 working conditions and actual external characteristics of the battery.

Table 7.8 Error between the simulated operating voltage and the real operating voltage of the battery under 15 working conditions.

Error	Standard deviation (V)	Average deviation (V)	Maximum deviation (V)
Simulation result	0.0226	0.0168	0.0274

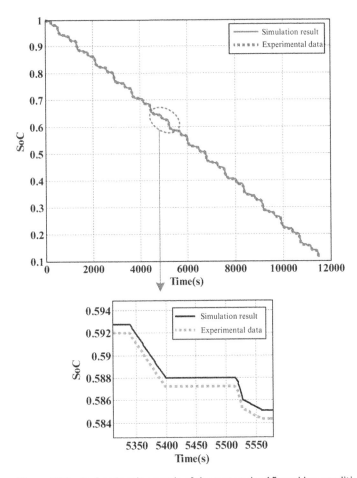

Figure 7.29 SoC estimation result of the test under 15 working conditions.

Battery performance simulation and SoC estimation under the UDDS working condition

The UDDS (Urban Dynamometer Driving Schedule) is designed for the lightweight vehicle by the Environmental Protection Agency. Although the UDDS is originally designed for traditional vehicles, it can also be used for testing the batteries of electric vehicles (e.g. the driving range test, etc.). Figure 7.30 shows the UDDS.

Figure 7.31 shows the simulation result of the battery obtained by model simulation verification under the UDDS and the experimental result. Due to the large rated capacity of the used battery and more cycles of the working condition and a limited picture size, the simulation results of some working condition cycles are provided here. The simulation results are represented by the solid line and the experimental results are represented by the dotted line in the figure.

Table 7.9 SoC estimation error of the test under 15 working conditions.

Error	Standard deviation	Average deviation	Maximum deviation
Simulation result	1.186e-004	7.7220e-005	6.22e-004

From the voltage curve shown in Figure 7.31, it can be seen that the external characteristics of the battery obtained from the simulation test of the established battery model under UDDS have a high degree of approximation compared with the external characteristics of the actual battery. Table 7.10 provides the error between the simulated operating voltage and the real operating voltage of the battery.

Figure 7.32 shows the SoC estimation result of the battery obtained by the model simulation verification under the UDDS. The simulation results are represented by the solid line and the experimental results are represented by the dotted line in the figure.

Table 7.11 shows the SoC estimation error shown in Figure 7.32.

Based on the above simulation results, the established battery model shows the external characteristics of the battery with a high degree of approximation with the real external characteristics of the battery during verification under the UDDS, and the estimated SoC value is accurate.

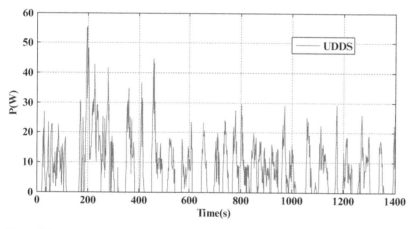

Figure 7.30 UDDS.

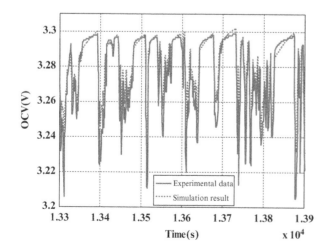

Figure 7.31 Simulated external characteristics under the UDDS and the actual external characteristics of the battery.

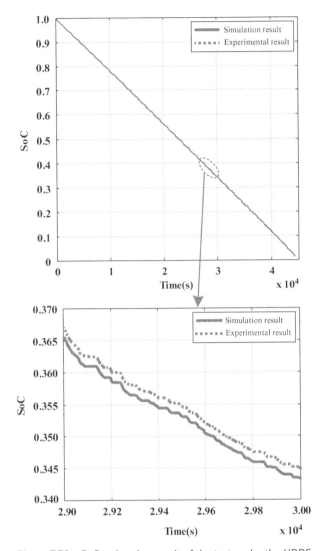

Figure 7.32 SoC estimation result of the test under the UDDS.

Table 7.10 Error between the simulated operating voltage under the UDDS and the real operating voltage of the battery.

Error	Standard deviation (V)	Average deviation (V)	Maximum deviation (V)
Simulation result	0.0198	0.0202	0.0496

Table 7.11 SoC estimation error of the test under the UDDS.

Error	Standard deviation	Average deviation	Maximum deviation
Simulation result	2.6125e-005	2.2296e-005	1.8988e-004

References

1 He, H., Xiong, R., Guo, H., et al., "Comparison study on the battery models used for the energy management of batteries in electric vehicles," *Energy Conversion & Management*, 2012, 64(4): 113–121.

2 Bruzzone, L. and Prieto, D. F., "A technique for the selection of kernel-function parameters in RBF neural networks for classification of remote-sensing images," *IEEE Transactions on Geoscience and Remote Sensing*, 1999, 37(2): 1179–1184.

3 Sankarasubramanian, S. and Krishnamurthy, B., "A capacity fade model for lithium-ion batteries including diffusion and kinetics," *Electrochimica Acta*, 2012, 70: 248–254.

4 Chen, Z. H., Qiu, S. Q., Masrur, M. A., et al., "Battery state of charge estimation based on a combined model of extended Kalman filter and neural networks," in *The 2011 International Joint Conference on Neural Networks*, IEEE, 2011, pp. 2156–2163.

5 Charkhgard, M. and Farrokhi, M., "State-of-charge estimation for lithium-ion batteries using neural networks and EKF," *Industrial Electronics*, 2010, 57: 4178–4187.

6 Sadli, I., Urbain, M., Hinaje, M., et al., "Contributions of fractional differentiation to the modelling of electric double layer capacitance," *Energy Conversion & Management*, 2010, 51(12): 2993–2999.

7 Moo, C. S., Hsieh, Y. C., Tsai, I. S., et al., "Dynamic charge equalisation for series-connected batteries," *Electric Power Applications*, 2003, 150(5): 501–505.

8 Unnewehr, L. E. and Naser, S. A., *Electric Vehicle Technology*, US: John Wiley, 1982, pp. 81–91.

9 Meethong, N., Kao, Y. H., and Tang, M., "Electrochemically induced phase transformation in nanoscale olivines $Li1-xMPO_4$ (M = Fe, Mn)," *Chemistry of Materials*, 2008, 20: 6189–6198.

10 Koltypin, M., Aurbach, D., Nazar, L., et al., "More on the performance of $LiFePO_4$ electrodes – the effect of synthesis route, solution composition, aging, and temperature," *Journal of Power Sources*, 2007, 174: 1241–1250.

11 Liu, P., Wang, J., Hicks-Garner, J., et al., "Aging mechanisms of $LiFePO_4$ batteries deduced by electrochemical and structural analyses," *Journal of the Electrochemical Society*, 2010, 157(4): A499–A507.

12 Liao, C., Li, H., and Wang, L., "A dynamic equivalent circuit model of $LiFePO_4$ cathode material for lithium ion batteries on hybrid electric vehicles," in *5th IEEE Vehicle Power and Propulsion Conference*, IEEE, 2009, pp. 1662–1665.

13 Moo, C. S., Hsien, Y. C., Tsai, I. S., et al., "Dynamic charge equalization for series-connected batteries," *Electric Power Applications*, 2003, 150(5): 501–505.

14 Dubarry, M., Truchot, C., Liaw, B. Y., et al., "Evaluation of commercial lithium-ion cells based on composite positive electrode for plug-in hybrid electric vehicle applications. Part II. Degradation mechanism under 2C cycle aging," *Journal of Power Sources*, 2011, 196: 10336–10343.

15 Belt, J., Utgikar, V., and Bloom, I., "Calendar and PHEV cycle life aging of high-energy, lithium-ion cells containing blended spinel and layered-oxide cathodes," *Journal of Power Sources*, 2011, 196(23): 10213–10221.

16 Seaman, A., Dao, T.S. and Mcphee, J., "A survey of mathematics-based equivalent-circuit and electrochemical battery models for hybrid and electric vehicle simulation," *Journal of Power Sources*, 2014, 256(3): 410–423.

17 Barsoukov, E., Kim, J. H., Yoon, C. O., et al., "Universal battery parameterization to yield a non-linear equivalent circuit valid for battery simulation at arbitrary load," *Journal of Power Sources*, 1999, 83: 61–70.

18 Bhangu, B. S., Bentley, P., Stone, D. A., et al., "Nonlinear observers for predicting state-of-charge and state-of-health of lead-acid batteries for hybrid-electric vehicles," *Vehicular Technology*, 2005, 54: 783–794.

19 Szumanowski, A. and Chang, Y. H., "Battery management system based on battery nonlinear dynamics modeling," *IEEE Transaction on Vehicular Technology*, 2008, 57(3): 1425–1432.

20 Date, C. J., *An Introduction to Database Systems*, Reading, Massachusetts: Addison-Wesley, 1983.

21 Kiran, M. S., "TSA: Tree-seed algorithm for continuous optimization," *Expert Systems with Applications*, 2015, 42(19): 6686–6698.

22 Wu, C.-H., Tzeng, G,-H., and Lin, R.-H., "A novel hybrid genetic algorithm for kernel function and parameter optimization in support vector regression," *Expert Systems with Applications*, 2009, 36(3): 4725–4735.

23 Daowd, M., Omar, N., Bossche, P. V. D., et al., "A review of passive and active battery balancing based on MATLAB/Simulink," *International Review of Electrical Engineering*, 2011, 6(7): 2974–2984.

Part III

Functions of BMS

8

Battery Monitoring

8.1 Discussion on Real Time and Synchronization

During battery state monitoring, the state information acquisition, the information transmission, and the information process are more or less lagging, while the "real-time" is relative. According to the factors that may cause delay in the battery management system, this section intends to analyze the reasons for non-real-time and non-synchronous conditions in the state monitoring process and describe their negative impacts in order to propose some solutions to solve the problems.

8.1.1 Factors Causing Delay

During power battery state monitoring, the state information delay may be caused by the information acquisition of the battery monitoring circuit (BMC), the information transmission of the communication network, and the battery control unit (BCU) responsible for whole decision-making.

1. **Delay caused by BMC**

BMC consists of a chip closest to the acquired physical quantities and its auxiliary circuits. According to different applications, the front-end chip comprises a single-chip microcomputer, an analog–digital converter and some chips specially designed for the battery management system, which are responsible for converting analog signals, such as battery voltage, into digital information. The delay is mainly caused by the analog/digital conversion time. Generally, it takes about 100 μs to complete an 8-bit analog/digital conversion of a signal. The time delay of the voltage acquisition increases with the increase of the conversion bits.

2. **Delay caused by the communication network**

If a bus network is used to transmit the information in the battery management system, the delay caused by the communication network cannot be ignored. The delay of the communication network is related to the communication control mode and the communication baud rate. The following provides a discussion by exampling the CAN bus network, which is generally used in the battery management system of the electric vehicle. When a baud rate of 250 kbps is used for the CAN bus, the inquiry frame sent by the BCU to the BMC is 2 bytes and the answer frame sent by the BMC to the BCU is 8 bytes. The time required for an inquiry and answer is equal to

$$(2 + 8) \times 8/250000 = 0.32 \text{ ms} \tag{8.1}$$

Battery Management System and its Applications, First Edition. Xiaojun Tan, Andrea Vezzini, Yuqian Fan, Neeta Khare, You Xu, and Liangliang Wei.

If there are other nodes in the communication bus, a longer delay may be caused by a bus contention.

3. Delay caused by BCU

The BCU comprises the chip that carries out the highest decision in the battery management system, and all functions, including safety management, energy management, and equilibrium management, are completed by the master chip. The BCU is obviously not often differentiated from the BMC, but sometimes a chip is responsible for both state monitoring and security and energy management. However, in the practical application of the electric vehicle, due to the large number of batteries and their scattered locations, even hierarchical management (i.e. separation of the BCU from the BMC) is adopted. It is required to solve a coordination problem between the BCU and the BMC. The causes for the delay by the BCU are complicated, maybe including the following causes.

First, a one-to-many relationship between the BCU and BMC. For example, the battery pack used in an electric vehicle consists of 96 cells, which are managed by eight daughter boards (the front-end chip on each daughter board is responsible for monitoring the information of 12 cells). A one-to-eight relationship is formed between the BCU and the BMC. In this case, the BCU often manages the cells by polling each BMC daughter board in order, and during an information exchange between the BCU and one of the BMC daughter boards, the real-time data from the other seven BMC daughter boards are delayed.

Second, delay by process scheduling. The master chip not only exchanges the information with each front-end chip, but also deals with many other functions of the battery management system, such as the charge management function, equilibrium control function, information display function, information storage function, etc. Such a multitask system is usually implemented by multiprocess technology, and it is inevitable that a delay may be caused during process scheduling.

Third, delay by peripheral devices of the master chip in the BCU. It is usual to install some peripheral devices in the peripheral of the master chip of the BCU in order to execute all of the control functions of the master chip. It takes a certain amount of time to execute these functions, so it is possible to cause a delay. For example, during the historical information management of the battery described in Chapter 14, a memory device is required to store the historical information of the battery. If 8-byte data (including maximum cell voltage, minimum cell voltage, maximum temperature, minimum temperature, and an instantaneous charge and discharge current) of the battery pack in a certain moment are written on a flash card, it takes about 0.2 ms. During the process, in case of a no higher priority event, the master chip does not process other tasks until the data storage is completed.

8.1.2 Synchronization

The "real time" is relative due to the delay. The synchronization is also relative. For example, in the battery management system of an electric vehicle, eight BMC daughter boards are responsible for acquiring the voltage information from 96 cells (each BMC is responsible for 12 cells), and when transmitting the data to the BCU by the CAN bus, the bus rate is 250 kbps. Even if the BMC acquires the voltage information of each cell "at full power" and intermittently transmits the information, there is about a 90 ms time difference between the information of the first cell and the information of the last cell due to the delay (an analog-to-digital conversion delay and a communication delay need to be considered at this time). Thus, the "real-time" information of 96 cells that appear to be acquired "simultaneously" is actually neither simultaneous nor real-time. If the chip in the BMC is controllable and programmable, the synchronization problem can be improved to a certain extent by the acquisition timing. For example, the BCU can simultaneously send the "start acquisition" instruction to each BMC daughter board through the bus, and requires the BMC daughter board to start the information acquisition as soon as it receives the instruction, in order to minimize the negative impact caused by the out-of-sync acquisition. However, not all chips used in the BMC are programmable or synchronously controlled, so synchronization issues are not necessarily eliminated in the acquisition process.

8.1.3 Negative Impact of Non-real-time and Non-synchronous Problems

As mentioned above, many other functions of the battery management systems rely on real-time monitoring of the battery state information. If the information has a long time delay, or the information acquisition from different cells is not synchronized, it is possible to affect the implementation of other functions of the battery management system.

For example, the battery equilibrium management is based on the battery consistency, and voltage consistency is often one of the important bases to reflect battery consistency. However, due to an out-sync voltage acquisition, the "real-time" voltage of different batteries at the same "moment" varies greatly, which will affect the effect of battery equilibrium.

For instance, in the rapid charging process, the charging current is reduced or disconnected based on the voltage value of each cell in the battery pack. Due to the large charging current, the battery voltage changes rapidly in the late stage of rapid charging. If the data acquisition and processing are not "real-time" enough, the best time for current regulation is missed. It is possible to affect the charging effect of the battery and damage the battery, or even cause a safety accident in the worst case.

Here is another example. The coulomb counting method (CC method, sometimes called the current integration method), discussed in Chapter 9, is one of the most important methods commonly used to calculate the SoC of the battery. Based on the real-time current information acquired at each moment, the master chip accumulates and estimates the accumulated discharge capacity of the battery by integration or summation, so as to estimate the SoC. However, if the non-real-time nature is ignored, the SoC estimation accuracy may be affected by failure to sample the current according to the current change during the operation of the electric vehicle, or non-uniform current sampling time.

From the above examples, it can be seen that the non-real-time and non-synchronous problems affect the effect of battery state monitoring, and if their negative impact is ignored, it is possible to affect other functions of the battery management system.

8.1.4 Proposal on Solution

According to the above analysis, the non-real-time and non-synchronous problems cannot be ignored in the process of battery state monitoring; otherwise, it is possible to cause some unpredictable conditions. Necessity and feasibility are considered to solve the problems. First, for the necessity, according to the requirements of different applications, the tolerable range of information delay is analyzed, the real-time and synchronicity requirements of state data monitoring are clarified, and the design indexes are determined. Second, for the feasibility, according to the requirements of the design index, comprehensive cost, reliability, etc., the appropriate topology, core devices, network parameters, and so on are selected to get a reasonable solution [1, 2]. The following proposals for the solution for these problems are provided according to past work experience.

1. Analyze the feature of state signal and select sampling frequency

According to Nyquist's sampling theorem, when the sampling frequency is greater than two times the maximum frequency of the signal during the analog/digital signal conversion, the information in the original signal is completely retained in the sampled digital signals. In practice, the sampling frequency should be 5 ~ 10 times the maximum frequency of the signal.

For example, when the current signal sampling is determined as mentioned above, to determine the required sampling frequency, the characteristics of the current signals in the operation of the electric vehicle should be studied first. The possible working current "spectrum" can be obtained through actual on-site sampling or backward simulation analysis and the spectrum range of the signal is determined through the comprehensive analysis of the current curve, especially the frequency domain analysis of the instantaneous current upon stepping and releasing the "accelerator" pedal. The design index of the current sampling frequency is obtained according to the sampling theorem.

2. Set different sampling frequencies for different physical quantities

Current, voltage, and temperature are the three main battery state monitoring indicators.

Because the power batteries are usually in series and operate at the same current, therefore, at some point, only one current value is acquired for the entire battery pack of the electric vehicle. Furthermore, as mentioned above, the higher the current sampling frequency, the higher the accuracy of SoC estimation in the accumulative charge method will be, so the sampling frequency of the physical quantity "current" should be set to the maximum.

The voltage sampling frequency should be set according to needs. The safety management, charging management, and equilibrium management of the battery should be based on the voltage value of each cell. The lower limit of the voltage sampling frequency can be determined according to the requirements of different functions. Due to the large number of cells in the electric vehicle, the high voltage sampling frequency may consume more system resources (for example, it will occupy the processing time of the master chip). Moreover, since the voltage acquisition time of each cell may be out of sync, high-voltage sampling frequency is also unnecessary.

The temperature sampling frequency should be the lowest among the three physical quantities because the temperature measurement has a large hysteresis and error, so the high sampling frequency is of little significance. However, this does not mean that temperature monitoring is not important, because the battery is a chemical product and its operating temperature is often closely related to safety. Measures, such as reducing the alarm threshold and improving the sensitivity of early warning, can be taken to avoid the potential safety risks caused by the low sampling frequency.

3. Full consideration of the instrument display, history storage, communication load and other factors

Some real-time information of the battery pack should be notified to the driver through the instrument and the refresh rate of the state information on the instrument is also a factor that needs to be considered for real-time monitoring. Too high or too low a refresh rate is not appropriate. If the refresh rate is too high, the number on the screen will jump or the pointer will jitter, resulting in the failure of the drive to see the state information. On the contrary, if the refresh rate is too low, the displayed information will seriously lag behind the actual operation. For example, the driver sees the current rise on the instrument 2 s after the acceleration. Such an instrument display cannot assist the driver in the optimal control of the vehicle.

The storage of historical data is required for some systems. For example, the storage of the measured voltage and current for accurate SoC estimation or the storage of historical data to facilitate fault diagnosis or post analysis are factors that should be considered during selection of real-time monitoring and sampling speed. If the sampling frequency is too low, it is impossible to meet the storage requirement. Conversely, if the sampling frequency is too high, the data may not be timely stored, which is of no significance.

If the battery state information is transmitted by the bus network, it is important to consider the load of the communication network. Due to the large number of the cells in the power battery pack for the electric vehicle, many management daughter boards, and large amounts of information, a high burden may be caused to the network by the too-high sampling frequency during state monitoring, which may reduce the real-time performance of other functions of the battery management system. For example, for safety management, the load margin is often reserved on the communication network for such operations as "alarming," "emergency disconnection," etc. Therefore, the bandwidth resources occupied by state monitoring should not be too high. In some applications in the electric vehicle, the battery management system even shares the communication bus with the motor control system and the vehicle control system. In this case, it is necessary to control the network load caused by real-time state monitoring.

A common network load reduction strategy is hierarchical management, which means that some responsibilities of the battery management system's BCU are transferred to the daughter boards to reduce the amount of state information uploaded from the daughter board to the master chip. In such a system, the daughter board only uploads the relevant information necessary for SoC estimation, instrument display, and historical storage at an appropriate rate, and processes the real-time data within the scope of its authority. Of course, this strategy sometimes deliberately sets a certain level of information redundancy, in order to improve system security and information reliability. The redundant design of the battery management system is a more in-depth research topic, which is not described here due to limited space.

8.2 Battery Voltage Monitoring

The monitoring of voltage, current, temperature and other state quantities is an important function of the battery management system. However, due to the fact that the high-power battery is usually formed by series connection of multiple cells, the voltage monitoring is more challenging compared with other two physical quantities. Several current mainstream voltage monitoring schemes are analyzed and compared in this section.

8.2.1 Voltage Monitoring Based on a Photocoupler Relay Switch Array (PhotoMOS)

In the BMS, the voltage monitoring based on photocoupler relay switch array is one of the common solutions. The principle is that each cell is selectively connected by the relay switch array, the voltage of the selectively connected cell is sent to the A/D converter by an isolated operational amplifier, and the voltage of each cell is acquired and monitored in proper order by polling, as shown in Figure 8.1.

In the figure, the relay switch array can be either a solid-state relay or an electronic switch. Due to the short life and large volume of the solid-state relay, it is preferable to use a high-speed PhotoMOS relay with a photoelectric isolation function. Compared with other FET type semiconductor relays, the PhotoMOS relay has advantages such as a high isolation voltage up to 1500 V, a high response speed, high sensitivity, no need for a special driving circuit, a stable on-resistance, a small open-circuit leakage current, and strong interference immunity.

Of PhotoMOS relays available on the market, the AQW214 is preferred, the electrical principle of which is given in Figure 8.2.

The voltage monitoring based on the PhotoMOS is adopted to solve such problems as digital signal isolation and a voltage signal floating ground during voltage measurements taken of the battery packs connected in series that effectively resist external interference. In addition, it can protect the MCU and other hardware circuits from damage when an external input is connected to an instantaneous high voltage or a high current due to a wrong operation or for other reasons.

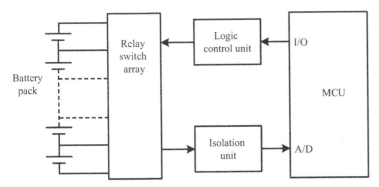

Figure 8.1 Functional block diagram for voltage monitoring based on the relay switch array.

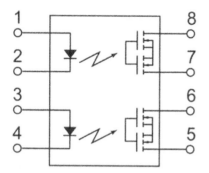

Figure 8.2 Electrical schematic diagram of the AQW214.

This solution has the following disadvantages: first, polling voltage sampling has a lower information acquisition speed and a poor real-time performance; second, a lower voltage acquisition accuracy; third, many relay array devices and a complicated circuit.

8.2.2 Voltage Monitoring Based on a Differential Operational Amplifier

In a large capacity and high-power DC power supply system, the battery pack is usually made by a series connection of many cells, and the voltage of the battery pack is up to tens of or more than one hundred volts. If each cell is directly connected to the conventional differential analog channel, a high common mode voltage exists at both ends of the cell, which exceeds the common mode voltage input range of the general electronic analog switches (such as CD4051, MAX358, etc.). A solution based on a PhotoMOS isolated differential operational amplifier can be used to eliminate the impact of the common mode voltage, the functional block diagram of which is shown in Figure 8.3.

The voltage acquisition circuit is mainly composed of three parts, including a cell channel selection module based on a PhotoMOS switch and its logic control unit, a high-voltage common mode differential amplifier and an absolute value circuit, and an MCU circuit with an A/D converter.

In the implementation of this scheme, the high common-mode resistance differential operational amplifier LT1990 from Linear Technology may be selected, which has features such as supporting high-voltage common-mode input voltage, low power consumption, high precision, rail-to-rail output, etc. When using a 5 V single power supply, its common mode voltage range is 85 V; when using a ±15 V power supply, the LT1990 is capable of operating within a range of ±250 V common mode voltage, thereby providing fault protection for the input by withstanding transients of ±350 V common mode voltages and differential voltages up to ±500 V. The schematic diagram of the differential operational amplifier and absolute value circuit used in this scheme is shown in Figure 8.4.

The output voltage of the LT1990 differential operational amplifier is calculated by the following equation:

$$V_{OUT} = G(V_{+IN} - V_{-IN}) + V_{REF} \tag{8.2}$$

where G is the optional gain coefficient, V_{+IN} and V_{-IN} are the input differential voltages, and VREF is the reference voltage. The common mode voltage range is calculated by the following equations:

$$V_{CM+} \leq 27 \times V^+ - 26 \times V_{REF} - 23 \tag{8.3}$$

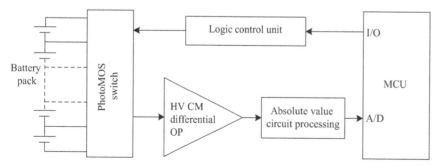

Figure 8.3 Functional block diagram of voltage monitoring based on a differential operational amplifier.

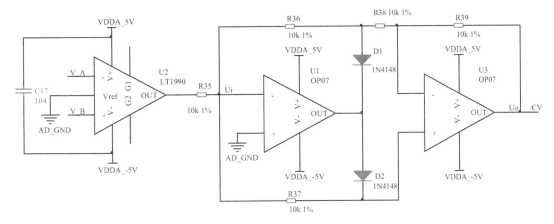

Figure 8.4 Differential operational amplifier and absolute value circuit.

$$V_{CM-} \geq 27 \times V^- - 26 \times V_{REF} + 27 \tag{8.4}$$

where V_{CM+} and V_{CM-} are the common mode voltages, V^+ and V^- are the supply voltages, and V_{REF} is the reference voltage.

In Figure 8.4, a precise absolute circuit is composed of a precision resistance divider circuit and an operational amplifier (OP07). When $U_i > 0$, the diode D1 is connected and the D2 is disconnected, with $U_o = U_i$; when $U_i < 0$, the diode D1 is disconnected and the D2 is connected, with $U_o = -U_i$. The connection and disconnection of PhotoMOS is controlled by the I/O of MCU. Since the cells are connected in series, the battery voltage needs to be switched simultaneously by positive and negative electrodes. The voltage signal is first transmitted to the differential amplifier and the absolute value circuit, and then to the A/D channel of MCU for processing. Absolute value resistance matching is one of the important factors affecting the precision of battery voltage monitoring. If the resistance is not matched properly, a large error may be caused. In the absolute value circuit, all resistances have a high precision resistance with a deviation of less than 1%.

The above voltage monitoring based on the differential operational amplifier has two disadvantages: first, the cost increase resulting from the use of more photoelectric relays at the front end; second, a higher circuit complexity resulting from the use of positive and negative power supplies for the operational amplifier.

8.2.3 Voltage Monitoring Based on a Special Integrated Chip

With the improvement of semiconductor process integration, many large semiconductor device manufacturers have developed special integrated chips for battery management systems, such as LTC6802, LTC6803, and LTC6804 chips from the Linear Technology Corporation, MAX14920 and MAX14921 chips from MAXIM, OZ8940 chips from O2Micro, and BQ series chips from TI. Compared with the previous schemes, the PhotoMOS photocoupler relay or isolator is not required for the circuit based on the special integrated chip. The circuit has been simplified to significantly reduce the area of the circuit board.

For example, the Linear Technology Corporation has developed three generations of special integrated chips for the battery management system, including LTC6802, LTC6803, and LTC6804. LTC6804 is a complete battery monitoring IC, each of which can monitor the battery pack formed by series connection of up to 12 cells, with less than 1.2 mV measurement error, and can complete the voltage measurement of all 12 cells connected in series in the system within 290 µs. It has a 0~5 V measurement range for the cell, can achieve the undervoltage and overvoltage monitoring of each cell, and can provide an associated MOSFET switch for battery equilibrium control.

Each LTC6804 is equipped with an isoSPI interface to achieve high speed and RF anti-interference LAN communication. The proprietary isoSPI design of the chip enables multiple LTC6804s to be used in series. The chip may be supplied directly by the monitored battery pack or through the power supply isolated with the battery. Battery voltage monitoring based on the LTC6804 chip is shown in Figure 8.5.

The battery management system based on the LTC6804 chip is described in more detail in the following sections.

8.2.4 Comparison of Various Voltage Monitoring Schemes

In the traditional BMS for a lead–acid battery, a classic and simple "precision resistance divider" method is used to monitor the voltage. The precision resistance divider and the above-mentioned three voltage monitoring methods are compared here to analyze and compare the voltage acquisition accuracy, circuit complexity, leakage current, resistance matching, and cost. The results are shown in Table 8.1.

8.2.5 Significance of Accurate Voltage Monitoring for Effective Capacity Utilization of the Battery Pack

In order to avoid over-charge or over-discharge of the battery, the battery management system normally controls the battery to operate in 5 ~ 95% of the SoC. As for the battery degradation factors, if the battery operates within 20–80% of the SoC, a very positive effect may be achieved on slowing down battery aging and extending the life of the whole battery pack. However, in order to optimize the charge and discharge control of the battery, it is necessary to carry out a relatively accurate SoC estimation for the battery, which depends on accurate voltage monitoring.

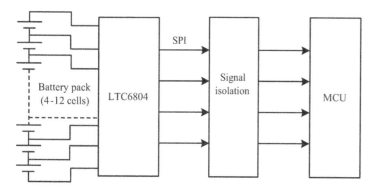

Figure 8.5 Battery voltage monitoring based on the LTC6804 chip.

Table 8.1 Comparison of voltage monitoring methods.

Voltage monitoring method	Acquisition accuracy	Circuit complexity	Leakage current	Resistance matching	Cost
Scheme based on precision resistance divider	Lower	Simple	Larger	Yes	Low
Scheme based on photocoupler relay switch array	Higher	Complicated switch array circuit	Smaller	No	Higher
Scheme based on differential operational amplifier	Higher	Complicated power circuit for operational amplifier	Smaller	Yes	High
Scheme based on special integrated chip	High	Simple	Larger	No	Depending on the chip

The necessity for accurate battery voltage state monitoring is illustrated by the electromotive force curve of a certain type of NCM lithium-ion battery and the change in the effective SoC range caused by the error, as shown in Figure 8.6.

In Figure 8.6, (a) shows the EMF–SoC curve of a ternary lithium-ion battery, (b) shows the amplification of the corresponding EMF value during 20–80% SoC, and (c) shows the possible misjudgment point at SoC = 50% and voltage measurement error of ±10 mV, which means that, when the voltage measurement error is +10 mV, the results that should have been judged during SoC = 50% will be wrongly judged as those during SoC = 52%. Similarly, when the voltage measurement error is –10 mV, the results that should have been judged during SoC = 50% will be wrongly judged as those during SoC = 48%.

The condition shown in Figure 8.6(c) can be generalized to other SoC values; that is, during different SoC values, the inaccurate voltage monitoring value may lead to the misjudgment of SoC values, which can be understood in combination with those in Figure 8.7.

In Figure 8.7, if the actual SoC value of a battery is 78% and the actual EMF is 3.75 V due to the error of the open-circuit voltage, the reading range of the voltage sensor is 3.74 ~ 3.76 V. If the SoC value is judged according to the reading of the sensor, the possible SoC range is 76–80%, that is, the inaccuracy of the voltage sensor may cause ±2% SoC error.

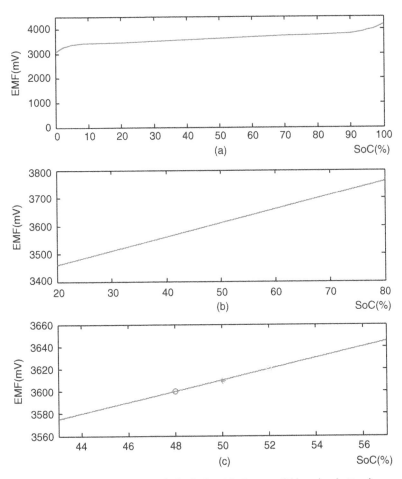

Figure 8.6 Voltage reading vs SoC relationship (ternary lithium-ion battery).

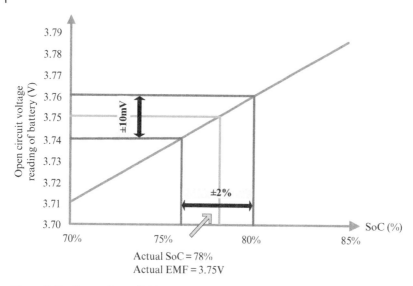

Figure 8.7 Change in available capacity resulting from error (ternary lithium-ion battery).

If an inaccurate voltage measurement has been confirmed, in order to ensure that the battery does not exceed the operating limit, the operating range of the battery must be limited by the "isolation zone." The principle is explained as follows:

1. Isolation zone for the upper charge voltage limit

According to the EMF–SoC curve (a), an 80% SoC corresponds to a voltage of 3.76 V in theory. If the voltage monitoring error is ±10 mV, when the actual SoC is 80%, the minimum measured voltage value may be 3.75 V. When the voltage reading is 3.75 V, the actual SoC may be 78%. In order to extend the battery life, SoC must be strictly controlled to < 80%, and it is preferable to stop the charge at 3.75 V. As a result, the battery must not be charged during SoC = 78%, which corresponds to a decrease of the effective battery capacity by 2% (2% corresponds to an isolation zone set to maintain the battery health).

2. Isolation zone for the lower discharge voltage limit

Similarly, the lower discharge voltage limit may be calculated, where 20% SoC corresponds to a voltage of 3.46 V in theory. If the voltage monitoring error is ±10 mV, when the actual SoC is 20%, the maximum measured voltage value may be 3.47 V. When the voltage reading is 3.47 V, the actual SoC may be 22%. In order to extend the battery life, SoC must be strictly controlled to >20%, and it is preferable to stop the discharge at 3.47 V. As a result, the battery must not be discharged during SoC = 22%, which corresponds to a decrease of the effective battery capacity by 2% (2% corresponds to an isolation zone set to maintain the battery health).

In the above examples, if an inaccurate voltage measurement has been confirmed, in order to ensure the health of the battery pack, the operating range of the battery is not 20–80%, but may be 22–78%, so it can be seen that the effective capacity of the battery pack is reduced by 4%.

The above data are calculated according to the characteristics of the ternary lithium-ion battery. Since the EMF–SoC curve of the lithium iron phosphate battery is gentler, according to a similar calculation, the capacity loss of the lithium iron phosphate battery may be more serious. Table 8.2 lists the available SoC range of ternary lithium-ion battery and lithium iron phosphate battery at different voltage measurement errors.

As can be seen from the table, for the ternary lithium-ion battery, if the measurement error of the sensor is reduced from ±10 to ±1 mV, the effective capacity of the battery pack will increase from 56 to 59.6%. For the

Table 8.2 Available SoC range of ternary lithium-ion battery and lithium iron phosphate battery at different voltage measurement errors.

Measurement error (Mv)	Available SoC range of ternary lithium-ion battery	Available SoC range of lithium iron phosphate battery
±1	20.2–79.8%	20.725–79.275%
±5	21–79%	23.625–76.375%
±10	22–78%	27.350–72.650%

lithium iron phosphate battery, the effective capacity of the battery pack will increase from 45.3 to 58.55% under the same conditions. It can be seen that the increase of voltage monitoring accuracy is of great significance for the effective utilization of the battery capacity.

8.3 Battery Current Monitoring

The real-time performance has been discussed above for battery current monitoring. Compared with other physical quantities such as voltage and temperature, current monitoring has the following characteristics. First are the limited current sampling channels. In the battery pack, there are many voltage sampling points and temperature sampling points due to the large number of cells. However, multiple cells are often used in series, so the working current of each cell is the same. Therefore, it is a basic requirement to monitor the total current after a series connection, with a few sampling channels. Second is the high current sampling frequency. As mentioned above, the current sampling frequency has an important impact on the SoC estimation accuracy and the system security, so the current sampling frequency is higher than the sampling frequency of other physical quantities.

8.3.1 Accuracy

Before designing the software and hardware for the system, the accuracy index of current monitoring should be determined first according to the following three aspects:

1. **Safety**

Although safety is very important for the whole battery management system, a high accuracy of current monitoring is not required for safety. Generally, in order to ensure the safety of the electric vehicle, a threshold is set in the battery management system for charge and discharge currents. Once the operating current of the battery pack exceeds the threshold value, over-current protection measures will be activated. In this case, a certain amount of error is allowed in current monitoring, because it is always assumed that the abnormal operating current is much larger than the normal operating current. For example, the normal discharge current range of an electric vehicle is 0 ~ 300 A and the protection threshold value is set to 400 A (because there are some inductive and capacitive loads in the circuit, it is necessary to reserve a certain margin for the threshold value). At this time, even if the error of current monitoring is 10 A, it is impossible to cause too much of an impact on the over-current protection function.

2. **Instrument display**

There are two possible scenarios. First, in the operation process of the electric vehicle, because the working current of the electric vehicle is usually large, the current value displayed by the instrument is allowed to have a large error. For example, when the actual working current is 50 A, 55 A is displayed on the instrument, and in this case, even though the current error is 10%, no large negative impact is caused on the driver's experience. Second, when the car is parked, the driver is sensitive to the absolute value of the error displayed on the instrument. For example, when a car is parked, almost all the electrical loads are disconnected and the actual working current should be less than 1 A. However, if a 6 A working current is displayed on the instrument, with an error of 5 A, it will make the driver feel uncomfortable, because of no high-power consumption during parking. More seriously, if during parking, the actual discharge current is 1 A and −1 A is displayed on the instrument (the minus sign is often used to indicate charging), this will cause a significant negative impact, because the car is obviously not in the charging state, but the instrument shows that it is in the state of small current charging. Although the absolute error is small, this case is unacceptable.

3. SoC estimation demand

If more consideration is given to the absolute error in the above discussion of the current monitoring accuracy based on safety and the instrument display, more consideration is given to the relative error in a discussion of the current monitoring accuracy based on the SoC estimation demand. It can be considered that, on the premise of a high enough current sampling frequency (satisfying the Nyquist sampling theorem), the SoC estimation accuracy evaluated using the current integration method (called the coulomb counting method or CC method) directly depends on the accuracy of the current monitoring. For example, if the average relative error of current monitoring is 5% in the past hour, the error of the estimated SoC in the current integral in the past hour is also 5%. If there is a systematic error in the current monitoring, that is, fixed too large or too small, then the estimated SoC value will be correspondingly too large or too small.

8.3.2 Current Monitoring Based on Series Resistance

The voltage is the most direct measured value. Generally, the A/D conversion chips are mostly used for voltage signals, so it is often a requirement to convert the current signal into a voltage signal in current monitoring. One of the conversion methods is a series connection of a shunt on the main circuit of an electric vehicle, as shown in Figure 8.8.

The shunt is actually a resistor with a small resistance value, high precision, and small temperature drift. When the current flows through the shunt, the current can be calculated by measuring the voltage drop U_r at both ends, i.e.

$$I = \frac{U_r}{r} \tag{8.5}$$

The resistance value of the shunt is mainly selected based on the working range of the electric vehicle current. For example, the working current range of the electric vehicle is 0 ~ 300 A. If it is required to produce a 75 mV maximum pressure drop on the shunt, the resistance value of the shunt is 0.25 mΩ. Of course, the voltage value of 75 mV is relatively small and an appropriate amplifier is usually added before the A/D conversion. The disadvantages of the series resistance method are as follows.

1. Thermal loss

It is well known that a certain amount of heat can be generated when the current flows through the resistor. If the resistance value of the shunt is $r\,(\Omega)$ and the current is $I\,(A)$, the heat loss in the shunt is $I^2 r\,(W)$. Since the working current of the electric vehicle is usually large, the heat loss on the shunt cannot be ignored.

2. Isolation

The method of using series resistance is strictly a contact measurement, so the problems, such as common ground and isolation, must be considered in the circuit design. It was originally thought that the greatest advantage of this current monitoring method was the low cost. However, due to the increase in the isolation circuit, the circuit complexity is increased, so the cost advantage is not obvious.

8.3.3 Current Monitoring Based on a Hall Sensor

A Hall current sensor is an electronic device that detects the current according to the Hall effect (discovered by the American physicist, E. H. Hall). It can be used to measure all kinds of current, from DC to tens of kilohertz of AC. A typical Hall sensor for current detection in the electric vehicle is shown in Figure 8.9.

As shown in the figure, the Hall sensor has three pins: the positive and negative power E^+ and E^- of the operational amplifier and the output V_{out} of the sensor (the voltage value of the output V_{out} is denoted as U_{out}). When a current-carried wire passes through the center hole of the Hall sensor, the positive, negative, and magnitude of the current are directly reflected in U_{out}. Figure 8.10 shows the relationship between the current I passing through the wire and the output voltage U_{out}.

The following three considerations should be given to the current monitoring scheme based on the Hall sensor.

1. Amplifying circuit

The Hall sensor measures current signals using the principle of electromagnetic induction. The voltage signal "induced" by the electromagnetic field is usually small, only several mV, while the general A/D converter requires several V at the input end, so it is necessary to add an amplifying circuit to solve this problem. In order to facilitate the use and improve the anti-interference capability, many Hall sensors embedded with an amplifying circuit are available on the market, so that the output signal of the sensor can be used directly. For this reason, the sensor pin is often equipped with the power supply input pin of the operational amplifier to supply positive and negative power for the embedded amplifier. The E^+ and E^- pins shown in Figure 8.9 are the power supply end of the embedded amplifier. In addition, the embedded amplifier is one of the factors affecting the accuracy of the current measurement, so it needs to be calibrated.

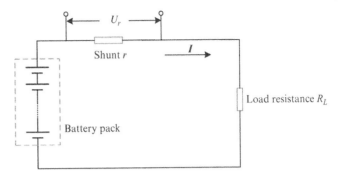

Figure 8.8 Current monitoring scheme based on a shunt.

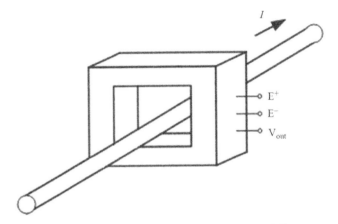

Figure 8.9 Current monitoring scheme based on the Hall sensor.

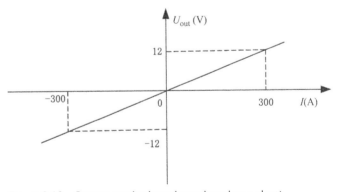

Figure 8.10 Current monitoring scheme based on a shunt.

2. Calibration

The I–U_{out} curve shown in Figure 8.9 is an ideal output curve. The actual product may have non-ideal conditions in two aspects, as shown in Figure 8.11. As shown in Figure 8.11(a), the sensor may have a systematic error, resulting in a condition where the zero output voltage V_{out} does not correspond to the zero working current I.

This is a sensitive situation during parking. As mentioned above, if a car is parked, the actual discharge current is 1 A and –1 A is displayed on the instrument (the minus sign is often used to indicate charging), this may cause the driver to become uncomfortable. On the other hand, the ideal I–V_{out} curve is a straight line, but it may have a certain non-linearity in fact, as shown in Figure 8.11(b). For the above two cases, the Hall current sensor should be calibrated before use.

3. Selection of the measuring range

The measuring range of the sensor is selected according to the actual operating conditions of the electric vehicle. For example, if the current monitoring range of an electric vehicle is –400 ~ 400 A and the input range of the A/D chip is 0 ~ 4 V, when selecting the Hall current sensor and designing the current monitoring circuit, it is best to correspond the maximum working current and the maximum voltage output value as far as possible. In other words, when I is 400 A, U_{out} is 4 V and when I is –400 A, U_{out} is 0 V. At this point, the resolution of current monitoring is at its highest. For example, if the resolution of the A/D conversion is 10 bits, then the maximum resolution of the current monitoring is

$$\frac{400-(-400)}{2^{10}} = \frac{800}{1024} \approx 0.78 \, \text{A/bit} \tag{8.6}$$

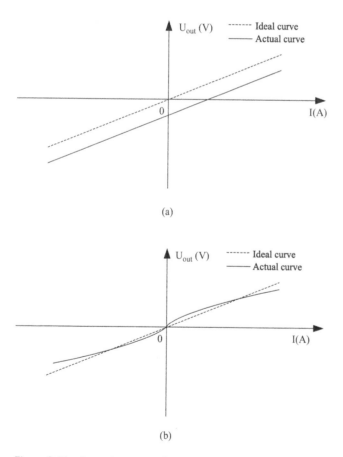

(a)

(b)

Figure 8.11 Operating curve of the Hall current sensor.

During the SoC estimation in the current integration method, this resolution is, to some extent, of decisive significance for estimating the accuracy.

8.3.4 A Compromised Method

It can be seen from the above discussion that the current monitoring based on a Hall sensor avoids the heat loss problem of the "series resistance method" and the sensor has a high measurement accuracy after calibration, which seems to be an ideal solution. However, the electric vehicles may have the following problems, which cannot be solved by the Hall current sensor. For example, during parking, the vehicular audio system is turned on, in which case, the motor driving system has been turned off, but the weak current system (equivalent to the 12 V system of a traditional car) continues to supply the power and the output current of the battery pack may be less than 1 A. Two problems may be caused during the current monitoring by the above-mentioned Hall current sensor: first, the relative error of the current monitoring is too large for the current integral method to use for an accurate SoC estimation; second, due to the limited resolution, it is even difficult to judge whether there is a working current. As mentioned above, if a 10-bit A/D

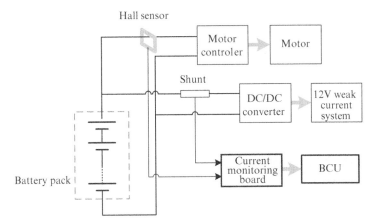

Figure 8.12 A compromised current monitoring scheme.

chip is used, the current value obtained at this time is between 0 bit and 1 bit, which cannot accurately reflect the real value of the current, and the relative measurement error is large. In this case, a compromised method can be adopted according to the characteristics of the two current monitoring methods: the Hall sensor is used to monitor the large current during the driving process of the electric vehicle, while the series resistance is used to monitor the working current of the weak current system (the input of low-power DC/DC), as shown in Figure 8.12. The larger current supplied by the battery pack to the motor and its controller is monitored by the Hall sensor, while the smaller current supplied by the DC/DC converter to the 12 V weak current system is monitored by the series resistance (shunt). In this case, because the power of the DC/DC converter is smaller than that of the motor control system, the heat loss on the shunt can be tolerated.

8.4 Temperature Monitoring

Careful readers may have noticed that the previous two sections are entitled "battery voltage monitoring" and "battery current monitoring," while this section is entitled "temperature monitoring," meaning that the temperature information to be monitored is not limited to the battery. In fact, in the battery management system, in addition to battery temperature monitoring, the ambient temperature and the battery box temperature should also be monitored, which are of great significance for the SoC estimation and safety protection of the battery. Temperature monitoring is discussed in this section.

8.4.1 Importance of Temperature Monitoring

The battery is extremely sensitive to its operating temperature. If the temperature is too high, it is possible to cause shell breakage, electrolyte leakage, explosion, and other safety accidents. If the temperature is too low, it is possible to cause electrolyte solidification and failure to charge or discharge the battery. The importance of temperature monitoring is analyzed in the following three aspects.

1. **Significance of temperature monitoring to safety protection**

Any power battery can be used in the allowable operating temperature range that corresponds to an upper operating temperature limit and a lower operating temperature limit. If the operating temperature of the battery exceeds the allowable upper limit, it is possible to cause the following adverse condition:

- The expansion of active chemicals, which leads to swelling of the cell.
- Mechanical deformation of the battery assembly, which causes a short circuit or open circuit of the battery.
- Irreversible chemical reaction, which reduces perpetually the active chemicals while reducing the battery capacity.
- Partial plastic decomposition resulting from operation at high temperature for a long time.
- Gas leakage.
- Pressure rises in the battery.
- Battery damage or explosion.

If the battery's operating temperature is below the allowable lower limit, the electrolyte may be frozen. For example, at temperatures below the freezing point of the electrolyte, the battery performance may deteriorate due to the decrease in the chemical reaction rate. Generally, the LiFePO$_4$ batteries do not work normally at less than –30°C temperatures.

2. Significance of temperature monitoring to SoC estimation

All batteries work by electrochemical reactions regardless of the charge or discharge process. We know that these chemical reactions are temperature-dependent to some extent. The nominal performance of a battery usually refers to the performance within a specified operating temperature range, such as +20 to +30°C. If the battery operates at a high or low temperature, the actual performance may be reduced substantially. The temperature must be fully taken into account during estimation of the SoC, because at different temperatures the battery can release different charges and provide different amounts of energy.

3. Thermal management of the battery and its significance

The thermal management of the battery packs is generally considered as an optional feature of the battery management system, but more and more attention has been given to its research. The thermal management of the battery pack involves electrochemistry and thermochemistry, and the thermal management methods also vary greatly with the specific application of the electric vehicle. The thermal management of the battery is not temporarily detailed here due to limited space. However, since temperature monitoring of the battery pack provided in this section is closely related to thermal management, thermal management is briefly discussed here.

In short, thermal management for the battery is conducted mainly to ensure the safety and high-efficiency operation of the battery within the temperature range. The difference between thermal management and normal temperature protection is that normal temperature protection only controls or protects the circuit when the battery may fail or has failed, while thermal management focuses on maintaining the more suitable temperature for the battery by various measures. For example, in the battery management system only with temperature protection, the common strategy is to set up the upper and lower temperature limits for the battery. The lithium iron phosphate battery, for example, can be set to +50 and –30°C respectively, and when the temperature is close to the limit, the working current of the battery pack is controlled or directly disconnected and the driver is notified by the instrument. For the battery management system with a thermal management system, the battery is made to work between 20 and 30°C as far as possible, and if the temperature is lower than 20°C, certain heating measures should be taken to heat the battery pack; if the temperature is higher than 30°C, cooling measures should be taken to cool the battery pack.

8.4.2 Common Implementation Schemes

Common temperature acquisition schemes are given below.

1. Thermistor

The thermistor is the most common temperature acquisition scheme and its resistance varies almost linearly with the temperature. If the thermistor is connected in series with another resistance whose value is known, the temperature can be determined by detecting the voltage difference between the two resistances. Some chips also support thermistor connection, such as the LTC6802 described above from LT, which has pins that connect directly to the specified type of thermistor and the temperature values to be monitored can be obtained directly through the chip's internal design.

2. 18B20

18B20 is a common chip-level temperature sensor. The MCU can be connected to multiple sensors by the bus, thus reducing the number of MCU pins and connection complexity.

3. Special integrated chip

Some chips specially designed for the battery management also integrate the voltage, current, and temperature acquisition functions, such as the DS2782 special chip of the MAXIM Company, which can sense the temperature in a certain area of the outer wall of the chip and save it into the register of the chip for reading by the upper computer MCU.

8.4.3 Setting of the Temperature Sensor

The temperature sensor is an electronic device used to acquire temperature. It can be classified into a thermal resistance type, a thermocouple type, etc., according to the acquisition principle, and can also be classified into an analog or digital type. In the battery management system, the temperature information to be monitored mainly includes the ambient temperature, the temperature of the battery box, and the battery temperature. In the actual design, how to set the temperature sensor is equivalent to the following problems, which are often difficult to solve and involve contradictions between the cost and performance.

1. Environmental temperature monitoring and battery temperature monitoring

The temperature information finally acquired by the battery management system is the battery temperature rather than the ambient temperature information. However, ambient temperature monitoring is also important, for the following reasons.

First, the ambient temperature information is simple and easier to acquire. In the battery pack, the temperature varies with the cells and the temperature varies with the positions of the same cell in the charging and discharging process. Therefore, the battery temperature can be estimated by monitoring the ambient temperature.

Second is hysteresis. In the charging and discharging process, with the increase of the ambient temperature, the cooling speed of the battery surface slows down, and the battery temperature will gradually increase. Since the heat of the battery is emitted from inside to outside, if the ambient temperature changes can be monitored in advance, the temperature increase of the battery surface after a period of time can be predicted. On the contrary, if only the external temperature of the battery is monitored, when the external temperature of the battery is found to be high, the temperature inside the battery may actually have reached a higher dangerous value.

2. Whether to monitor the temperature of each cell

From the perspective of safety, it is necessary to monitor the temperature of each cell, because any cell may exceed the temperature threshold, causing battery damage or a serious safety accident. However, if every cell is equipped with a temperature sensor, on the one hand, the cost of the battery management system will increase; on the other hand, many wires will be added inside the battery box due to the configuration of the temperature sensor, thus reducing the actual maintainability of the battery pack.

Table 8.3 Temperature change at different locations of a 100 Ah square battery discharging at 1 C for 20 minutes.

Sensor location	Temperature
Positive terminal of battery	33°C
Negative terminal of battery	36°C
Upper surface of battery (middle position)	37°C
Front surface of battery (middle position)	43°C
Side surface of battery (middle position)	39°C

Note: ambient temperature of 25°C.

3. Selection of the sensor installation location on the battery surface

The chemical reactions take place inside each cell, but in practice, the temperature sensor cannot be placed inside the battery but only on the surface. Where is the sensor placed on the surface of the battery? Table 8.3 shows the temperature change of a 100 Ah square battery discharging at 25°C and 1 C for 20 minutes.

It can be seen from the table that the temperature is higher at the middle position of the front surface of the battery, due to slow heat dissipation and a most obvious temperature rise in the middle part of the large-capacity square battery. Therefore, the sensor should be arranged in the middle of the front surface of the square battery to monitor its ultimate temperature.

8.4.4 Accuracy

The power battery management system has a high tolerance of the temperature monitoring error. From the perspective of an SoC estimation, however, according to the tests provided in Section 4.2, the effective capacity of the battery varies with the temperature at the same discharge rate and the temperature measuring error will also directly affect the SoC estimation accuracy. However, the impact of a 1 ~ 2°C error is basically small, and the error does not accumulate over time. From the perspective of safety protection, the temperature monitoring error does not cause a serious impact.

It is worth mentioning that temperature monitoring is vulnerable to electromagnetic interference, because the voltage difference on the analog thermistor is relatively small, while the strict error verification mechanism is not available for a data message in the digital 18B20 chip. For such interference, the best solution is to add a digital filter to filter out a high-frequency signal change, because for the electric vehicle, the temperature is a gradual variable without a "step" change.

References

1 Podlubny, I., "Fractional differential equations: An introduction to fractional derivatives, fractional differential equations, to methods of their solution and some of their applications," *Mathematics in Science & Engineering*, 2013, 3: 553–563.
2 Monje, C. A., Chen, Y. Q., Vinagre, B. M., et al., *Fractional-Order Systems and Controls*, London: Springer, 2010, pp. 35–57.

9

SoC Estimation of a Battery

9.1 Different Understandings of the SoC Definition

The SoC estimation is an important function of the BMS and is the basis for the implementation of the control strategies. For example, many battery equilibrium control methods are based on SoC estimation results. If the same BMS development team has different understandings of the SoC definition, it is detrimental to the efficiency and robustness of the system.

9.1.1 Difference on the Understanding of SoC

It is found that the SoC is an interesting term: people acknowledge its importance and propose a variety of different ways to implement it, but there is little literature to strictly define the SoC, with the subtext "you should understand such a simple problem without explanation." However, in fact, if the SoC is not strictly defined, the definition may vary with the people, and it is difficult to verify a certain SoC estimation method.

However, in any case, most people still have some basic consensus to the definition of SoC. The SoC is often defined often by the following equation:

$$\mathrm{SoC} = \frac{\text{Remaining dischargeable}}{\text{Battery capacity}} \times 100\% \tag{9.1}$$

where the numerator and the denominator refer to the electric charge quantity (rather than energy), in coulombs (C) or ampere-hours (A h or Ah), and

$$1\,\mathrm{Ah} = 3600\,\mathrm{C}$$

According to Equation (9.1), the SoC value is in the range of [0, 1] or 0 ~ 100%.

The difference in the understanding of the SoC is caused by the different definition of the numerator and the denominator.

1. How do we define the remaining dischargeable charge (Q_{remain})?

The remaining dischargeable charge, i.e. the remaining charge, has generalized and narrow definitions.

The generalized remaining charge refers to the amount of charge released by all possible chemical reactions, i.e. the maximum charges dischargeable at a suitable temperature and discharge rate without damage to the battery.

Battery Management System and its Applications, First Edition. Xiaojun Tan, Andrea Vezzini, Yuqian Fan, Neeta Khare, You Xu, and Liangliang Wei.

The narrow remaining charge refers to the charges dischargeable from the battery at a limited temperature and discharge rate. In the case of a discharge at room temperature and a small discharge rate, the generalized remaining charge is equal to the narrow remaining charge. However, for the power battery for the electric vehicle, due to a larger change in the operating ambient temperature, and the higher discharge rate of the electric vehicle, the difference on the understanding of the remaining charge may cause different SoC estimation results with more than a 20% error.

2. How do we define the battery capacity (Q_c)?

In Equation (9.1), the denominator is the "battery capacity," which is not defined. The different definition of the "capacity" may lead to a different understanding of the SoC. The mainstream opinions are given below.

9.1.1.1 Use of Nominal Battery Capacity

Many engineers define the SoC by the rated capacity (Q_{rated}). However, the author considers that the rated capacity is not suitable to define the SoC for the following two reasons:

First, the rated capacity is different from the actual capacity (Q_{true}). Table 9.1 shows the comparison of the actual capacity and rated capacity among the new batteries provided by the manufacturers A, B, and C. It can be seen from the table that the actual capacity is not fully equal to the rated capacity.

Second, the rated capacity does not reflect battery aging. The maximum Q_{true} is reduced with the battery aging. This is a good explanation for why the SoC can be accurately estimated for the new battery in some estimation methods and a large deviation is caused with the increase in the charge and discharge cycle of the battery.

9.1.1.2 Regard the Battery Capacity as a Function of the "Degradation Degree" and "Discharge Condition"

Many engineers understand the "battery capacity" used in Equation (9.1) as

$$Q_c = kQ_{rated} \tag{9.2}$$

where k is the compensation factor added outside the rated capacity. Then

$$k = f(\sigma, \rho, \mu) \tag{9.3}$$

where σ, ρ, and μ are respectively the apparent temperature, discharge rate, and aging degree of the battery.

In fact, for a fully-charged battery, the maximum dischargeable charge is affected by the discharge rate, the ambient temperature, and other factors and it is not a constant value. Table 9.2 shows the discharge capacity of a 100 Ah new battery provided by the manufacturer A at different temperatures and different discharge rates.

The SoC is estimated using Equation (9.2) for the following benefits. The denominator of Equation (9.1) is no longer regarded as a static value, but is determined by considering the dynamic factors, such as the capacity loss after battery aging, coulomb efficiency, and the impact of temperature on the discharge rate.

Table 9.1 Comparison of the actual capacity and the rated capacity of three new batteries (test temperature 25°C and discharge rate 0.02 C).

	Sample of manufacturer A	Sampler of manufacturer B	Sampler of manufacturer C
Q rated	100 Ah	100Ah	80 Ah
Maximum Q true	115 Ah	103 Ah	83 Ah

Table 9.2 Discharge capacity of the same battery sample at different temperatures and different discharge rates.

Ambient temperature	Discharge rate (current)	0.2 C (20 A)	0.5 C (50 A)	1.0 C (100 A)
25°C		110 Ah	105 Ah	100 Ah
40°C		112 Ah	108 Ah	103 Ah

Note: the rated capacity of the battery sample is 100 Ah.

Therefore, the estimated value of the SoC becomes a function of the degradation degree and "discharge condition," which is more realistic.

However, according to Equation (9.3), used to define the compensation factor k, although the ageing factor μ can be used to compensate for the battery capacity to some extent, it is not easy to identify the function relationship of σ, ρ, and k. Furthermore, the independent variables σ and ρ are variable during the operation of the electric vehicle, and it is more difficult to forecast the ambient temperature and the discharge rate in advance.

3. The "maximum capacity" of the battery is defined and corrected with battery ageing

Before an in-depth discussion of the battery capacity, consider the working principle of the lithium-ion battery, which can be analyzed by referring to Figure 9.1.

In general, during the charging, under the influence of the external electric field, the lithium ions escape from the positive electrode of the battery and reach and "engage" with the negative electrode of the battery through the electrolyte. During the discharging process, the negative electrode loses a lot of electrons, the positive electrode gains the electrons, and the lithium ions engaged with the negative electrode go back to the positive electrode through the electrolyte.

As shown in Figure 9.1, in a sense, there is actually a more objective measurement index for the battery capacity, i.e. the size of the space of the negative electrode for engagement with the lithium ions.

Let us think about it this way: the total lithium-ions engaging with the negative electrode after full charging the battery may be regarded as the battery capacity. Although the "total" may vary slightly due to some microscopic factors, we can still understand at the macro level that the total lithium ions engaging with the negative electrode are limited, and the maximum of the limit is defined as the battery capacity.

Of course, this "maximum value" decreases continuously with battery ageing, which macroscopically means the battery capacity decreases. However, this "maximum value" is not affected by the discharge rate and the ambient temperature upon discharging. That is to say, compared with the above understanding (2), this value is not affected by the "future" and is an objective value.

In contrast, the author prefers to use the last definition of Q_c.

9.1.2 Difference and Relation Between SoC and SoP as Well as SoE

In the practical work of the electric vehicle, the concepts of SoP (State of Power) and SoE (State of Energy) are close to that of SoC. This section intends to discuss the easily confused three concepts and provide their relation.

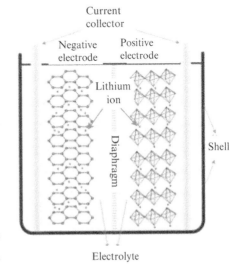

Figure 9.1 Structure diagram of the lithium-ion battery.

1. Definition of SoP

For the electric vehicle, SoP can be defined as the power provided by the battery pack to various electrical loads such as the motor at a specific moment. It can be simply considered that the SoP is a function of the SoC and the temperature, namely:

$$SoP = f(SoC, T) \tag{9.4}$$

The SoP is often used as a real-time parameter provided by BMS to the motor controller and other electrical control systems of the electric vehicle to represent the power provided by the battery at a specific moment. In actual work, the unit (W) of power or current (A) is used as the unit of SoP, mainly because the BMS provides the total voltage of the battery pack simultaneously, and the product of the total voltage and the current is the total power that the battery pack can provide, so it can also be expressed only by the current.

In fact, for the powertrain of many electric vehicles, BMS not only estimates the external output power SoP of the battery pack at a particular time, but also provides the allowable maximum charge power SoP2 of the battery pack. On the one hand, SoP2 is sent to the motor through the communication bus to tell the motor that it should not exceed a certain limit value during braking energy recovery. On the other hand, it is sent to the charger together with the charging strategy, so as to avoid damage to the battery resulting from the high charging current provided by the charger. To distinguish it from the SoP, the maximum discharge power provided by the battery pack is expressed by SoP1.

2. Definition of SoE

Just as its name implies, the SoE refers to the rest energy of the battery, which can be expressed by percentage (%) or as a unit of energy (J), but often by kWh for the electric vehicle, namely "how many kilowatt-hours are left." Their conversion equations are given below:

$$1\,Wh = 3600\,J \tag{9.5}$$

$$1\,kWh = 3\,600\,000\,J \tag{9.6}$$

However, detailed analysis shows that the SoE also has two levels of meaning. The SoE refers to the "chemical energy" of the battery pack at a certain moment, which is expressed as SoE0, and the energy that the battery pack supplies to the load, which is expressed in SoE1.

In fact, the SoR (state of range) of an electric vehicle is related to the external output energy SoE1 of the battery, rather than to the rest energy SoE0 of the battery. However, the SoE1 is not only an independent variable of SoC, but also changes with different temperatures and discharge currents, namely:

$$SoE_1 = f(SoC, T, I) \tag{9.7}$$

where I is the discharge current of the battery pack.

That is to say, in the case of the same SoC, during discharge at different currents, the power battery can supply different energies.

3. Relationship between SoC and SoP1 as well as SoE1

According to the above analyses, it is not hard to see that the user of an electric vehicle is actually more concerned about SoP1 and SoE1 than SoC. Because the SoP1 reflects the maximum power the battery pack can provide at this point in time, which means whether an electric vehicle with the battery system can be started on a ramp or is driven at a certain speed (e.g. 100 km/h) on a highway. The SoE1 generally corresponds to the kilometers that the electric vehicle travels.

The relationship between SoC and SoP1 as well as SoE1 can be identified by the above Equations (9.4) and (9.7), as shown in Figure 9.2.

It can be seen from the figure that the SoP1 can be derived from the SoC and the temperature T. Specifically, according to Equation (9.4):

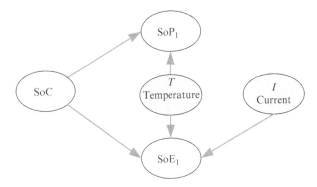

Figure 9.2 Relationship between SOC and SOP$_1$ as well as SoE$_1$.

$$SoP_1 = f(SoC, T) = u_p I_p \qquad (9.8)$$

where u_p is the operating voltage of the battery at maximum power, I_p is the operating current of the battery at maximum power, and they are mutually constrained to decide the maximum power SoP of the battery. Based on voltage analysis,

$$u_l = EMF(SoC) - 1r(SoC, T) \qquad (9.9)$$

or

$$1 = [EMF(SoC) - u_l] / r(SoC, T) \qquad (9.10)$$

where $r(SoC, T)$ is the internal resistance of the battery, which is affected by SoC and temperature T, u_l is the discharge cut-off voltage of the battery, and l is the current at u_l. In the case of $I < I_{max}$, where I_{max} is the allowable maximum discharge current of the battery, then $u_p = u_l$ and $I_p = l$:

$$SoP_1 = f(SoC, T) = u_l I = u_l \left[EMF(SoC) - u_l \right] / r(SoC, T) \qquad (9.11)$$

Based on the current analysis,

$$u = EMF(SoC) - I_{max} r(SoC, T) \qquad (9.12)$$

In the case of $u > u_l$, $u_p = u$, and $I_p = I_{max}$,

$$SoC_1 = f(SoC, T) = u I_{max} = I_{max} EMF(SoC) - I_{max^2} r(SoC, T) \qquad (9.13)$$

SoE$_1$ can be derived from the SoC, the temperature, and the operating current. Specifically, the specific form of Equation (9.7) is

$$SoE_1 = f(SoE, T, I) = \frac{\int_{t_0}^{t} u(\tau) i(\tau) d\tau}{E_N} \qquad (9.14)$$

where the battery starts discharging at t_0 and is fully discharged at t, $u(\tau)$ and $i(\tau)$ are respectively the voltage and current change during discharging, and E_N is the nominal energy of the battery. The operating voltage $u(\tau)$ can be expressed by

$$u(\tau) = EMF(SoC(\tau)) - i(\tau) r(SoC(\tau), T(\tau)) \qquad (9.15)$$

In case of discharge at constant current I,

$$\text{SoE}_1 = f(\text{SoC},T,I) = \frac{\int_{t_0}^{t} I\,\text{EMF}(\text{SoC}(\tau)) - I^2 r(\text{SoC}(\tau),T(\tau))\mathrm{d}\tau}{E_N} \tag{9.16}$$

According to the above analysis, the SoP and the SoE are different from SoC and can be deduced by the SoC, the temperature, the current, and other factors. The temperature (T) is a key factor affecting the output power (SoP) of the battery pack and for the SoE, in addition to the temperature, the current (I) during operation of the electric vehicle should be considered.

9.2 Classical Estimation Methods

In past 50 years, scholars have proposed a number of classical methods for estimation of battery level or SoC in their literatures, all of which have their own scope of application. This section provides several classical estimation methods applicable to the lithium-ion batteries, such as the coulomb counting method, the open-circuit voltage method, and their combination. This section also gives a brief description of some classical methods that are not applicable to the lithium-ion batteries, such as the internal resistance method, the load voltage method, etc.

9.2.1 Coulomb Counting Method

The coulomb counting method (CC method) refers to calculation of the current SoC based on known SOC at a previous moment, and the charge and discharge capacity of the battery recorded during a certain period.

Suppose that the SoC is Q_{t_1} at a previous moment t_1 and the SoC is Q_{t_2} at the current moment t_2; then the cumulative charge and discharge capacity during the period from t_1 to t_2 is

$$Q_{t_1}^{t_2} = \int_{t_1}^{t_2} i(\tau)\mathrm{d}\tau \tag{9.17}$$

Then

$$Q_{t_2} = Q_{t_1} - Q_{t_1}^{t_2} \tag{9.18}$$

In Equation (9.17), $i(\tau)$ can be either positive or negative. In the case of $i(\tau) > 0$, the battery is discharging and in the case of $i(\tau) < 0$, the battery is charging. Similarly, in Equation (9.18), in the case of $Q_{t_1}^{t_2} > 0$, during the period from t_1 to t_2, the discharge capacity is higher than the charge capacity and, on the contrary, in the case of $Q_{t_1}^{t_2} < 0$, this means that during the period from t_1 to t_2, the charge capacity is higher than the discharge capacity.

After calculation of Q_{t_2} by Equation (9.18), the SoC at the current moment can be calculated by proportional calculation.

However, the coulomb counting method has following three disadvantages:

1. Dependence on the initial value

In fact, the coulomb counting method is used only to solve the capacity change $Q_{t_1}^{t_2}$ over a period of time, and we are ultimately concerned with the SoC Q_{t_2}, which depends on Q_{t_1} accuracy. If there is an error in the initial value, it is impossible to make corrections during use of Equations (9.17) and (9.18).

2. Accumulative error

For causes such as insufficient accuracy of the current sensor, a low sampling frequency, and signal interference, the current i_r used for the integral has some errors compared with the real value. The cumulative charge estimated by Equation (9.17) for a period of time has an error, which will be accumulated to Q_{t_i} at the next moment by Equation (9.18), so that the SoC estimation error is increased.

In order to eliminate the cumulative error, it is necessary to correct the estimated SoC. The effective correction method is to fully charge or fully discharge the battery. Of course, such an operation cannot be carried out frequently in the practical application of the electric vehicle, thus reducing the practicability of this method.

3. Failure to cope with the self-discharge of the battery

Almost all storage batteries have a self-discharge problem, in which the battery is discharged at an extremely low rate. It is difficult to cope with the self-discharge in the coulomb counting method for the following reasons. First, the equivalent current of the self-discharge is very small, so the general current sensor cannot measure it accurately. Second, a considerable part of the self-discharge current does not flow through the operating current circuit and cannot be detected by the sensor installed in the operating current circuit. Third, the self-discharge may occur when the battery management system is not working, for example, when the electric vehicle is parked in the garage after "turning off the engine," so the BMS is not required to work and naturally cannot monitor the self-discharge of the battery.

In addition to above three considerations, some other details are considered during use of the coulomb counting method. For example, the BMS is required to record the SoC Q_{t_i} at the last moment. Otherwise, when the driver starts the electric vehicle the next day, no initial value is available for an SoC estimation. In addition, after the power pack is replaced, the SoC must be corrected once by a full charge of the battery pack.

9.2.2 Open Circuit Voltage Method

The open circuit voltage (OCV) method refers to an estimation of the SoC by measuring the OCV of the battery when the battery is neither charged nor discharged, that is, when the operating current is zero.

The OCV method is generally used based on three preconditions.

First, it is considered that the SoC has a one-to-one correspondence with the electromotive force (EMF) of the battery, that is, any given SoC value between 0 and 100% corresponds to the unique EMF value.

Second, it is considered that the OCV is equal to the EMF of the battery when the operating current is zero.

Third, the factors, such as the temperature and aging degree, are not considered, that is to say, it is considered that the batteries with different aging degrees have the same SoC–EMF curve at different temperatures.

Figure 9.3 shows the SoC–EMF curve obtained through the battery test procedure described in Chapter 5.

In practical applications of the electric vehicle, the battery voltage UO is measured usually at zero operating current in the OCV method and then the SoC of the battery is reversely calculated according to the curve shown in Figure 9.3.

However, in the specific application of the electric vehicle, the OCV method also has shortcomings:

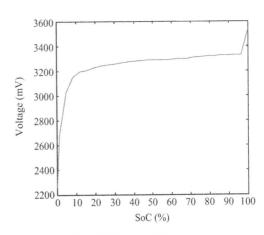

Figure 9.3 SoC–EMF curve of the battery.

1. Problem with the zero operating current

First, the above-mentioned OCV method is used based on the precondition of a zero operating current. However, it is often necessary to know the SoC when driving the electric vehicle or charging the battery, namely when the operating current is not zero. In this case, the OCV method is obviously not applicable.

Second, even if the electric vehicle is in static state without operation or charging, it cannot be considered that the operating current of the battery is zero, since the weak current system of the car is still operating. For example, the controller is not turned off, the communication network is still in a working state, the instrument desk may be on, and at least the BMS is still working, which means the operating current of the electric vehicle is not necessarily absolutely zero. In practice, a threshold value of the current can be set. When the current value is less than the threshold value, the OCV method can be used to estimate the SoC.

2. Acquisition and use of the EMF–SoC curve

The hysteretic characteristic of battery voltages discussed in Chapter 7 undoubtedly has a significant impact on the representation of the EMF in OCV. Although the difference between the charging and discharging equilibrium EMF of the battery in the voltage platform area is a few millivolts and the total voltage drop is not more than 200 millivolts when the SoC is reduced from 90 to 10%, the estimated SoC may have an error of about 10% if the hysteretic characteristic is ignored. In a practical application, since the users pay more attention to the SoC of the battery during discharge, the discharge EMF–SoC curve is used as the standard for the SoC estimation. However, due to the braking energy recovery and the cell equilibrium control in the use of the electric vehicle, it is possible to alternately discharge and charge the battery and the impact of the hysteretic characteristic on the OCV method can only be reduced, but not removed.

3. Voltage rebound problem

The voltage rebound effect must be considered for the OCV method. When the electric vehicle is parked for a short time for some reason (such as waiting for a green light), it is also necessary to consider whether the SoC can be estimated using the OCV method. If the parking time is too short, due to the voltage rebound effect, the voltage has not rebounded to a stable state, so the estimated SoC value must be small, and the size of the error is related to the parking time. Two solutions are available for the voltage rebound. On the one hand, a time threshold can be set in the practical application, if the period during which the operating current is continuously less than the current threshold value is greater than the time threshold, the OCV method can be used to estimate the SoC. On the other hand, the battery model can be used to estimate the limit value of the voltage rebound through the voltage rebound trend within 10 to 30 seconds, and then the limit value can be used to estimate the SoC.

In addition, some problems are considered for the OCV method. For example, the precondition for use of the OCV method is that the OCV is equal to the battery EMF at a zero operating current, but in fact the OCV is not only related to the EMF, but is also related to the temperature, battery usage history, and other factors. In addition, because there is a platform area on the OCV–SoC curve, in order to obtain a higher SoC estimation accuracy, a higher resolution is required for the voltage sensor, which will also increase the cost of the SoC estimation.

To sum up, compared with the CC method, the OCV method also has many shortcomings. However, the biggest limitation of the OCV method is that it cannot be used when the battery is in normal operation, but can only be used after the battery is stopped for a certain period of time.

9.2.3 A Compromised Method

From the discussion provided in the above two sections, it can be seen that the CC method and the OCV method are obviously complementary in their advantages and disadvantages, so many researchers proposed to combine these two methods to estimate the SoC, which is a compromised method.

The compromised method is used as follows. When the battery is in an operating state (the operating current is greater than the set threshold value), the SoC value is updated in real time by the CC method. At the same time, in order to eliminate the accumulated errors caused by the CC method and solve the problems of the CC method, such as estimation of the initial SoC, the OCV method is used to calibrate the SoC when the battery is started, or when the battery pack is temporarily out of work.

Obviously, the deficiency of the CC method can be removed by the compromised method to a certain extent, such as by regularly eliminating the accumulated error and solving the initial SoC problem after there has been no use of the battery for a long time, and also by the self-discharge problem. At the same time, the compromised method also solves the problem where the SoC cannot be estimated in the OCV method when the battery pack is working normally.

Is the compromised method perfect? Sadly, the answer is "no." During the use of the compromised method, the SoC is calibrated by the OCV–SoC curve. However, the deficiencies of the OCV method can be removed by the compromised method, such as zero current, the voltage hysteresis effect, and the impact of temperature and usage history on the EMF. In addition, the combination of the two methods also causes a new problem: SoC fluctuation. In the process of an intermittent discharge, since the OCV method can eliminate the accumulated error resulting from the CC method, the SoC is likely to fluctuate, depending on the accumulated error resulting from the CC method between two calibrations and after the SoC is calibrated by the OCV method. For example, the SoC calculated by the current accumulation method is 70% after driving the electric vehicle for a period, and the driver may find that the SoC had fluctuated to 65% upon restarting the electric vehicle. This is the result of eliminating the accumulated error resulting from the CC method by using the OCV method for the SoC calibration. However, the driver does not know this principle and may think the battery or battery management system is out of order. That is obviously not what we want.

Therefore, although this compromised method has been widely used in the actual BMS, the improvement of the SoC estimation method is still a topic worth studying.

9.2.4 Estimation Methods Not Applicable for the Lithium-Ion Battery

In addition to the above methods, other common SoC estimation methods are disclosed in the literature, some of which do not apply to the electric vehicle and some of which do not apply to lithium-ion power batteries. The "internal resistance method" and the "load voltage method" are now briefly analyzed.

1. Internal resistance

The internal resistance of the battery is generally divided into AC impedance (also known as AC internal resistance) and DC internal resistance. The AC impedance is a complex variable, indicating the resistance of the battery against an AC current, which can be measured by an AC impedance meter. The DC resistance represents the resistance of the battery against a DC current, which can be calculated by the ratio of the voltage change to the current change in the same short period of time.

In some literature, a large number of experiments have proved that the AC impedance and the DC internal resistance of the battery are closely related to the SoC. If a certain functional relationship can be obtained from the battery sample, the SoC value can be estimated by monitoring the internal resistance of the battery in real time. This is the basic idea of the internal resistance method.

However, the internal resistance method is rarely applied to the electric vehicle for at least three reasons:

First, the relationship between the internal resistance and the SoC is very complex, and no accurate conclusion has been reached for its wide application. Because the internal resistance of the battery is not only related to the SoC, but also to the temperature, SoH, and other factors, it cannot be concluded that an internal resistance state corresponds to a certain SoC.

Second, the internal resistance of the battery is a milliohm level value. Because its value is too small, it is difficult to accurately measure the internal resistance of the battery using the conventional measuring circuit. Moreover, due to a complex electromagnetic interference in the electric vehicle, it is difficult to accurately measure the internal resistance of the battery.

Third, the internal resistance is quite different in the same manufacturer's model and the same batch of the batteries for the physical and chemical characteristics of the battery. The product tolerance cannot be eliminated, so it is difficult to obtain the "internal resistance SoC" characteristic curve of the battery through the battery sample.

In spite of this, the DC internal resistance of the lead–acid battery will increase significantly in the late stage of discharge, and the impact of the above three points can be ignored within a certain range of accuracy requirements. Therefore, some people also use the internal resistance method to estimate the SoC value of a lead–acid battery in the late stage of discharge. However, this method is not applicable to the lithium-ion batteries because the internal resistance is still small at the end of charging and discharging and the characteristics of the internal resistance are not clear.

2. Load voltage method

The load voltage method is based on the principle of the OCV method and an improved method proposed for failure to real-time estimate the SoC in the OCV method. The principle is as follows: according to the discharge equivalent circuit of the battery pack in Figure 8.8, when the internal resistance r and working current I of the battery are known, the EMF of the battery can be calculated by the following equation after measuring the operating voltage UL at both ends of the load RL:

$$EMF = U_L + Ir \tag{9.19}$$

The SoC can be calculated according to the EMF–SoC curve as shown in Figure 9.3.

This method is feasible in theory, but there are many difficulties in practical application: first, the internal resistance r of the lithium-ion battery is related to various factors and cannot be accurately measured due to large differences in the cells. Second, in the practical application of the electric vehicle, a large fluctuation in the current I causes a great negative impact on the accuracy of the calculation. Third, it is required for the load voltage method to estimate the OCV or EMF, so this method cannot avoid some defects of the OCV method, such as the large error of the platform area list. Therefore, the load voltage method is rarely applied to the electric vehicle, but it is often used to determine a stop in the charge and discharge process.

9.3 Difficulty in an SoC Estimation

It is difficult to accurately estimate the SoC for the following causes.

9.3.1 Difficulty in an Estimation Resulting from Inaccurate Battery State Monitoring

The SoC cannot be directly measured but is estimated indirectly through the measured value of the state quantity, such as voltage and current. Because the error of battery state monitoring is inevitable, the SoC estimation error is also inevitable. The inaccurate battery state monitoring is caused by the following two aspects.

1. Inaccurate battery state monitoring resulting from sensor accuracy

In the process of monitoring physical quantities such as voltage and current, the error is inevitable, which will lead to difficulty in estimating the SoC. Two typical examples are provided.

Example 1 The SoC is estimated in the CC method by reference to Equations (9.17) and (9.18). In this example, the accumulated charge during the period from t_1 to t_2 is calculated by $Q_{t_1}^{t_2} = \int_{t_1}^{t_2} i(\tau)d\tau$, where the error of $i(\tau)$ is directly reflected in $Q_{t_1}^{t_2}$ due to the error of the current sensor. Generally, the error of $i(\tau)$ includes systematic error and random error. For a longer operation time, the random error will be offset in the integration summation process, while the systematic error will be retained proportionally. Assuming that the systematic error i_τ caused by the current sensor is 2%, the error of calculated $Q_{t_1}^{t_2}$ is also 2%. This error will be transferred to the estimated SoC by Equation (9.18).

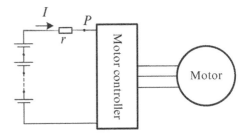

Figure 9.4 Schematic diagram of the driving platform for an electric vehicle.

As shown in Figure 9.4, during 10 ~ 95% SoC, the battery is in the platform area, the voltage is 3.18–3.32 V, that is to say, 85% SoC change corresponds to 0.14 V voltage change. Assuming that the platform area voltage changes approximately linearly, this means that a 0.01 V voltage error may cause about a 6% SoC estimation error or a 0.005 V voltage error may cause about a 3% SoC estimation error. The voltages of 0.01 V and 0.005 V are discussed because the mainstream voltage sensors used for the current battery management systems meet this sampling accuracy.

2. Inaccurate battery state monitoring resulting from electromagnetic interference

As analyzed above, the SoC estimation error is caused by the sensor accuracy, and it seems that the error can be removed by improving the sensor accuracy at a higher cost. In fact, that is not the case, because in the process of work, electromagnetic interference factors will also cause inaccurate state monitoring, resulting in the SoC estimation error. Figure 9.4 shows a simple schematic diagram of the driving platform for the electric vehicle. As shown in the figure, the driving platform consists of a battery pack, a motor, and a motor controller matching the motor, and r is the equivalent resistance produced by the wire connection. In the operating process, the interference from the motor and the motor controller is transferred to the point P by the bus, so the operating voltage and operating current of point P contain rich AC components. As a result, the monitored voltage and current values are not accurate enough.

Figures 9.5 and 9.6 show the waveform of the voltage at the point P and the current flowing through r. Due to the rich working frequency spectrum of the motor for an electric vehicle, many high-frequency harmonics are superimposed (as shown in Figure 9.7). Although the resistance-capacitance filter circuit is generally added at the sampling end before A/D conversion, it is impossible to completely eliminate the impact of electromagnetic interference on the sampling accuracy. Moreover, in an electric vehicle, the sampling circuit will be subject to various non-conductive forms of radiation interference, resulting in errors in the monitoring values of state quantities such as the voltage and current.

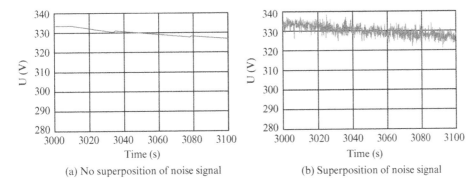

(a) No superposition of noise signal   (b) Superposition of noise signal

Figure 9.5 Voltage waveform at point P.

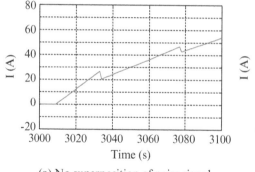

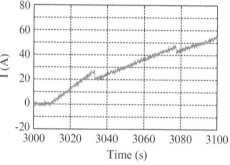

(a) No superposition of noise signal   (b) Superposition of noise signal

Figure 9.6   Current waveform flowing through *r*.

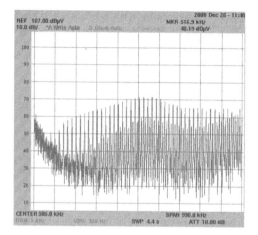

Figure 9.7   Spectrogram of voltage at point P.

9.3.2   Difficulty in an Estimation Resulting from Battery Difference

In the manufacturing process of the power batteries, due to the differences in materials, technology, and other aspects, a great difference is caused between different batches of batteries and even between the different batteries of the same batch. The difference causes difficulty in the SoC estimation for the following two reasons:

1. **Difference between the battery sample and the actual battery**

The current SoC estimation methods are based on the characteristics of the battery samples. The characteristics of the sample are compared with those of the battery used practically. The difference between the battery sample and the actual battery imposes an impact on the SoC estimation accuracy. Two common conditions are provided below.

First, when the SoC is estimated by the CC method, capacity non-uniformity may cause an inaccurate estimation. For example, if the capacity of the sample battery is 100 Ah, while the capacity of the actual battery is 105 Ah, when the accumulated capacity discharged in a certain actual process calculated by the CC method is 95 Ah, according to the capacity of the sample, the SoC should be 5 Ah, while the SoC of the actual battery is 10 Ah.

Second, the SoC is estimated by the OCV method based on the SoC–EMF curve of the battery sample, that is, after measuring the OCV of the actual battery, the SoC is reversely estimated for the actual battery through the curve. However, due to the difference in the characteristic curves of the actual battery and the sample battery, the SoC estimated by the OCV method will have errors.

2. **Difference in cells of a battery pack**

In practical work, attention should be given to the following three conditions:

First, in the battery management system, it takes a lot of time to evaluate each cell. In addition, if more precise voltage sampling is performed for each cell, the device required is more expensive. In order to solve these problems, in some battery management systems, voltage sampling is performed only for some cells in the battery pack, in order to estimate the SoC of the power pack. This method greatly saves the SoC estimation time and cost. However, due to the difference in the cells of the battery pack, certain errors are inevitable in the estimation results.

Second, in the working process of the electric vehicle, due to the different position of the cells in the battery box, the heat absorption and heat dissipation vary with the cells, and the temperature has a greater impact on the SoC and even the cycle life of the battery. Therefore, at a certain time, there are some differences in the SoC and SoH of the cells in the battery pack. If the SoC is estimated using a uniform method, the error may be large.

Third, after the battery pack of an electric vehicle has been used for a period of time, it is sometimes necessary to replace a particularly poor performing cell. However, once replaced, the difference between the new cells and the original cells may cause more difficulty in the SoC estimation.

9.3.3 Difficulty in an Estimation Resulting from an Uncertain Future Working Condition

In the working process of the electric vehicle, the possible working conditions are ever changing and so the driver cannot predict the working condition at the next moment, which causes certain difficulties in the battery level or SoC estimation.

On the one hand, the SoC is affected by many factors and it is impossible to fully discharge the battery. For example, in practical work, if it is required to discharge a larger operating current at any time in the future, the dischargeable capacity of the battery pack is lower; on the contrary, if the electric vehicle requires a small operating current in the future and is used under stable working conditions, the dischargeable capacity of the battery pack is higher. In addition to the operating current, the operating temperature of the battery pack will also affect the maximum dischargeable capacity of the battery.

On the other hand, in the case of a given SoC, the dischargeable capacity varies with the battery packs. Qualitatively, at a constant operating temperature and a constant internal resistance, the larger the operating current, the less the dischargeable capacity is and the shorter the endurance mileage of the electric vehicle will be. At the same time, in the case of a given working current, the higher the operating temperature, the smaller the internal resistance is, the larger the dischargeable capacity is, and the longer the endurance mileage will be.

9.3.4 Difficulty in an Estimation Resulting from an Uncertain Battery Usage History

The battery usage history has an impact on assessing its current capacity. However, it is difficult to understand the battery usage history. Without an understanding of the battery usage history, it is difficult to assess its current state.

1. Importance of understanding battery usage history to an SoC estimation

The power battery is a chemical product. Many characteristic quantities of the battery are process quantities, which are related to the battery usage history. Understanding the battery usage history is important to estimate the SoC for three reasons.

First, the EMF of the power battery has a hysteretic effect. The EMF is of great significance to the estimation of the SoC. However, due to the hysteresis effect of the EMF of the power battery, the EMF of the power battery at a certain moment depends on the charging and discharging operation for a long time before that. If these historical data are insufficient, it is difficult to estimate the current EMF of the battery and thus it is difficult to estimate the SoC by the operating voltage or the open-circuit voltage.

Second, due to the polarization capacitance effect, the open-circuit voltage of the power battery often has rebound resilience. It can be seen from the previous chapters that when an electric vehicle is parked, the operating current is close to zero, while the battery voltage is not necessarily zero. Due to the existence of the polarization capacitance, the battery voltage will rebound over a period of time. Without historical data, it is difficult to determine whether the battery's voltage rebound has ended and it is difficult to estimate the current SoC of the battery from the voltage. Let us imagine the following situation: a driver parks the electric vehicle and shuts down the engine while the BMS stops its operation, but he/she remembers that there is something unfinished and immediately restarts the electric vehicle while the BMS restarts its operation and the interval time between the shutdown and the restart is not more than 10 s. In this case, if BMS does not know the working history of the electric vehicle and will think that the battery voltage at this time has no rebound action. Therefore, it will correct the SoC of the battery by using the open-circuit voltage at the time, which will cause a large error. Therefore, it is suggested that for the BMS of the electric vehicle, it is necessary to record the last parking time in order to correct the SoC in the open-circuit voltage method.

Third, the battery's SoH is related to the battery usage history. After the battery leaves the factory, its performance will decline monotonously with time, that is to say, the SoH of the battery will decrease continuously and the SoH is of great significance for the estimation of the SoC. It can be seen that the accurate SoH can be estimated by understanding the battery usage history, which is beneficial to an accurate estimation of the SoC.

2. **Difficulty in a full understanding of the power battery history**

As mentioned above, it is important to obtain as much historical information of the battery as possible for an accurate SoC estimation. However, it is difficult to fully understand the historical information of the power battery for three reasons.

First, it is difficult to obtain complete historical information about each cell. To obtain historical information about each cell, the voltage, current, and temperature of each cell must be monitored at every moment, and not every battery management system has sufficient BMC to accomplish this task. For example, as mentioned in the previous sections, many battery management systems cannot collect temperature information of each cell in real time, making it impossible to obtain the historical information of each cell.

Second, it is difficult to store all historical information of each cell. On the one hand, resources of the BCU are occupied by the stored data, and the BCU of the general battery management system is implemented by the embedded system, so the system resources are scarce. Occupation of a large amount of BCU resources by the stored data may affect other functions of the battery management system. On the other hand, a large amount of memory banks is required to store all the historical data, which is not practical for the embedded systems.

Third, even if all historical information is recorded, a large number of operations is required to process the historical data in the actual work. Since it is required to estimate the SoC in real time, such operations are an almost impossible task for the embedded battery management system.

9.4 Actual Problems to Be Considered During an SoC Estimation

Due to the difficulties in an SoC estimation of the battery for the electric vehicle detailed in Section 9.3, it is almost impossible to estimate the SoC of the battery pack 100% accurately. What SoC error is acceptable in practice? This section will attempt to discuss some practical considerations relating to the electric vehicle.

9.4.1 Safety of the Electric Vehicle

For electric vehicles, safety is the first consideration, so the SoC estimation error of the remaining capacity must not threaten the safety of people and electric vehicles. Let us hypothesize the following two conditions. First, for the electric vehicle, the battery of which has been fully charged, the actual SoC is 100% and the estimated SoC is 90% due to an estimation error. In such a case, the battery is not fully charged according to the estimated SoC and energy recovery is allowed when the electric vehicle slides down a long slope. Therefore, the BMS sends out the "allow charging" message to other control systems and a part of the mechanical energy is converted into electrical energy to charge the battery, resulting in the battery becoming overcharged, which may cause irreversible damage to the battery, battery performance degradation, and even battery explosion and other safety accidents. Second, for the electric vehicle with 5% actual SoC, the estimated SoC is 10%, which is twice as much as the actual SoC. If the driver knows the mistaken information of 10% from the instrument, according to his or her experience, he or she may consider that the SoC is enough for him or her to go home, and the battery may be charged at home. However, in fact, due to an SoC estimation error, the battery may be fully discharged several hundred meters from home.

According to the analysis provided in the previous section, although the SoC is important, its estimation error is inevitable. Therefore, the following two suggestions are given for practical working of the electric vehicle.

First, redundancy is set for vehicle safety. The redundancy design is often used in a vehicle design. For example, for the performance degradation after the motor system is used for a certain period of time, it is often necessary to provide a certain degree of power redundancy when designing the motor for the electric vehicle [1]. Some redundancy should be considered for the battery management system, for example, the linear transformation is performed for the estimated SoC and the transformed result is displayed in the instrument. That is, when the estimated SoC is 10%, 5% SoC is displayed, and when the estimated SoC is 95%, 100% SoC is displayed. In this way, redundancy is provided not only to ensure the safety of the electric vehicle but also to reduce the charge and discharge depth of the battery, which is beneficial to extend the battery cycle life. Of course, the specific linear transformation algorithm is calibrated according to the SoC evaluation error of the battery management system by a large number of reliable experimental data.

Second, a protection mechanism independent of the SoC must be set up in the BMS. Although the SoC estimation has an important guiding significance for many other functions of the battery management system, it is also often an important reference index for the safety protection function of the BMS. However, in the process of battery management system design, a set of safety protection mechanisms independent of the SoC estimation must be set up for two reasons. First, if the SoC is used as the sole safety protection criterion, the estimation error will affect the vehicle safety. However, as analyzed in the previous section, since the SoC estimation error is inevitable, the vehicles must have safety risks. Second, some faults or accidents happen so fast that the system has not yet had time to finish the state monitoring and information calculations. Therefore, it is of great significance for the safety of an electric vehicle to additionally set up a set of simple and reliable protection mechanisms. For example, the simplest voltage comparator analog circuit can be used to reliably and effectively protect the battery voltage.

9.4.2 Feasibility

It is difficult to 100% accurately estimate the SoC, but the estimation accuracy can be improved under certain conditions. However, during the accuracy improvement, it is important to take the system feasibility into account, including technical feasibility and cost feasibility.

1. Technical feasibility from the perspective of the software algorithm

With the development of digital signal processing technology and artificial intelligence, many complex computing methods have been proposed in the past few decades, in order to improve the accuracy of numerical processing. However, for a special application, the electric vehicle, the SoC estimation algorithm should not be too complex, because the vehicle battery management system basically relies on the embedded system, with limited computing capacity and memory space. This requires the battery management system designers to select the appropriate algorithm for the embedded system so that the system can effectively estimate the SoC in real time.

2. Cost feasibility from the perspective of hardware

No matter how sophisticated the software algorithm is, the SoC estimation still relies on accurate battery state monitoring. The monitoring accuracy mainly depends on the hardware. Admittedly, the choice of high-precision hardware and the increase in the number of sensors have a positive significance on the SoC estimation. However, the cost inevitably increases with the increase in the quantity and quality of the hardware. Therefore, it is necessary to make a trade-off between the hardware cost and estimation accuracy.

9.4.3 Actual Requirements of Drivers

One of the main purposes of the battery level or SoC estimation is to report the state of the battery pack to the driver through the dashboard, in order to help the driver determine the endurance mileage of the electric vehicle and choose the appropriate charging time. Therefore, the following actual requirements of the drivers must be considered when selecting the SoC estimation algorithm.

1. Information content and accuracy

As mentioned above, the SoC and the estimated endurance mileage are the information drivers care about. However, as described in Section 9.1, the remaining charge of the battery pack may be shown to the driver in the form of the battery level (Ah) or the SoC (%). Should the remaining charge be displayed in Ah or a percentage? Let us imagine the following two situations. First, since the driver may not understand the definition of Ah, if told that "the SoC of the electric vehicle is 10 Ah", the driver will not be able to judge how much energy corresponds to 10 Ah. Second, for the battery pack operating in the platform area, if the driver is not sensitive to the SoC error, for example, he or she may consider that "70%" SoC and "75%" SoC displayed during driving have a similar meaning.

Therefore, when designing the motormeter for the electric vehicle, many large vehicle manufacturers use a relatively intuitive graphic scale to display the SoC of the battery so that the SoC is displayed in the same display mode as the fuel gauge of the orthodox car to meet the driving habit. Although the accurate value becomes fuzzy by the use of a graphic scale to display the SoC, the graphic scale is not inappropriate because, in fact, the graphic scale is used to display the battery level in many electronic products, such as mobile phones, and people do not feel it is inconvenient to use.

2. Consideration of the driver's subjective feeling

An SoC estimation error is inevitable as the driver cannot judge whether the SoC estimation algorithm used in the BMS is accurate, but will judge whether the estimated SoC is reliable based on his or her own subjective feelings. The following four aspects are considered.

First, there should be no great fluctuation in the data shown to the driver. For example, the error of a certain SoC estimation algorithm, a high-accuracy estimation algorithm, is not more than ±3% of the full range. However, if the actual SoC is 70%, the SoC estimated by the algorithm evaluation is 72% at a certain moment and 68% after 10 s in theory, the error of two estimated values is not large, but if two values are displayed on the instrument, the driver is surprised by a 4% loss of SoC within 10 s. Thus, for an electric vehicle, the accuracy is not the only criterion by which to judge the SoC algorithm.

Second, an abnormal numerical fluctuation may affect the driver's trust in the system. For example, the error of a certain SoC estimation algorithm, a high-accuracy estimation algorithm, is not more than ±3% of the full range. When the electric vehicle is normally driven without the braking energy recovery, if the actual SoC is 70%, the SoC estimated by the algorithm evaluation is 69% at a certain moment and 71% after 10 s, so, in theory, the error of two estimated values is not large and the fluctuation is small. However, a 2% fluctuation of the SoC within 10 s may cause distrust of the driver in the system, because the SoC may not be increased when the electric vehicle is being driven.

Third, the driver's sensitivity to the error increases at the end of a battery discharge. The driver may not be sensitive to the SoC estimation error when the battery is operating in the platform area (15 ~ 90%), but the driver would be very concerned about the SoC estimation error at the end of a battery discharge, because of the direct impact on the driver in judging whether the battery capacity is enough to arrive at the destination.

Fourth, information uniformity. The following facts are considered: for the battery pack with 100 Ah actual capacity, the dischargeable capacity during the first 50 Ah (corresponding to 50 ~ 100 SoC) is not equal to that during the second 50 Ah (corresponding to 0 ~ 50% SoC). As known from the SoC–EMF curve, at the same operating current, the dischargeable capacity is higher during the first 50 Ah corresponding to the higher electromotive force and lower during the second 50 Ah. Therefore, the same SoC reduction may correspond to different travel distances. For example, if the travel distance is 100 km during the reduction from 100 to 50%, it does not mean that the travel distance is 100 km during reduction from 50% to 0. However, the driver may not have the relevant experience and may think that the whole process is uniformly linear. Therefore, the battery management system designers must consider this actual condition.

9.5 Estimation Method Based on the Battery Model and the Extended Kalman Filter

9.5.1 Common Complicated Estimation Method

Different technologies (such as fuzzy logic, Kalman filtering, neural network, recursion, the self-learning method, etc.) are often used to improve the accuracy of the algorithm for difficulties in the SoC estimation. Several representative methods are described below.

1. Fuzzy logic algorithm

Fuzzy logic is a simple method used to extracting exact conclusions from vague, ambiguous, or inaccurate information. Its ability to extract exact answers from approximate data is similar to a human decision.

Unlike classical logic, which requires a deep understanding of the system, precise equations, and precise numerical values, fuzzy logic allows modeling with higher abstractions derived from our knowledge and experience. It allows expression of such knowledge by subjective concepts such as big, small, very hot, bright red, long time, fast, or slow. This qualitative linguistic description of professional knowledge is more like nature than a numerical description of a system, and its algorithm development is relatively simple compared with a numerical system. The system output can then be mapped to a precise range of values to characterize the system. The fuzzy logic is widely used in automatic control systems. Using this technique, all available information about battery performance can be used to estimate the SoC or SoH of a battery more accurately.

2. Kalman filter algorithm

The Kalman filter solves an age-old problem: how to get accurate information from inaccurate data? More precisely, how to select a "best" value as the latest data of the system state to update the system data when the input data are still inaccurate? Hybrid vehicles are an example of this condition. The SoC of the battery is influenced by many factors and will change with the change of the user's driving mode. The purpose of a Kalman filter is to remove noise from the data stream [2]. It is achieved by predicting the new state and its uncertainty, and then calibrating the predicted value with the new measured value. It is suitable for multi-input systems and is widely used in predictive control circuits of navigation and targeting systems. The accuracy of the SoC estimation model with the Kalman filter can be improved, and it has been reported that the system accuracy is now improved by 1% [3, 4]. As with the fuzzy logic, standard software packages are available to assist the implementation of Kalman filter algorithms [5, 6].

3. Artificial neural network algorithm

The neural network, a computer architecture model based on the interconnection of the human brain nerve system, mimics the processes of information processing, memory, and learning in the human brain. It mimics the brain's classification patterns and the ability to learn from trial and error to distinguish and extract the relationship between the presented data and the hidden information. Each neuron in a neural network has one or more inputs and produces an output. Each input has a weighting factor that modifies the input value of the neuron. The neuron precisely manipulates the input and then outputs the result. The neural network is made by a simple combination of the neurons, with the output of one neuron acting as the input of the next neuron until the final result is obtained. The neural network completes the learning phase by the samples submitted to it (with known results). The final output of the system is adjusted on the basis of the sample by human intervention or programming algorithm weighting factors to approximate the known result. In other words, the neural network "learns" from the sample (as a child learns to recognize a dog from a given example of a dog) and then gets some ability to generalize beyond the sample data. Thus, the neural network resembles the human brain in two ways: first, the neural network acquires knowledge through learning; second, the knowledge of the neural network is stored in the cross-neuron connection strength of synaptic weights.

When the battery performance is affected by many quantized parameters and the mathematical accuracy of most of the parameters cannot be defined, the neural network technology is very useful in the SoC estimation. The algorithm accuracy can be improved by the experience in a similar battery performance.

9.5.2 Advantages of a Kalman Filter in an SoC Estimation

A variety of more complicated estimation methods are described in the previous section to solve some problems that cannot be solved by classical estimation methods. A special SoC estimation method based on the Kalman filter is described in this section. This method is based on the lithium-ion battery model proposed in Chapter 7 and the extended Kalman filter (EKF), and is applicable to application in the electric vehicle.

It is not difficult to imagine that the application of a Kalman filter will increase the algorithm complexity and the system cost. However, it must be hoped that a certain result will be obtained by such efforts. Let us first identify the advantages of the Kalman filter and the reasons for its use for an SoC estimation in an electric vehicle. Compared with some classical estimation methods described in Section 9.2, the Kalman filter has the following advantages.

First, it is usable at any time. The OCV method mentioned in Section 9.2.2 is used only when the battery is idle for a period of time and the voltage rebound is sufficient. However, the OCV method cannot be used when the battery is in normal operation. The Kalman filter method is applicable to the battery in any state ("idle," "discharge," or "braking energy recovery"), and can be used to estimate the SoC of the battery.

Second, it helps to correct the initial value. The CC method mentioned in Section 9.2.1 has fatal disadvantages, such as dependence on the initial value and error accumulation. In other words, if there are errors in the initial value of the SoC, such an error will continue to accumulate. Although Section 9.2.3 proposes a compromised method to correct the accumulated error by an open circuit voltage, certain conditions, for example stopping the battery for a period of time (or enough small working current), must be met during the correction. The error may not be corrected for an hour for a vehicle that has been running. The Kalman filter can overcome this deficiency to correct such an error at any time, and even if the error of the initial SoC is very large, after a period of time the filter will eliminate such an error.

Third, it is helpful in solving insufficient sensor accuracy. Since the vehicle-mounted BMS is restricted by many factors such as cost and reliability, the sensor accuracy is often limited, resulting in a large SoC estimation error. The Kalman filter is helpful in removing this defect. The principle is analogous to removal of the random errors by multiple observations of a measured object. For example, if the voltage measurement resolution of 5 mV is achieved by a 10-bit A/D chip, less than a 5 mV measured value may have a certain random error. The measured value may have a larger random error for single sampling, and the use of the Kalman filter is equivalent to the measurement of the same test object many times in order to eliminate random error and obtain higher precision.

Fourth, it is helpful in the removal of the impact of electromagnetic interference. As described in Section 9.3, inaccurate battery state monitoring may be caused by electromagnetic interference to the sensor. Such electromagnetic interference is particularly pronounced in the electric vehicle. It is helpful in the removal of the impact of electromagnetic interference with the use of the Kalman filter. Of course, according to theoretical analysis, the Kalman filter is only helpful in eliminating the noise that follows normal distribution.

9.5.3 Combination of an EKF and a Lithium-Ion Battery Model

This section intends to describe the basic framework of the complicated algorithm. It is required to answer the following questions: What is the EKF? What is difference between the EKF and the classical Kalman filter? How can the EKF be combined with the lithium iron phosphate battery model described in the previous chapter? Why are EKFs chosen?

1. **Kalman filter and EKF algorithm**

The classical Kalman filter needs to satisfy the following linear model:

$$x_k = Ax_{k-1} + Bu_{k-1} + w_{k-1} \tag{9.20}$$

$$z_k = Hx_k + v_k \tag{9.21}$$

where $x_k \in R^n$ is the state variable of the system, $z_k \in R^m$ is the observational variable of the system, the random signals w_k and v_k are respectively the process excitation noise and observed noise of the system, and u_k is a control function, or so-called system excitation. Equation (9.20) is called the random state difference equation (hereinafter referred to as the state equation or the difference equation), where A is an $n \times n$ matrix reflecting the mapping of the state $(k - 1)$ at the last time to the current state (k) and B is an $n \times l$ matrix representing the impact of the system excitation on the state vector. Equation (9.21) is called the measurement equation, where H is an $m \times n$ matrix reflecting the impact of the state variable x_k on the measured variable z_k.

It can be seen from the above relationship that, in the classical Kalman filter, the relationship between state vectors, system excitation, and observed variables is all linear. However, if their relationship is not linear, that is, at least one of Equations (9.20) and (9.21) is not valid, the classical Kalman filter is no longer applicable and the EKF may be used [7–9].

The state equation and measurement equation of the EKF are as follows:

$$x_k = f(x_{k-1}, u_{k-1}, w_{k-1}) \tag{9.22}$$

$$z_k = h(x_k, v_k) \tag{9.23}$$

The variables used in two equations have same meaning with those for the classical Kalman filter, but the two equations are no longer formally linear, but can be non-linear.

According to the forms of the above two equations, the big difference between the EKF and classical Kalman filters cannot be seen. In fact, their differences are reflected in the specific algorithm. If two equations are non-linear, some additional processes are introduced and some approximations are made at some point during the calculation. In short, the algorithm of the EKF is more complex than that of the classical Kalman filter. If the relationship between the state variable and the observational variable is linear, the EKF is not used very much.

2.   Selection of main variables for the EKF

After the EKF has been determined to solve the problem of SoC estimation, it is necessary to clarify the specific forms of the state variable, observational variable, and system excitation [10, 11]. After the analysis of recent literature, these values are determined by the following ideas generally.

First, the operating current of the battery is used as a system excitation. In this case, it is generally accepted that the current is the factor that drives the "state" change of the whole system, because the change of the current affects the change in other parameters of the battery, including the battery level or SoC, the EMF, and of course, the operating voltage at both ends of the battery. In fact, the ambient temperature can be used as another excitation factor, but the relationship between the temperature and other state variables is not very clear, and in a certain working process of an electric vehicle, it is considered that the temperature does not change very much, so people often abandon this excitation factor. As an introduction to the principles, the temperature is not used as the excitation herein. A more rigorous demonstration should be performed to determine whether this factor is discarded.

Second, the operating voltage of the battery is used as an observational variable. A close inspection of the battery's working process shows that the operating voltage at two ends of the battery is a physical quantity that can be accurately measured. Other variables, such as the SoC and the EMF, cannot be accurately measured from the outside during the actual operation of the electric vehicle. Therefore, the working voltage of the positive and negative electrodes of the battery is selected as the observational variable.

Third, the state parameters are determined for the variables changing over time under the system excitation and are combined to form the state variables in vector form. The state parameters are various in form. The equivalent ohmic resistance of the battery can be used as the state parameter or the voltage at both ends of the polarization capacitance of the battery can be used as the state parameter. The parameter, the SoC to be estimated, is basically indispensable.

3.   Establishment of the state equation based on the lithium iron phosphate battery model

However, after the above variables are determined, the Kalman filter algorithm cannot start to operate because the specific form has not been determined for a state equation and measurement equation, that is, the specific relationship is determined between variables in Equations (9.22) and (9.23). The algorithm discussed in this section is based on the lithium iron phosphate battery model presented in Chapter 7, so two equations are also established based on this model.

According to the lithium iron phosphate battery model, the following several elements are determined for the Kalman filter: the system excitation is i_k (the operating current of the battery), the observational variable is u_k (the working voltage at both ends of the battery), and the state variable is $x_k = \begin{bmatrix} u_k^\Omega & u_k^s & u_k^m & u_k^l & \mathrm{SoC}_k \end{bmatrix}^T$. The first four terms of the state variable are based on the model described in Chapter 7, which respectively represent the voltage at both ends of the equivalent ohmic resistance and the voltage at both ends of the three RC networks. The last item of the state variable is the SoC value of the battery that we are most concerned about.

The state equation is determined below for the Kalman filter. According to the battery model described in Chapter 7, under the action of current excitation, all state parameters have the following relationship:

$$\begin{cases} u_k^{\Omega} = i_{k-1} R_{\Omega} \\ u_k^s = i_{k-1} \dfrac{R_s}{1 + R_s C_s} + \dfrac{R_s C_s}{1 + R_s C_s} u_{k-1}^s \\ u_k^m = i_{k-1} \dfrac{R_m}{1 + R_m C_m} + \dfrac{R_m C_m}{1 + R_m C_m} u_{k-1}^m \\ u_k^l = i_{k-1} \dfrac{R_l}{1 + R_l C_l} + \dfrac{R_l C_l}{1 + R_l C_l} u_{k-1}^l \\ \mathrm{SoC}_k = i_{k-1} \dfrac{1}{C_{cap}} + \mathrm{SoC}_{k-1} \end{cases} \tag{9.24}$$

where C_{cap} is the battery capacity, in ampere seconds.

The above equation is modified as a matrix, as follows:

$$x_k = i_{k-1} \begin{bmatrix} R_{\Omega} \\ \dfrac{R_s}{1 + R_s C_s} \\ \dfrac{R_m}{1 + R_m C_m} \\ \dfrac{R_l}{1 + R_l C_l} \\ \dfrac{1}{C_{cca}} \end{bmatrix} + \begin{bmatrix} 0 & 0 & 0 & 0 & 0 \\ 0 & \dfrac{R_s C_s}{1 + R_s C_s} & 0 & 0 & 0 \\ 0 & 0 & \dfrac{R_m C_m}{1 + R_m C_m} & 0 & 0 \\ 0 & 0 & 0 & \dfrac{R_l C_l}{1 + R_l C_l} & 0 \\ 0 & 0 & 0 & 0 & 1 \end{bmatrix} x_{k-1} \tag{9.25}$$

Assuming that

$$A = \begin{bmatrix} 0 & 0 & 0 & 0 & 0 \\ 0 & \dfrac{R_s C_s}{1 + R_s C_s} & 0 & 0 & 0 \\ 0 & 0 & \dfrac{R_m C_m}{1 + R_m C_m} & 0 & 0 \\ 0 & 0 & 0 & \dfrac{R_l C_l}{1 + R_l C_l} & 0 \\ 0 & 0 & 0 & 0 & 1 \end{bmatrix}$$

then

$$B = \begin{bmatrix} R_{\Omega} & \dfrac{R_s}{1 + R_s C_s} & \dfrac{R_m}{1 + R_m C_m} & \dfrac{R_l}{1 + R_l C_l} & \dfrac{1}{C_{cap}} \end{bmatrix}$$

After consideration of the noise factor, the following state equation can be obtained:

$$x_k = A x_{k-1} + B i_{k-1} + w_{k-1} \tag{9.26}$$

where the random signal w_k represents the process excitation noise, which is related to the measured current noise, and can be ignored sometimes and set to zero directly. The characteristics of w_k may be analyzed according to specific problems.

4. Establishment of the measurement equation based on the lithium iron phosphate battery model

From Equation (9.26), it can be seen that the state equation is linear. If the measurement equation is also linear, the classical Kalman filter can be used to avoid the use of EKF. In fact, the measurement equation is not linear, because according to the battery model described in Chapter 7, when the battery is in a discharge state, the operating voltage of the battery is related to the equilibrium EMF of the battery, the voltage at both ends of the three RC networks, and the voltage at both ends of the ohm resistance, and has the following circuit relations equation:

$$z_k = u_k = U_k^{EMF} - u_k^{\Omega} - u_k^s - u_k^m - u_k^l \tag{9.27}$$

where U_k^{EMF} is the equilibrium EMF of the battery, which has a non-linear functional relationship with the SoC of the battery, namely

$$U_k^{EMF} = g(\mathrm{SoC}_k) \tag{9.28}$$

Since the function $g(\cdot)$ is non-linear according to the model, Equation (9.27) is also non-linear. Equation (9.27) is just the observational equation reflecting the relationship between the state variables and the observational variables. Therefore, the EKF, rather than the classical Kalman filter, is used to obtain a variable estimation [12].

9.5.4 Implementation Rule of the EKF Algorithm

In the previous section, the framework of the SoC estimation algorithm is established according to the basic principle of the EKF and the LiFePO$_4$ power battery model described in Chapter 7. It is worth discussing a few details.

1. Hysteretic voltage of the model

In Chapter 7, the hysteretic voltage is described and it is proposed that there is no clear functional relationship between the hysteretic voltage and the SoC, and if not handled properly, it is possible to cause an SoC estimation error. For pure electric vehicles, however, it is relatively easy to solve the problem. Since the pure electric vehicles are in the state of "charging" or "discharging" for a long time, that is, charging or discharging always lasts for quite long periods of time, the value of the hysteretic voltage in the model can be fixed as the value of the "charging" or "discharging" state. However, for the hybrid vehicles, since the battery pack often is switched between "charging" and "discharging," it is not easy to solve the hysteretic voltage problem, and the resulting SoC estimation error is often 10%. Fortunately, hybrid vehicles have a much higher tolerance to SoC estimation errors than pure electric vehicles, because there is no need to worry about misjudging the endurance mileage due to the SoC estimation errors.

2. Function g(•) and its applicable scope

Function $g(\cdot)$, a non-linear function according to the model in Chapter 7, reflects the relationship between the SoC and the electromotive force U^{EMF} of the battery and can be approximately expressed by higher-order polynomials. It can be seen from the previous contents of this book that the EMF of the lithium iron phosphate battery has an obvious three-stage characteristic, that is, the curve is relatively flat and approaches linear during SoC 10 ~ 90%, and is significantly changeable during more than 90% SoC and less than 10% SoC. Function $g(\cdot)$ has two understandings. First, the function applicable to 0 ~ 100 SoC is established so that the order is higher and estimation fluctuation is larger during function derivation. Second, the function applicable to 10 ~ 90% SoC is established to achieve a relatively simple operation and more accurate estimation. It is required to identify the estimation method for more than 90% SoC and less than 10% SoC.

3. Initial value

It is unavoidable for the Kalman filter to identify the initial value for each operation. The initial value is set for each element of state variable x_k at the beginning of a recursive operation. In this algorithm, X_k includes five elements, i.e. u_k^Ω, u_k^s, u_k^m, u_k^l, and SoC_k, which cannot be directly measured. In fact, the Kalman filter has a good adaptability. Even if the initial value is set incorrectly in the operation process, the estimated value close to the real value of the estimated object can be obtained after several steps of calculation. Therefore, the following solution is given for the initial value of the state variables

First, the initial values of the four state variables, i.e. u_k^Ω, u_k^s, u_k^m, and u_k^l, are set as 0. Second, the initial value of SoC_k is set by two methods. First, the initial value is estimated in the OCV method according to the initial open-circuit voltage of the battery (if available). Second, the last estimated SoC, namely the SoC obtained by the BMS before shutdown of the electric vehicle, is called from the stored historical information (if the BMS has a historical information storage function).

4. Specific operation process

After all details are determined, the EKF algorithm can be completed according to the classical recursive method, specifically as follows [13].

First, according to Equation (9.26), the estimated value of the state variable at moment k can be obtained as follows:

$$x_k^* = Ax_{k-1} + Bi_{k-1} + w_{k-1} \tag{9.29}$$

The "*" is marked above the state vector to indicate that this is a value estimated by the state equation. Accordingly, the covariance matrix after completing the state recursion is obtained as

$$P_k^* = A_k P_{k-1} A^T + Q_{k-1} \tag{9.30}$$

Second, this is a crucial step, the gain K_k of the Kalman is solved by

$$K_k = P_k^* H_k^T (H_k P_k^* H_k^T + R_k)^{-1} \tag{9.31}$$

where R_k is the covariance matrix of the observation noise. In addition, H_k is the Jacobian matrix obtained according to the non-linear relation of Equation (9.23), and the elements in the matrix H_k meet the following equation:

$$H_{k[i,j]} = \frac{\partial h_{[i]}}{\partial x_{[j]}} [x_k^*, 0] \tag{9.32}$$

Third, the estimated value of the state vector and the corresponding covariance matrix are modified according to the gain of the Kalman:

$$x_k = x_k^* + K_k(z_k - H_k x_k^*) = x_k^* + K_k(u_k - H_k x_k^*) \tag{9.33}$$

$$P_k = (I - K_k H_k) P_k^* \tag{9.34}$$

where I is an element matrix. In addition, since the operating voltage at both ends of the battery is used as the observational variable, $z_k = u_k$.

After the third step is completed, the time indicator k is increased by 1 and the steps are repeated to continue the calculation. So far, a recursive algorithm has been achieved for the estimation method based on the battery model and the E [14].

9.5.5 Experimental Verification

Experiments are conducted to verify the above SoC estimation methods. The experimental sample is a lithium iron phosphate battery with a rated capacity of 100 Ah. The charging and discharging of the battery under "10–15" conditions are simulated by the high-power battery tester. The step length is 1 second.

The experiment result is shown in Figure 9.8. During the whole experiment, the SoC of the battery is estimated by two methods simultaneously. The combination of the OCV method and the CC method are described in Section 9.2.3, which is represented by a dash–dotted line in the figure, and the estimation method based on the battery model and EKF described in this section, which is represented by a solid line in the figure. Both methods have a common premise that voltage and current sampling values are not accurate or precise enough. The resolution of the voltage sampling value is 0.005 V and the average error of the current sensor is 1%. This arrangement is designed to simulate the actual situation that may occur in the onboard sensor. At the same time, the true value of the SoC is obtained as a reference by verification of charge and discharge data, as shown by the dotted line in the figure.

Figure 9.8(a) shows the whole process. It can be seen that the SoC estimation result obtained by the EKF fluctuates to a certain degree, but is basically close to the real value all the way, and the maximum error is not more than 5%. The estimation result obtained by a combination of the OCV

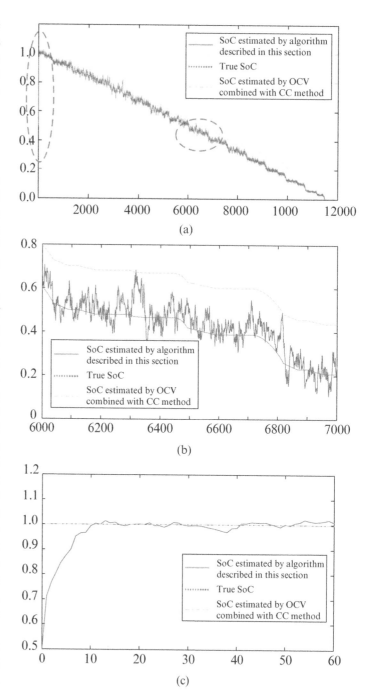

Figure 9.8 Validity verification of the algorithm described in this section by experimental results. *Note:* in the above figures, the horizontal axis is time, *s*, and the vertical axis is SoC.

method and the CC method fluctuates slightly, and has a higher accuracy in the beginning, but over time, the error of this method becomes bigger and bigger for the sensor, and at the end of the discharge, the accumulated error is more than 10%.

Figure 9.8(b) is a partial enlargement of Figure 9.8(a), which reflects the situation from the 6000th to the 7000th second. It shows the comparison result of two algorithms: the value estimated in the method described in this section fluctuates but has no accumulated error; the value estimated in the method described in Section 9.3.2 is stable but has a higher accumulated error.

In order to test the sensitivity of the EKF algorithm to the initial value, we deliberately set the initial value of the algorithm in this section as 0.5 (in fact, it is 1.0). The result shows that, under the effect of the EKF recursive algorithm, the estimated SoC value rapidly converged from 0.5 to 1.0 after about 10 seconds, as shown in Figure 9.8(c).

9.6 Error Spectrum of the SoC Estimation Based on the EKF

The error is inevitable for any SoC estimation algorithm, including the SoC estimation method based on the extended Kalman filter (EKF). In the process of estimation, the errors of the voltage and current sensors and the errors of the model affect the algorithm accuracy to different extents. Moreover, even if all other factors are the same, the battery level also has an impact on the SoC estimation accuracy. For example, when the SoC is estimated for a battery, if the reading of the voltage sensor has a 0.05 V error, the SoC estimated in the same algorithm at a 50% battery level and a 10% battery level has different errors. Therefore, it is necessary to make a quantitative analysis of the SoC estimation errors caused by various factors, which is to study the "error spectrum" of the algorithm.

The error of the algorithm is analyzed in this section from two aspects. First is an analysis of the estimation error caused by the inaccurate battery model. Second is an analysis of the estimation error caused by the measurement error of the sensor.

9.6.1 Estimation Error Caused by the Inaccurate Battery Model

As described in Chapter 7 of this book, the model includes an "equivalent voltage source" and an "equivalent impedance." The "equivalent voltage source" is mainly used to describe the relationship between the equilibrium potential (E_B) and the SoC of the battery. The equilibrium potential (E_B) comprises the electromotive force (EMF) and the hysteretic voltage (V_h). The former is controlled by the V_{SOC}, while the latter is controlled by the V_{Lh}. The "equivalent impedance" is mainly used to simulate the voltage rebound characteristic of the lithium iron phosphate battery with a third-order resistance-capacitance network, where R_Ω is mainly used to describe the ohmic resistance R_s, C_s, R_m, C_m, R_l, C_l, and so on, which reflect the polarization resistance characteristic of the battery. However, such a battery model may have an error. Three causes for an SoC estimation error are analyzed below.

1. SoC estimation error resulting from negligence of the hysteretic voltage

First, it is necessary to clarify that the EMF of the battery is conceptually different from the equilibrium potential (E_B) of the battery. The former is only related to the current SoC and temperature of the battery, while the latter is required to consider the voltage hysteresis characteristics. In general,

$$E_B = \text{EMF} + V_h \tag{9.35}$$

The definitions of the EMF, the equilibrium potential (E_B), and the OCV) are analyzed below:

First, the EMF is the quantity value of chemical energy carried in the battery. It is related to the SoC and the temperature of the battery, but is not related to the charging or discharging state of the battery.

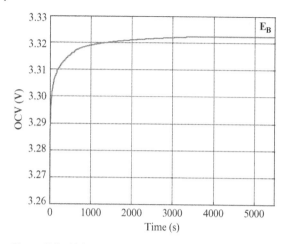

Figure 9.9 Voltage rebound curve of a battery at room temperature after discharge at 0.5 C.

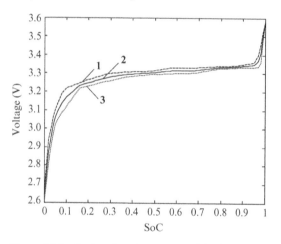

Figure 9.10 Hysteretic voltage curve of the lithium iron phosphate battery. *Note:* In the figure, 1 is the charge equilibrium potential (E_{charge}) curve, 2 is the EMF curve, and 3 is the discharge equilibrium potential ($E_{discharge}$) curve.

Second, the equilibrium potential (E_B) refers to a steady-state potential of the battery after it has been charged and discharged. It has a relational expression with the EMF and the hysteretic voltage: $E_B = \text{EMF} + V_h$, where V_h is less than 0 when the battery is charged, more than 0 when the battery is discharged, and is an uncertain value when the battery is in the process of switching between the charging and discharging states. It can be seen that the magnitude of E_B depends on the EMF and V_h, and is indirectly related to the SoC and the temperature of the battery. Since E_B is defined as the "equilibrium potential," its magnitude will not change in the case of no change in the "charging" and "discharging" states of the battery. That is, if the temperature is constant and the battery is not loaded (equivalent to no use of the battery), then the E_B value remains unchanged.

Third, the OCV is the voltage measured between the two electrodes of the battery without any load. Because the battery is made of chemical materials, the voltage at the two ends of the battery is not a constant value, even without a load. Figure 9.9 shows the open-circuit voltage curve of the battery at room temperature after discharging at a discharge rate of 0.5 C and disconnection of the load and standing for some time. It can be seen from the figure that, although the battery is not loaded, the OCV at both ends of the battery is still changing (rebound) for more than an hour. It can also be seen from the figure that after standing for some time, the OCV of the battery will rebound to the equilibrium potential (E_B) of the battery. Therefore, it can be understood that at a certain temperature, if the battery is not loaded, the OCV of the battery converges to the equilibrium potential (E_B) of the battery.

For convenience of expression, the equilibrium potential of the battery during charging is expressed by E_{charge} and the equilibrium potential of the battery during discharging is expressed by $E_{discharge}$, that is:

$$E_B = \begin{cases} E_{\text{charge}}, & \text{During charging} \\ E_{\text{discharge}}, & \text{During discharging} \end{cases} \tag{9.36}$$

In general, for the lithium iron phosphate battery, the charge equilibrium potential E_{charge} is slightly higher than the EMF, while the discharge equilibrium potential $E_{discharge}$ is slightly lower than the EMF due to the hysteretic voltage characteristic of the lithium iron phosphate battery, as shown in Figure 9.10.

When the SoC is estimated using the battery model, if the hysteretic voltage (V_h) is ignored and it is considered that $\text{EMF} = E_{charge} = E_{discharge}$ is possible, a large estimation error will be caused. Figure 9.11 shows an

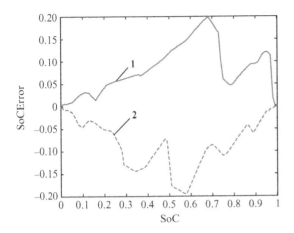

Figure 9.11 Spectrum of the SoC estimation error caused by ignoring the hysteretic voltage during charging and discharging. *Note:* 1 is the SoC estimation error caused by considering the EMF curve as the charge equilibrium potential (E_{charge}) curve during charging and 2 is the SoC estimation error caused by considering the EMF curve as the discharge equilibrium potential ($E_{discharge}$) curve during discharging.

error spectrum, which suggests that the SoC estimation error is caused by ignoring the hysteretic voltage at different actual SoC values during charging and discharging.

Table 9.3 shows the SoC estimation error caused by ignoring the hysteretic voltage during charging and discharging. As shown in the table, the standard error of the SoC estimation error caused by ignoring the hysteretic voltage during charging and discharging is more than 5% and the maximum error may be up to 20%.

For Table 9.3, a further understanding is required.

First, it can be seen from the Figure 9.11 and Table 9.3 that, in general, if the hysteretic voltage characteristic of the battery is ignored, the estimated SoC is larger during charging and smaller during discharging.

Second, the above data are obtained from the lithium iron phosphate battery. The EMF–SoC curve of the lithium iron phosphate battery is relatively flat and the estimation error is relatively large. For the ternary lithium-ion battery, each value listed in the table can be reduced by more than half. Therefore, it is considered that it is easier to manage the ternary lithium-ion batteries compared with the lithium iron phosphate batteries. This view is reasonable from the point of view of the above error spectrum.

Table 9.3 SoC estimation error caused by ignoring the hysteretic voltage during charging and discharging.

Error cause	Standard error	Maximum error	Mean error
Negligence of V_h during charging	0.0540	0.1955	0.0848
Negligence of V_h during discharging	0.0538	-0.1982	-0.0832

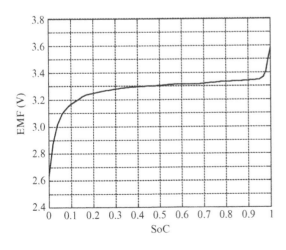

Figure 9.12 The EMF–SoC curve of the lithium iron phosphate battery at 20°C.

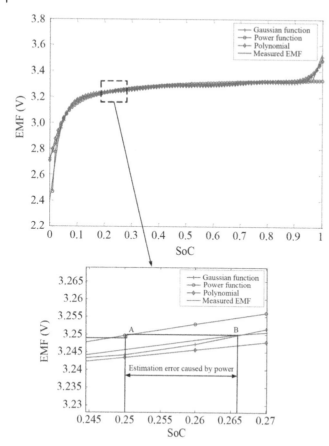

Figure 9.13 Fitting result of the EMF–SoC curve obtained in different fitting methods.

2. SoC estimation error resulting from an inaccurate EMF–SoC curve

Even if the hysteretic voltage is taken into account, if the inaccurate EMF-SoC curve is used in the model, it is possible to cause a certain SoC estimation error. The EMF refers to the electromotive force at two ends of a battery, which is related only to the material type and the operating temperature of the battery. For the lithium iron phosphate battery, the EMF value is unique at the same temperature. Figure 9.12 shows the EMF–SoC curve of the lithium iron phosphate battery at 20°C.

Since the EMF–SoC curve has good stability and can be used for accurate measurement, why would it cause the SoC estimation errors? The reasons are as follows.

First, due to the error of the temperature sensor for the battery management system and the different positions of the temperature sensor, it is impossible to accurately obtain the true operating temperature of the battery in the actual work. In this case, the EMF–SoC curves at various temperatures can be measured and an analysis is performed by comparing the curves. The following second reason is detailed.

Second, due to the limitation of the storage unit of the embedded system, the look-up table is replaced by the fitting function method in the actual control unit, in order to reduce the huge EMF–SoC look-up table. The author fit the measured EMF-SoC curves respectively by different functional forms, such as the Gauss function, power function, and higher-order polynomials, as well as the MATLAB fitting tool [15]. Figure 9.13 shows the error of the EMF–SoC curves fitted by the several common functions, and the obtained composition error is shown in Table 9.4.

Table 9.4 Composition error of the EMF–SoC fit in different fitting methods (unit: V).

Fitting method	Gaussian function	Power function	Polynomial
Corresponding function	$f(x) = a_1^* e^{-\left((x-b_1)/c_1\right)^2} + \cdots$ $+ a_6 *^* e^{-\left((x-b_6)/c_6\right)^2}$	$f(x) = a^* x^b + c$	$f(x) = p_1^* x^7 + p_2^* x^6$ $+ p_3^* x^5 + p_4^* x^4$ $+ p_5^* x^3 + p_6^* x^2$ $+ p_7^* x + p_8$
Standard error	0.0449	0.0761	0.0216
Maximum error	0.1480	-0.1643	0.0623
Average error	0.0129	-0.0573	-0.0082

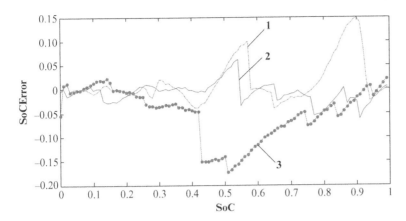

Figure 9.14 SoC estimation error spectrum of different fit EMF–SoC curves.

Notes:
1. SoC estimation error spectrum of the Gaussian function-fit EMF–SoC curve.
2. SoC estimation error spectrum of the polynomial function-fit EMF–SoC curve.
3. SoC estimation error spectrum of the power function-fit EMF–SoC curve.

It can be seen from Figure 9.13 and Table 9.4 that it is better to use higher-order polynomials to fit the EMF–SoC curves in terms of comprehensive indexes. Even so, the errors are inevitable. When the fitted function is used to estimate the SoC, a large error may be caused. In order to quantitatively describe such errors, we assume that the voltage sensor used in the battery management system does not have any errors, and the SoC is estimated by all fitted EMF–SoC curves. The relationship between the obtained estimation error (SoCError) and the SoC is shown in Figure 9.14.

It can be seen from the figure that, in the platform area of the EMF-SoC curve ($0.1 < SoC < 0.9$), three fitting methods have a large fitting error for the EMF–SoC curve characteristic of the lithium iron phosphate battery. When the SoC is in the platform area, the EMF of the lithium iron phosphate battery changes relatively gently. In this case, if a slight EMF error is generated during fitting, the large SoC error can be caused. It is noted that the above error is obtained by the assumption that the measurement error of the voltage sensor is zero. If the voltage measurement error exists, the SoC estimation error will be larger.

In order to reduce the fitting error, the piecewise fitting method can be used. The EMF–SoC curve is divided into three sections ($0 < SoC1 \leq 0.1$ is the first section, $0.1 < SoC2 \leq 0.9$ is the second section, and $0.9 < SoC3 \leq 1$ is the third section), which are respectively fit by different polynomials. The fitting results are shown in Figure 9.15.

It is not difficult to find that the SoC estimation error can be effectively reduced in the EMF–SoC curve fitted by a piecewise polynomial. Table 9.5 shows the comparison of the error statistics between piecewise fitting and non-piecewise fitting.

It can be seen from the table that the maximum error and the standard error of the SoC are reduced to about one-third of the original ones when the EMF–SoC curve fitted by a piecewise polynomial is used for the SoC estimation. Figure 9.16 shows the estimation error spectrum of an EMF–SoC curve fitted by a piecewise polynomial.

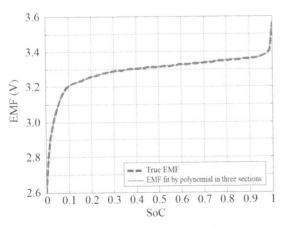

Figure 9.15 Fitting result of the EMF–SoC curve obtained by a polynomial in three sections. (*Note:* two curves in the figure almost coincide.)

Table 9.5 Error statistics of different EMF–SoC fitting methods.

Fitting method	Standard error	Maximum error	Average error
Whole fitting of polynomial (non-piecewise)	0.0216	0.0623	-0.0082
Fitting of polynomial in three sections	0.0068	-0.0229	-0.0011

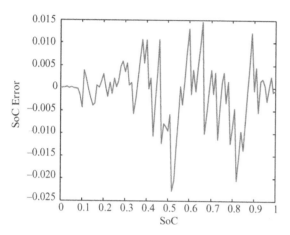

Figure 9.16 SoC estimation error spectrum of the EMF–SoC curve fitted by a polynomial in three sections.

Therefore, for lithium iron phosphate batteries, it is advocated for the EMF curves to be fitted using the polynomial in three sections in order to reduce the SoC estimation error.

3. Error resulting from inaccurate equivalent impedance of the model

In the equivalent circuit model of the lithium iron phosphate battery, the resistance capacitance network is often used to show the external characteristic relationship between the battery voltage and current. However, in practice, the set parameter of the model may not be completely close to the real value for the following reasons.

First, due to the manufacturing process of the battery and other reasons, it is impossible for the current power battery to guarantee exactly the same equivalent internal resistance of all the cells, so the equivalent impedance of the cells has some errors. If all cells are calculated with the same parameters, the error can be caused in a real situation. The errors were inevitable during the SoC estimation.

Second, the SoC estimation error may be caused due to battery degradation (aging) and the gradual change in the internal parameters of the battery during use. Failure to 100% accurately evaluates the battery degradation and obtains the true impedance parameter of the battery after degradation.

Third, if the battery management system has an error in the temperature (T) monitoring fails to accurately estimate the SoC at the current moment, and the impedance of the battery is a partial function of T and the SoC, it is difficult to accurately set the impedance value for the battery model.

The reasons for the inaccurate equivalent impedance of the battery model are summarized as above, and the inaccurate SoC estimation caused thereby is analyzed qualitatively. Then the SoC estimation error caused by the inaccurate impedance value is analyzed quantitatively.

The equivalent circuit model of the battery has multiple values of resistance and capacitance, including R_Ω, R_s, C_s, R_m, C_m, R_l, C_l, etc. Since the impedance values have an equivalent impact on the relationship between the voltage and the current, the impact on the estimation error of these parameter values is not analyzed respectively, but as a whole. For example, the overall equivalent error of impedance is set to +5% manually, that is, R_Ω, R_s, C_s, R_m, C_m, R_l, and C_l of the model are respectively increased by 5%.

Simulation experiments can be designed to analyze the impact of the model parameter error on the estimation results. The following simulation experiments can be similarly generalized to the following analysis of an "estimation error resulting from inaccurate measurement of the sensor."

1) Apply NEDC operating conditions to the power battery: the corresponding current spectrum is generated according to the parameters of an electric vehicle under the NEDC operating condition (see Section 5.5.3 of this book for specific methods), and the current spectrum can be applied to a battery model. The simulation step size (namely the current sampling step size) can be set to 1 second.

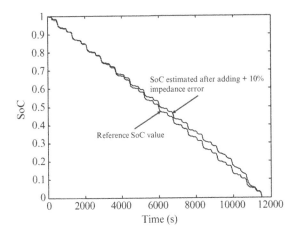

Figure 9.17 Comparison of SoC estimated by the EKF at a +10% impedance error added in the model and reference SoC.

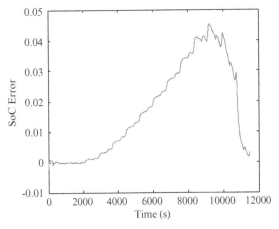

Figure 9.18 SoC estimation error spectrum by the EKF at a +10% impedance error added in the model.

2) A set of "reference data" can be obtained using the above simulation steps, and includes the output voltage, working current, and SoC value of the battery per second.

3) The errors are manually added to the various impedance values of the battery model and the SoC of the battery is estimated by the EKF (extended Kalman filter). The step size and implementation period of the algorithm are consistent with those used in the above simulation steps, and thus a set of "estimated data" can be obtained.

4) The "estimated data" generated in Step 3 is compared with the "reference data" obtained in Step 2 (contrapuntal subtraction), in order to obtain the estimation error of the algorithm.

If a +10% impedance error is manually added to the battery model and the SoC is estimated by the EKF, the SoC estimation error spectrum obtained based on the above steps is shown in Figure 9.17.

Based on Figure 9.17, the SoC estimation error spectrum can be obtained at a +10% impedance error, as shown in Figure 9.18.

The impedance error can be manually set to more values in order to analyze further its impact on the SoC estimation (e.g. +20%, –10%, –20%), and the SoC is estimated by the EKF once every modification of the model parameters, which is equivalent to estimation of the SoC by the EKF many times in the same process. Combined with the above +10% error, four operation results can be obtained and compared with the reference SoC value, as shown in Figure 9.19.

The estimation error spectrum is obtained by subtracting the true SoC from the four operation results, as shown in Figure 9.20. At the same time, the statistical value of the error spectrum of the four operations is tabulated, as shown in Table 9.6.

It can be seen from the table that the estimation error is approximately in direct proportion to the impedance error, that is:

1) If the impedance error is doubled, the SoC estimation error will approximately double.

2) The estimation error caused by an impedance error of +10% is digitally equal to the estimation error caused by an impedance error of +10% approximately with a difference in signs (+ and –). The estimation error caused by the impedance error of +20% is digitally equal to the estimation error caused by the impedance error of –20% approximately with a difference in signs (+ and –).

It is not difficult to understand the above rule, because, if the impedance value in the model is smaller than its real value, the equilibrium potential E_B (EMF + V_h) of the battery is smaller at the same measured current and voltage value, and the estimated SoC value of the battery is also smaller. On the contrary, if the impedance value of the model is larger than its real value, the estimated SoC of the derived battery is also larger. It can be

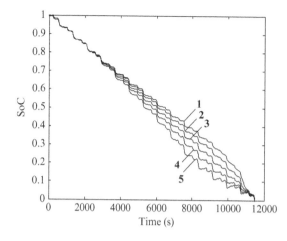

Figure 9.19 Comparison of the SoC estimated by the EKF after modifying the impedance parameter of the model and the reference SoC value.

Note:
1. Value estimated by the EKF at the +20% impedance error
2. Value estimated by the EKF at the +10% impedance error
3. Reference SoC value
4. Value estimated by the EKF at the −10% impedance error
5. Value estimated by the EKF at the −20% impedance error

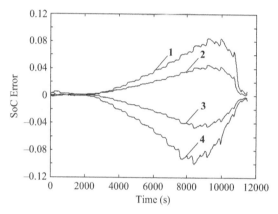

Figure 9.20 Error spectrum of the SoC estimated by the EKF after modification of the impedance parameter in the model.

Note:
1. Estimation error spectrum by the EKF at the +20% impedance error
2. Estimation error spectrum by the EKF at the +10% impedance error
3. Estimation error spectrum by the EKF at the −10% impedance error
4. Estimation error spectrum by the EKF at the −20% impedance error

Table 9.6 Error caused by the inaccurate equivalent impedance of the model.

Impedance error	Standard error	Average error	Maximum error
−20%	0.0332	−0.0375	−0.1003
−10%	0.0162	−0.0174	−0.0473
+10%	0.0146	0.0183	0.0455
+20%	0.0292	0.0320	0.0848

seen that the sign of the impedance error is the same as the sign of the SoC estimation error, and the SoC estimation error also increases with the increase of the impedance error. From this point of view, the internal resistance error has a similar rule for the impact on an SoC estimation with the systematic error of the current sensor mentioned later.

Meanwhile, since most of the EMF–SoC curve (10 ~ 90%) is quasi-linear, if the impedance error is increased by 10%, the error of the estimated voltage value is increased by 10% and the error of the estimated SoC is also increased by 10%, approximately showing the direct proportional relation.

9.6.2 Estimation Error Resulting from a Measurement Error of the Sensor

The SoC estimation error caused by an inaccurate battery model has been analyzed in the previous section. Another factor causing an SoC estimation error, namely the inaccurate measurements of the sensor used in the battery management system, is analyzed in this section. The SoC estimation errors caused by the inaccurate measurements of the sensor are discussed based on the current and voltage measurement errors respectively.

9.6.2.1 SoC Estimation Error Resulting from the Measurement Error of the Operating Current

In practice, the measurement error of the current sensor can be divided into systematic error and random error. Among them, the systematic error is the fixed error of the current sensor. The fixed current error may be caused when the current is measured at any moment. The random error is caused by inaccuracy of the current sensor, and can be considered as the superposition of the measured current value with a Gaussian noise.

9.6.2.1.1 Estimation Error Resulting from the Systemic Error of the Operating Current Measurement

The Hall sensor is often used to measure the operating current in the application of a high-power battery pack. As described in Section 8.3.3, due to the defects or inaccurate calibration of the sensor, the systematic error may be caused while the actual measurement is being taken. Such systematic errors can be positive or negative. When the EKF is used to estimate the SoC in the discharge process of the battery, a +5 A systematic error may be artificially added during the estimation process. The SoC estimation curve before and after the addition of the systematic error is shown in Figure 9.21.

The reference SoC value is subtracted from the value estimated after introducing the sensor error to obtain the estimation error per second, as shown in Figure 9.22. It can be seen that when the current sensor has a +5 A systemic error, the maximum SoC estimation error can exceed 7%.

The following rule can be found according to the above two figures:

First, the systemic error of the current sensor has the opposite sign to the SoC estimation error.
Second, the error spectrum curve features "large in the middle and small at both ends".

The above rules can be understood as follows:

First, it can be seen that, if the error of the current sensor is "positive," this means that the charge "discharged" from the battery increases and the SoC estimation error is "negative." Although the error may be modified by the observed value of voltage (U) in this process, since the state equation is also acting at this time, the systemic error can be partially offset by the correction of the measurement equation but cannot be eliminated completely.

Second, the shape of the error spectrum curve is as shown in Figure 9.12, where the EMF–SoC curve of the lithium iron phosphate battery is a "monotone increasing" function curve featuring "steep at two ends and flat in the middle." Since the SoC estimation algorithm using EKF is based on the EMF–SoC curve of the battery, in the case of the same systemic error of the current, the estimation error is "big in the middle and small at both ends." In other words, when the discharge depth is less than 60%, due to the systematic error of the current measurement and the existence of the state equation, the accumulated systematic error will increase until the late discharge period. The accumulated error is gradually eliminated

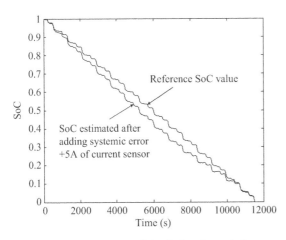

Figure 9.21 Comparison of the SoC estimated after adding a systemic error of +5 A of the current sensor and reference value.

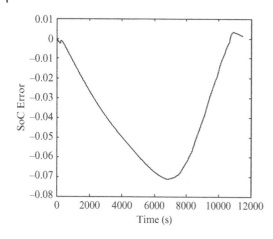

Figure 9.22 Error spectrum of the SoC estimated by the EKF after adding a systemic error of +5 A of the current sensor.

Table 9.7 Impact of the systemic error of the operating current on the SoC estimation error.

Systemic error of current	Standard error	Average error	Maximum error
−5 A	0.0275	0.0498	0.0839
−3 A	0.0163	0.0288	0.0493
−1 A	0.0054	0.0092	0.0161
+1 A	0.0053	-0.0088	-0.0157
+3 A	0.0154	-0.0253	-0.0455
+5 A	0.0243	-0.0386	-0.0712

under the effect of the EKF with the continuous decrease of the SoC, the EMF curve becoming steeper, a Kalman gain increase, and the function improvement of the measurement equation.

The above provides an analysis of the SoC estimation error spectrum at a +5 A systemic error of the current sensor. In addition, other systemic errors of the current sensor can be set manually, and the same method can be used for a simulation analysis in order to obtain the statistical value of the SoC estimation error at different systemic errors of current measurement, as shown in Table 9.7.

It can be seen from Table 9.7 that the SoC estimation error has the opposite sign to the systemic error of the current sensor, but their absolute values are approximately directly proportional.

9.6.2.1.2 *Estimation Error Resulting from a Random Error of the Operating Current Measurement*

While measuring the operating current, the sensor is often found to have a random error, whose feature is that the magnitude of the current measurement error is random at every moment, but the average error is zero. The reason for this is the measurement noise of the sensor and the resolution of the A/D converter at the back end of the sensor. The measurement error of the Hall current sensors available on the market is 1%, 3%, 5%, etc.

First, the maximum measurement error condition is analyzed, the current sensor is assumed to have a 5% random error, and the EKF is used to estimate the SoC in the discharge process of the battery. The SoC estimation curve before and after addition of the random error is shown in Figure 9.23.

It can be seen from Figure 9.23 that, even if the current sensor has a 5% random error, the error of the SoC estimated by the EKF is negligible. Therefore, it can be predicted that, if the random error is less than 5%, the SoC estimation error can be ignored.

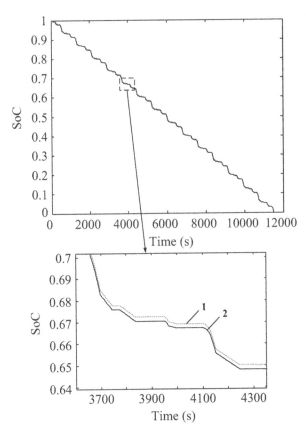

Figure 9.23 Comparison of the SoC estimated by the EKF at a 5% random error of the current sensor and the reference value.

Note:
1. The SoC estimated after addition of a 5% random error.
2. Reference SoC.

Table 9.8 Impact of a random error of a current measurement on the SoC estimation error.

Random error of current	Standard error	Average error	Maximum error
1%	0.0001	0.0002	0.0011
3%	0.0002	0.0002	0.0020
5%	0.0008	0.0015	0.0051

In order to verify the above opinion, according to the accuracy grade of the current sensors available on the market, different random errors (1% and 3% respectively) of current measurement are set in the process of simulation to get the statistical value of the SoC estimation error at various random errors. This is shown in Table 9.8, together with the error caused by a random error of 5% shown in Figure 9.23.

According to the analysis of the above two cases, (1) and (2), the systemic error has a major impact on the SoC estimation accuracy, while the random error has a small impact. It is not difficult to get such a conclusion, because the Kalman filter itself can suppress the random noise.

9.6.2.2 SoC Estimation Error Resulting from Measurement of the Error of the Operating Voltage

In practice, the measurement error of the voltage sensor can be divided into a systematic error and a random error, where the systematic error is the fixed error of the voltage sensor. The fixed voltage error may be caused when the voltage is measured at any moment. The random error is caused by the resolution ratio of the voltage sensor and can be considered as the superposition of the measured voltage value with a Gaussian noise.

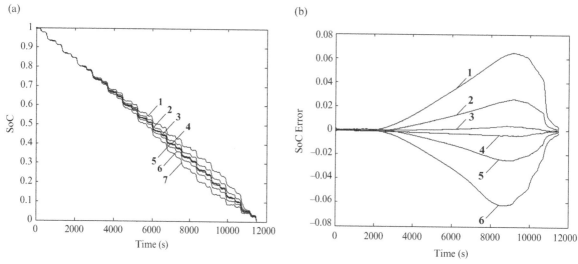

Figure 9.24 SoC estimated by the EKF at different systemic errors of voltage sensor and reference values.

(a) *Note:*
1. Value estimated by EKF at +20 mV systemic error of voltage
2. Value estimated by EKF at +8 mV systemic error of voltage
3. Value estimated by EKF at +1 mV systemic error of voltage
4. Reference SoC
5. Value estimated by EKF at −1 mV systemic error of voltage
6. Value estimated by EKF at −8 mV systemic error of voltage
7. Value estimated by EKF at −20 mV systemic error of voltage

(b) *Note:*
1. Estimation error spectrum at +20 mV systemic error of voltage
2. Estimation error spectrum at +8 mV systemic error of voltage
3. Estimation error spectrum at +1 mV systemic error of voltage
4. Estimation error spectrum at −1 mV systemic error of voltage
5. Estimation error spectrum at −8 mV systemic error of voltage
6. Estimation error spectrum at −20 mV systemic error of voltage

Table 9.9 Impact of the system error of the voltage measurement on the SoC estimation error.

Systemic error of voltage	Standard error	Average error	Maximum error
+20 mV	0.0235	0.0269	0.0648
+8 mV	0.0093	0.0100	0.0261
+1 mV	0.0013	0.0010	0.0041
−1 mV	0.0013	−0.0022	−0.0045
−8 mV	0.0089	−0.0097	−0.0248
−20 mV	0.0220	−0.0247	−0.0623

9.6.2.2.1 Estimation Error Resulting from a Systemic Error of the Operating Voltage Measurement

Generally, for any voltage monitoring scheme, the measured voltage must have the systematic error due to the limitation of chip precision. For example, in several schemes described here, the systemic error of the measured voltage may be ±8 mV for the LTC6802 chip scheme, ±1 mV for the LTC6804 chip scheme (although the maximum error given in the datasheet of LTC6804 is ±1.2 mV, the absolute error of the measured voltage of LTC6804 is more than 1 mV at 0 ~ 40°C), and ±20 mV for the poor voltage monitoring scheme used previously. Based on the above typical systematic error values, the NEDC operating condition is applied to perform the same simulation experiment as described above, where the systematic error is manually added to the measured voltage value and the SoC is estimated by the EKF. The value estimated in each simulation is compared with the reference value, as shown in Figure 9.24(a). Meanwhile, the SoC estimation error spectrum is obtained by subtracting the estimated value from the reference value, as shown in Figure 9.24(b).

The data shown in Figure 9.24(b) is tabulated as shown in Table 9.9.

It can be seen from Table 9.9 that the SoC estimation error has the same sign and is approximately proportional to the systemic error of the voltage sensor. This phenomenon is analyzed similarly to the analysis of "the estimation error caused by an inaccurate equivalent impedance." Readers may understand this error to be according to the analysis provided after Table 9.6, which is not described here.

9.6.2.2.2 Estimation Error Resulting from a Random Error of the Operating Voltage Measurement

The resolution ratio of the voltage sensor is always limited, which may cause a random error in the voltage reading. According to the common voltage acquisition schemes available on the market, several typical values are selected, including 20 mV (corresponding to the resolution ratio of the poor accuracy schemes), 5 mV (corresponding to the resolution ratio of LTC6802), 1 mV (corresponding to the 12-bit ADC scheme), and 0.01 mV (corresponding to the resolution ratio of LTC6804). Such random noise is manually added to the SoC estimation algorithm based on the EKF.

First, a maximum measurement error condition is analyzed when the noise intensity of the voltage sensor is 20 mV. In such a case the EKF is used to estimate the SoC in the discharge process of the battery. The SoC estimation curve before and after adding the error is shown in Figure 9.25.

Figure 9.25 shows the comparison of the SoC estimated after the addition of voltage measurement noise and the reference value. It can be seen from the figure that, even if the voltage sensor has 20 mV of random measurement noise, the error of the SoC estimated by the EKF is negligible. It can be predicted that the SoC estimation error can be ignored when the random error of the voltage is less than 20 mV.

In order to verify the above opinion, according to the accuracy grade of the voltage sensors available on the market, different random errors of voltage measurement are set in the process of simulation to get the statistical value of the SoC estimation error at various random errors, as shown in Table 9.10.

It should be noted that the random errors set above are 20 mV, 5 mV, 1 mV, and 0.01 mV respectively, which correspond to the 8 bit, 10 bit, 12 bit, and 16 bit resolution ratios respectively of the A/D converter at a 5 V reference voltage. These are exactly the parameter indexes of most A/D converters available on the market at present. For example, when the 10-bit A/D converter is used for voltage acquisition, there are $2^{10} = 1024$ possible readings. If the reference voltage is 5 V,

$$\frac{5V}{1024} \approx 5mV$$

Therefore, when the 10-bit A/D converter is used for voltage acquisition with a resolution of 5 mV, it corresponds to the second line of Table 9.10.

Similar to the above estimation error rule of the inaccurate current measurement, the analysis of the above two cases (1 and 2) shows that the systemic error has a greater impact on the SoC estimation accuracy, while the random error has a smaller impact. Such a conclusion can be expected, because the Kalman filter can suppress the random noise.

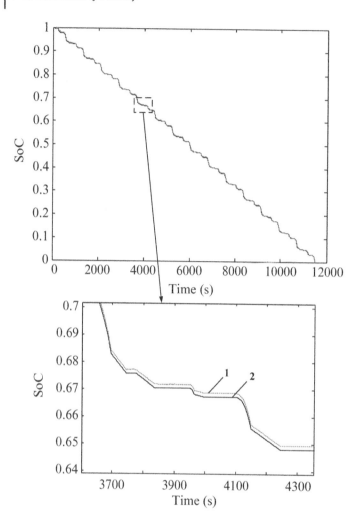

Figure 9.25 Comparison of the SoC estimated by the EKF after the addition of a 20 mV voltage measurement noise intensity and reference value.

Note:
1. SoC estimated after the addition of a 20 mV random error of voltage
2. Reference SoC value

Table 9.10 Impact of the random error of voltage measurement on the SoC estimation error.

Resolution ratio (bit)	Random error of voltage measurement	Standard error	Average error	Maximum error
8	20 mV	0.0006	0.0013	0.0054
10	5 mV	0.0005	0.0009	0.0028
12	1 mV	0.0003	0.0004	0.0011
16	0.01 mV	0.0002	0.0001	0.0009

9.6.3 Factors Affecting SoC Estimation Accuracy

This section intends to analyze the SoC estimation error spectrum of a Kalman filter caused by various factors. The error spectrum is distinguished from the isolated error after considering the SoC error distribution.

In addition, in order to compare the impact of various factors on the SoC estimation accuracy, the data of a lithium iron phosphate battery are tabulated, as shown in Table 9.11.

Table 9.11 Impact of various factors on the SoC estimation error.

Error sources	Specific causes		Approximate conversion relation	Typical maximum value
Inaccurate battery model	Hysteretic voltage neglected		General 5~ 10 error	±20% error in an extreme case
	Inaccurate EMF–SoC curve fitting		Error controllable to 1 ~ 2% by piecewise fitting	3 ~ 5% error for non-piecewise fitting
	Inaccurate equivalent impedance model		Error increased by 1% every impedance error increase by 5%	±4% error when the impedance error is ±20%
Inaccurate sensor measurement	Operating current	Systemic error	1 A systemic error convertible to 0.53% estimation error	5 A systemic error of current corresponding to 7 ~ 8% error
		Random error	Basically ignorable	5% random error corresponding to 0.15% error
	Operating voltage	Systemic error	1 mV systemic error convertible to 0.13% estimation error	20 mV systemic error corresponding to 6.48% error
		Random error	Basically ignorable	20 mV random error corresponding to 0.54% error

1) The "approximate conversion relation" is obtained from the statistical standard deviation. The "typical maximum value" refers to the statistical maximum. (Please refer to the "standard error" and "maximum error" columns of the relevant above-mentioned tables.)
2) The extended Kalman filter can remove the impact of the random error caused by an insufficient sensor resolution and fails to remove the estimation error caused by the systemic error.
3) The data shown is based on the lithium iron phosphate batteries. The error value in the table can be roughly divided by 4 for the lithium-ion batteries made of other materials, such as lithium manganate and ternary materials, in order to get their approximate error distribution. In this sense, the SoC of the lithium iron phosphate batteries estimated by the EKF has a maximum error. In other words, it is most difficult to estimate the SoC of the lithium iron phosphate battery because the EMF curve of the lithium iron phosphate battery is relatively flatter.

According to the analysis in this section, the SoC estimation accuracy depends on the parameter accuracy of the battery model and the measurement accuracy of the sensor.

References

1 Yu, W., Luo, Y., and Pi, Y. G., "Fractional order modeling and control for permanent magnet synchronous motor velocity servo system," *Mechatronics*, 2013, 23(7): 813–820.
2 Welch, G. and Bishop, G., "An introduction to the Kalman filter," University of North Carolina, 2004.
3 Plett, G. L., "Sigma-point Kalman filtering for battery management systems of LiPB-based HEV battery packs: Part 1. Introduction and state estimation," *Journal of Power Sources*, 2006, 161: 1356–1368.
4 Plett, G. L., "Sigma-point Kalman filtering for battery management systems of LiPB-based HEV battery packs: Part 2. Simultaneous state and parameter estimation," *Journal of Power Sources*, 2006, 161: 1369–1384.
5 Sierociuk, D., Macias, M., Malesza, W., et al., "Dual estimation of fractional variable order based on the unscented fractional order Kalman filter for direct and networked measurements," *Circuits Systems & Signal Processing*, 2016, 35(6): 2055–2082.
6 Zhou, D., Zhang, K., Ravey, A., et al., "Parameter sensitivity analysis for fractional-order modeling of lithium-ion batteries," *Energies*, 2016, 9(3): 123.

7 Plett, G. L., "Extended Kalman filtering for battery management systems of LiPB-based HEV battery packs: Part 1. Background," *Journal of Power Sources*, 2004, 134: 252–261.

8 Plett, G. L., "Extended Kalman filtering for battery management systems of LiPB-based HEV battery packs: Part 2. Modeling and identification," *Journal of Power Sources*, 2004, 134: 262–276.

9 Plett, G. L., "Extended Kalman filtering for battery management systems of LiPB-based HEV battery packs: Part 3. State and parameter estimation," *Journal of Power Sources*, 2004, 134: 277–292.

10 Windarko, N. A., Choi, J., and Chung, G. B., "SOC estimation of LiPB batteries using extended Kalman filter based on high accuracy electrical model," in *8th International Conference on Power Electronics*, IEEE, 2011, pp. 2015–2022.

11 Lee, J., Nam, O., and Cho, B. H., "Li-ion battery SOC estimation method based on the reduced order extended Kalman filtering," *Journal of Power Sources*, 2007, 174: 9–15.

12 Hu, X. S., Sun, F. C., and Cheng, X. M., "Recursive calibration for a lithium iron phosphate battery for electric vehicles using extended Kalman filtering," *Journal of Zhejiang University SCIENCE A*, 2011, 12: 818–825.

13 Hu, C., Youn, B. D., and Chung, J., "A multiscale framework with extended Kalman filter for lithium-ion battery SOC and capacity estimation," *Applied Energy*, 2012, 92: 694–704.

14 Zhang, L., Wang, Z., Sun, F., et al., "Online parameter identification of ultracapacitor models using the extended Kalman filter," *Energies*, 2014, 7(5): 3204–3217.

15 Grewal, M. S. and Andrews, A. P., *Kalman Filtering – Theory and Practice Sing MATLAB*, New Jersey: John Wiley & Sons, Inc., 2001.

10

Charge Control

10.1 Introduction

The huge demand placed on a rechargeable battery creates a need for efficient battery chargers. An objective of a battery charger is to put the energy into the battery by forcing current through. The charging process should not damage or affect the battery's capabilities. A battery charger should be efficient in order to minimize any loss of energy during the charging, should be reliable to help the battery charge when needed, and should have a high-power density to avoid a longer charging time [1]. In addition, a charger should be of low cost and portable with a lower volume and weight. Designers and manufacturers are trying hard to make a universal charger that can serve all of these items; however, it is hard to achieve a compact design with all the necessary features. The charger operation depends on the components used, switching strategies, and its controller design. Chargers also depend on the size and type of the battery being charged. The control design includes the charging algorithms. These define the charging steps by combining current, voltage, and pulse charging methods with termination conditions at various levels.

A simple battery charger is either a current source or a voltage source that can charge the battery with more tolerance to over-charging. Over-charging is when the battery continues to receive current even after being fully charged – in most battery chemistries, over-charging damages the battery components. Simple chargers do not have efficient control of over-charging or intelligent termination conditions from charging. Charging ends at the end of the charging cycle by a manual disconnection or with a fixed timer setting. The simple charger is inexpensive and carefully designed for a lower charging rate that takes longer to charge a battery than otherwise. Even so, many batteries left on a simple charger for too long will shorten the life and be subject to over-charging.

On the other hand, complex chargers are used for sensitive battery chemistry with efficient control over the charging steps and on the termination of charging are able to avoid any damage from over-charging. Over-charging reduces battery life by over-heating, increasing the risk of thermal runaway, excessive gassing, and even fire. In such cases, the charger must have an inbuilt temperature, voltage, and current sensors to collect information about the battery voltage, temperature, and the input current while charging. With the information, the charger can calculate the state of charge in the battery and employ safe termination conditions to stop the charging. For example, a heavy-duty automatic has "intelligent charging" systems that can be programmed with complex charging cycles specified by the battery manufacturers. With the configurable system, there is a universal charger available, i.e. it can charge all battery types under limited conditions.

The battery needs an efficient charger, not only to save the charging time but also to increase its performance and expected life. The charging methods help in preventing over-charging by appropriate terminations and control heating by making the charge current high and low. For the lithium-ion (Li-ion) battery, specific charging methods have been developed to suit the chemistry that can prevent metal plating at the electrode surface, provide a limited SoC operating window for extending the life of the battery, and avoid excessive heating.

Battery Management System and its Applications, First Edition. Xiaojun Tan, Andrea Vezzini, Yuqian Fan, Neeta Khare, You Xu, and Liangliang Wei.

Designing the charging method in a battery charger varies with the chemistry, size of the battery, and with the application of the battery. For a small consumer, electronic products like smart phones, power bank, and laptop chargers are portable and lightweight. The available voltage varies from +5 V to +12 V. Other common battery charger applications are stationary battery plants: a grid-connected energy storage plant, a data warehouse power backup, a UPS power backup, and a telecommunication tower power backup. Such chargers are permanently equipped with temperature compensation and supervisory alarms for various system faults. Chargers for stationary battery plants should have adequate voltage regulation and filtration. These chargers should be capable of supporting a DC load while the battery plant is under maintenance.

This chapter mainly emphasizes the battery charger and charge controller used for automotive applications [2]. The other chargers are beyond the scope of the book.

In an automobile, a three-stage charger is used to recharge a fuel vehicle's starter battery, but the case is very different for recharging an electric vehicle (EV) or hybrid electric vehicle (HEV) battery pack [3]. The chargers for an EV and HEV are categorized according to power rating, charging methods, charging time, and suitability with infrastructures. The charging systems are also classified on technical grounds, such as on-board and off-board, conduction and induction chargers, and unidirectional and bidirectional power flows.

For an example, Figure 10.1 illustrates a basic electric drive of the EV and categories of charging systems. The figure shows the charging facilities outside the vehicle and the on–off board charging system inside the vehicle. Inside the vehicle, the on–off board battery charger system consists of an AC/DC inverter followed by a power factor correction or controlled rectifier and a DC–DC converter. The DC–DC converter output is directly connected to the DC bus that feeds into the battery pack and electric load and is connected to an electric motor through a DC–AC converter. The left side of on/off board battery charger is connected to the outside charging facility. The diagram shows three different options:

a) Vehicle can be plugged into a grid connected to an on-board charging system which is a level-2, three-phase 240 V facility available either in a public vehicle or a privately owned vehicle.
b) Vehicle can be plugged into a home garage level-1, single phase of a 120 V socket.
c) Alternately, it can be charged via a commercially off-board level-3 facility of a three-phase AC–DC system such as a gas filling station.

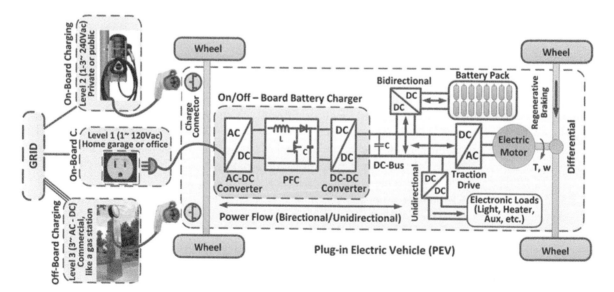

Figure 10.1 On/off board EV/ HEV battery charger.

On-board chargers are limited in power capabilities due to weight, space, and cost constraints. Thus, most on-board chargers are typically integrated into the electric drive to avoid limits on high power.

On-board charger systems can be conductive or inductive. Conductive charging systems use a direct physical contact between the connector and the charge inlet whereas an inductive charger uses inductive coupling to charge the vehicle battery without any physical connection.

An off-board battery charger is less constrained by size and weight and are at a commercial station where the vehicle battery can be charged.

The other most popular charger used is the intelligent charger. The intelligent charger communicates with the BMS to receive information about the battery characteristics and conditions to control and modify the charging actions. An intelligent charger may monitor the battery's voltage, temperature, or time under charge to determine the optimum charge current and termination condition. The intelligent chargers employ a combination of cut-off systems, which are intended to prevent over-charging.

Various charging methods can regulate the reaction rates in the battery in order to increase the system efficiency and extend the battery life. Control charging steps prevent over-charging and over-heating, which limit the damage in the battery electrodes and prolong the battery life. High-rate charging reduces the charging time and restores most of the capacity much faster. However, current density of the electrode material limits the maximum allowed charging rate. A high charging rate increases the overall heat in the battery. It also damages the electrode surface by generating a localized high current density point on the surface. High-rate charging is ideally needed for an EV with especially a public infrastructure charging facility. Very rapid charging rates, 1 hour or less, generally require the charger to carefully monitor battery parameters such as terminal voltage and temperature to prevent over-charging and damage to the cells.

The charge control method varies with the type of battery chemistry. For example, due to a high self-discharge rate, it is necessary to use trickle charging in a lead–acid battery. The trickle charge is a very slow charging step used in the last part of the charging process. Most batteries are left on trickle charging for a long time for maintenance. In contracts with a lead–acid battery, most of the Li-ion batteries may be damaged by trickle charging with the risk of over-charging. The nickel–cadmium cell is another exceptional chemistry that shows a significant voltage drop at the end of charging. The voltage drop is an indication for an NiC battery charger to stop charging at a ΔV drop, but the voltage drop at the end of charging is not a universal phenomenon for other battery chemistries. Each manufacturer defines a special method and safety limits for charging the battery in order to obtain the highest coulombic and energy efficiency. Usually, the range of charging rates, steps of charging, operating temperature window, end of the charging steps, and termination conditions are defined for a rated performance of the battery. Under the scope of the book, we will focus only on the Li-ion automotive battery charger.

An ideal charger should also have built-in safety features to avoid any consequences of over-charging and accidents. The safety standards recommend that battery charger design should include all the protection circuits to ensure common safety measures. These protection circuits are over-voltage protection, over-charge protection, over-current protection, over-heat protection, as well as short-circuit protection and over-discharge protection.

Isolation is another safety precaution needed between the EV charging board and the charging station. This includes isolation between the high-voltage battery, DC–DC converter, inverter for driving the electric motor, and also for a charger module connected to the grid. Therefore, the key component in the interface between the existing electrical system and the electric vehicle supply equipment (EVSE), i.e. the charging station, is the transformer. The EV body should be connected to the earth while being charged with on-board or off-board chargers. An inexpensive simple AC powered battery charger may suffer from a high ripple current and ripple voltage. Usually, the ripple current should remain within the battery manufacturer's recommendations.

With non-isolated DC/DC converters, a line-frequency transformer is needed to isolate the battery from the charger.

EVs and HEVs are slowly becoming popular, but still need to appeal at the mass level. High cost, battery life, complexity of chargers, and lack of charging infrastructure are the important barriers to gaining an acceptance for EVs/HEVs on a larger scale. Often, battery charger manufacturers struggle with minimizing charging time, cutting down the cost, increasing the power capacity and component ratings, reducing the volume and weight, and designing an appropriate charger according to location and infrastructure.

10.2 Charging Power Categories

Based on the infrastructure and power capacity, battery chargers are categorized in three levels.

Level 1 is a slow charging system for typically home garage overnight charging or work place charging, where the EV can be plugged into a convenience outlet, AC 120 V for the US and AC 230 V for the EU. It uses a single-phase grid outlet. It is suited for private domestic installations where extra authentication and billing are not required. Level 1 is an on-board charger that is equipped in the vehicle with most of the safety features built inside. A level 1 charging station, however, has very simple and limited safety features, like ground fault detection and disconnects. Figure 10.2 gives examples of the level 1 chargers. Figure 10.2(a) is an external cord set level 1 charger that users carry to charge the EV at a level 1 outlet. Figure 10.2(b) is a level 1 EVSE system that allows the user to connect a vehicle directly. Figure 10.2(c) is a level 1 charger mainly for workplaces offering the facility.

The level 2 charging system can be a facility for both private and public use (see Figure 10.3). It uses semi-fast charging methods through an AC 240 V outlet, provides enough power, and can be implemented in

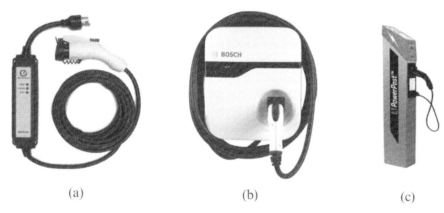

(a)   (b)   (c)

Figure 10.2 Level 1 chargers: (a) Nissan line-cord EVSE; notice the NEMA L6-30 plug on the cord. Credit: Nissan, (b) BOSCH power maximum level 1 EVSE. Credit: Bosch Automotive Service Solutions Inc, (c) L1 power post EVSE charger 120 V/16 A. Credit: Konnectronix, Inc.

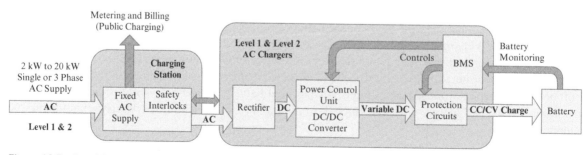

Figure 10.3 Level 2 charger.

most environments. The self-contained system gives the charger the flexibility to connect to different AC charging sources. The level 2 system, when installed for public use, is suited for parking lots, shopping centers, hotels, rest stops on highways, theaters, restaurants, etc. The system needs to incorporate intelligence features to communicate with the charging station. The charging station would verify and authorize the user to draw power from the source and allow the source to bill the customer for the energy transferred unless the charger is installed as a free service in the workplace or shopping mall. Figure 10.4 shows a Siemens level 2 charging station.

Figure 10.4 Siemens VersiCharge EVSE level 2 charging station with 30 A, a bottom inlet panel – VC30BLKB. Credit: SIEMENS

The level 3 charger uses high power. It is a three-phase, AC 208 V or DC 600 V connected outlet and has a fast-charging system (see Figure 10.5). It is intended for commercial and public applications, similar to a gas filling station. It can be installed in parking lots, shopping centers, hotels, rest stops, theaters, restaurants, etc. Having a high-power system design, AC/DC conversion, control circuits and power conditioning components it becomes very large and expensive. The level 3 system keeps all heavy-duty components and complex circuits at the charging station. This saves vehicle costs and reduces the on-board charger weight in the vehicle. In this system, an on-board BMS and charge controller must communicate with the power controller at a charging station to control the process of charging.

An example of a level 3 fast charging system is the CHAdeMO system developed by Japanese auto industries, as shown in Figure 10.6. The CHAdeMO is a direct current source that can deliver up to 62.5 kW. The voltage level varies from 50 to 500 V. CHAdeMO uses a proprietary electrical connector that is a global standard for a charger. The vehicle charger sends the battery information and set points of the voltage to the charging station through a CAN bus. The charging station then follows the CHAdeMO specified constant-current and constant-voltage charging profiles and terminates the charging process according to the set points. The charging station also

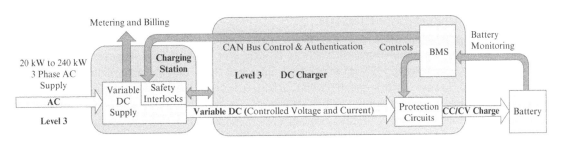

Figure 10.5 Level 3 charger schematic.

Figure 10.6 CHAdeMO connector. Credit: C-CarTom / Wikimedia Commons / CC BY-SA 3.0 and Nissan CHAdeMO level-3 Charger. Credit: Caxton & CTP Printers and Publishers Ltd.

Table 10.1 Categories of a charging system.

Power level types	Charger location	Typical use	Energy supply interface	Expected power level	Charging time	Vehicle technology
Level 1 (opportunity) AC 120 V (US) AC 230 V (EU)	On-board 1-phase	Charging at home or office	Convenience outlet	1.4 kW (12 A) 1.9 kW (20 A)	4–11 hours 11–36 hours	PHEVs (5–15 kWh) EVs (16–50 kWh)
Level 2 (Primary) AC 240V (US) AC 400V (EU)	On-board 1- or 3-phase	Charging at private or public outlets	Dedicated EVSE	4 kW (17 A) 8 kW (32 A) 19.2 kW (80 A)	1–4 hours 2–6 hours 2–3 hours	PHEVs (5–15 kWh) EVs (16–30 kWh) EVs (3–50 kWh)
Level 3 (Fast) (AC/DC 208-600 V)	Off-board 3-phase	Commercial analogous to a filling station	Dedicated EVSE	50 kW 100 kW	0.4–1 hours 0.2–0.5 hours	EVs (20–50 kWh)

ensures the initial safety conditions by checking the vehicle charger and battery fault conditions, such as a short circuit, over-voltage, over-heating, or any high leakage current.

Table 10.1 summarizes the charging system and their power levels.

10.3 Charge Control Methods

Battery charging methods include constant current (CC), constant voltage (CV), constant power (CP), pulse charging, and trickle and float charging. The constant current charging is the most efficient method for putting bulk capacity in the battery; however, it does not charge the battery completely. In addition, usually CC is combined with CV to charge the battery up to 100%. In a few applications of fast charging with specific chemistries, pulse and power charging methods are also used.

Often, in an automotive application, the Li-ion batteries are charged with a two-stage CC–CV profile, where the battery is charged with CC up to a certain threshold voltage (set as the end of the charge voltage) and moves to CV charging until charging current drops to 3–5% of the rated current. Often this is the acceptable definition for completion of the charge state. However, with some definitions, a battery is fully charged when the current becomes constant or levels off and does not drop further. The condition is often difficult to achieve within the limited time.

The charging methods affect the battery life significantly, even if an applied charge rate remains the same in each method. At the initial stage of charging while the SoC is below 1%, a low charging rate is desirable. Because, at a low SoC, resistance remains high, a high-rate charging can generate more heat. On the other hand, at a high SoC, due to electric polarization near the end of charging, the graphite anode potential falls below with respect to Li+/Li and allows metallic lithium plating. At this stage, continual intercalation into graphite generates the secondary reactions between Li-ions with an electrolyte surface. The loss of Li-ions in secondary reactions reduces charging efficiency and increases the resistive surface layer. Slow rate charging or CV charging can limit the metal plating at the end of charging and can help maintain a healthy battery life.

For most of the Li-ion batteries, the threshold voltage varies from 4.1 to 4.2 V/cell, but a 4.3 V/cell is typically for a high-capacity Li-ion. By contrast, LTO charges up to 2.7 V/cell. Commonly, for all chemistry, boosting the voltage increases the capacity, but going beyond the specified voltage may challenge the safety. Over-voltage results in over-heating and thermal runaway. In most of the battery packs, BMS does not allow voltage to go beyond the specified value. A continuous trickle charge can cause plating of metallic lithium on the electrode surface and compromise safety by short circuiting the electrodes. Thus, the charge current should be cut off as soon as charging ends.

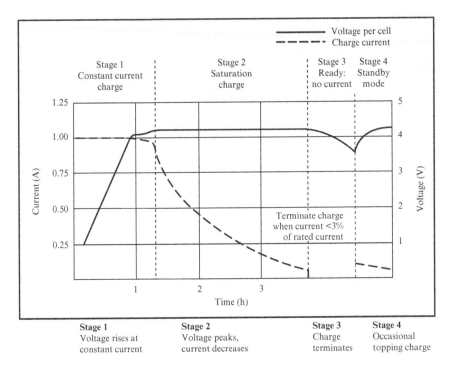

Figure 10.7 Li-ion battery charging steps.

Figure 10.7 shows four stages of charging an Li-ion battery. Stage 1 is a constant current of 1 C rate or a rated current where the voltage rises almost linearly with the input Ah capacity and then attains the threshold voltage. Stage 2 defines the starting of the CV step or saturation charge stage, where the voltage remains at the threshold voltage and the current starts dropping. A drop in the charging current continues until it reaches 3% of the rated current. At this stage the battery is considered fully charged. After removing the charge load at Stage 3, the voltage drops to the open circuit voltage (OCV). In some cases, Stage 4 is required where the battery charges push a little top-up current to bring the battery voltage up.

A higher charge current mostly fills up the battery faster, but it does only approximately 80% up to the voltage peak. In such a case, the saturation charge will take much longer to fill the battery completely. Some chargers follow only a bulk charging step without going to the saturation charge. This may fill the battery up to 80–85% of SoC, which is acceptable for a few applications. However, it affects the life of the battery.

The high rate of charging in the Li-ion battery is mainly limited due to heat generation as well as Li-ion intercalation. The charging stations and the chargers should limit the charge current to maintain the temperature within the specified limit. Most of the chemistries allow heating of 5–7°C at the rated charging current. The heating is further intensified in the older battery due to high internal resistance. The charger and charging station often disconnect or terminate charging if the temperature rises to higher than 10°C while charging and/or beyond the specified limit.

Table 10.2 lists the performance comparison of the most commonly used charging methods of the battery.

10.3.1 Semi-constant Current

The method uses impedance of the circuit to control the charging current. As the voltage of the battery increases with charge, the current starts to decease. As the name suggests, the charge current does not vary significantly, but it is not controlled accurately during the end of the charge, which increases the risk of over-charging.

Table 10.2 Battery charging methods.

Charging method	Advantage	Disadvantage
Semi-constant current charging	Fast	Not completely charge the battery Risk of over-charging
Constant current charging	Fast	High cost Protecting circuit is need to prevent over-charging
Constant voltage	Best suited for trickle charging	Slow Risk of over-charging
Constant current constant voltage	Controlled process No risk of over-voltage charging No risk of over-charging	
Two-step constant voltage charging	Reasonably fast Recovery step is possible	High cost
Pulse charging	Fast, low heat generation, increase in battery life	Expensive, needed more controls
Constant power charging	Fast	

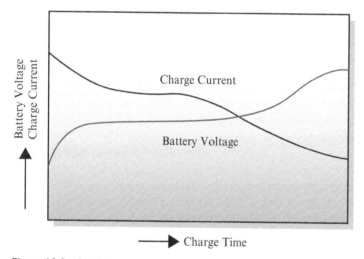

Figure 10.8 Semi-constant current charging characteristics.

This method, referred to as a simple method, is easy to adopt and is widely used for cycle service batteries. The charger consists of a transformer, diode, and resistor. Impedance from these elements ensures charging without an excessive charging current. With this method, the battery voltage increases while the charging current decreases as the charging proceeds. The problem occurs when at the end of the charging a large charging current flows and causes over-charging. Care should be taken to avoid over-charging by limiting the charging duration. Figure 10.8 provides the current and voltage profile of a cell while charging with semi-constant charging methods.

10.3.2 Constant Current (CC)

Constant current charging is an easy way to characterize the battery behavior and so it is a standard method applied during analysis, as shown in Figure 10.9. The battery behavior remains uniform as the coulombic charge is forced at a constant rate throughout the charging process. The process is comparatively fast, but needs an expensive charging circuit to control the constant current output. Circuits using voltage regulators and combinations of a diode and resistors are a simple design used to generate a constant current. Alternatively, accurate controllers and charging chip sets are available. Some examples are LT1510 or LT1510-5 from Linear Technologies, CCR-AND9031 from On Semiconductors, BQ 24650 from Texas Instruments, and IC7805 Electcircuit.

The basic circuit of a constant current regulator (CCR) is given in Figure 10.10. It uses a simple charging terminating circuit that compares battery voltage with reference voltage in the controller. The current control is a combination of transistors.

10.3.3 Constant Voltage (CV)

A constant voltage charging method is mostly used in conjunction with CC, as shown in Figure 10.11. It is used as a saturation charging step. The method mainly prevents over-charging with a carefully selected threshold voltage. A constant voltage unit applies voltage to the battery. The charging method basically applies the current based on the difference between the battery voltage and the threshold voltage. The charging unit should have a

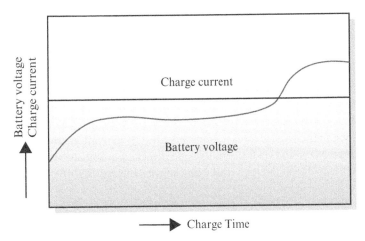

Figure 10.9 Constant current charging.

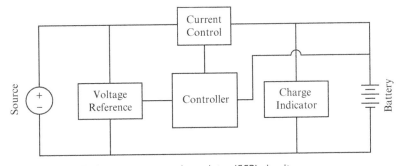

Figure 10.10 Basic constant control regulator (CCR) circuit.

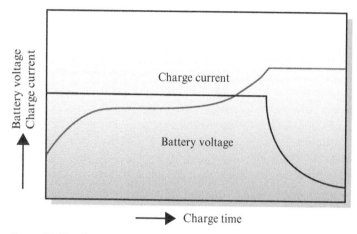

Figure 10.11 Constant voltage.

high capacity in order to control the initial high current flow. This method is slow and does not allow a uniform charge transfer rate for the electrochemical system. The termination of charging is mostly defined by achieving a very low current (3–5% of the rated current) state.

The charging threshold voltage is technology specific and varies with the operating temperature. An inaccurate voltage causes an over-charge or an under-charge.

CV is widely used in conjunction with CC for both cyclic and stand-by applications.

10.3.4 Constant Power (CP)

CP charging is suitable for a fast-charging requirement as the initial current is very high in order to maintain a constant power during the charging process. A high starting current gradually decreases with charging time in order to keep CP-correlation between the current and the voltage. Most commonly, the CV step follows the CP in a similar way as described for CC–CV. During the CP, the charge transfer resistance increases with aging. In the beginning of charging, the high current generates heat due to the high initial resistance. However, CP offers a low overall heating loss compared to CC.

The CP method accelerates aging in the battery by increasing resistance and losing Li-ions in secondary reactions due to metal plating at the end of charging.

10.3.5 Time-Based Charging

When pre-set, the time-based chargers are low cost and easy to implement and are mostly used for Ni-based batteries, for example Ni-Cd. Ni-Cd batteries are not very sensitive to over-charging. Therefore, it is safe to use a pre-set time to charge those technologies that are insensitive to over-charging, but, unfortunately, Li-ion batteries are not among them.

The charges often provide a large number of charging timers to match varying capacity batteries. However, it is difficult to avoid the possibility of under-charging and over-charging. For example, if the battery is in a pre-charge condition, the pre-set timer cannot adjust the time accordingly and results in over-charging.

10.3.6 Pulse Charging

Pulse charging is another fast-charging technique that uses controlled pulses. Pulses break the DC current by inserting a short relaxation time in between. Pulses have a strictly controlled rise time, pulse width, frequency, and amplitude. Pulse charging is a uniform method that suits all sizes, capacities and chemistries of batteries. This is widely implemented in the automotive applications. Pulse charging allows relaxation time intervals that prevent over-heating in the battery. Particularly, in a lead–acid battery, high current pulses break down the hard sulfate on the electrodes and extend the battery life significantly.

Pulse charging offers the advantage of fast charging by eliminating the need for the CV (saturation state) step during the charging. It also prevents the concentration polarization with relaxation time intervals that increase the power transfer rate by improving the Li-ion diffusion rate.

In some cases, pulse charges are designed by combining small discharging pulses with charging pulses. Short relaxation periods and short discharge pulses during charging help the Li metal plating back to the electrolyte and improve the active material utilization that provides the battery with a higher discharge capacity and a longer cycle life. Pulse charging overall maintains the stability of the electrode surface compared to DC charging, ensures reversibility of electrodes, and prevents an increase in the thickness of passive film on the electrode surface by combining the discharging pulse with charging. Figure 10.12 compares the charging time when used with CC + CV and Pulse + CV techniques. Pulse + CV saves half the charging time.

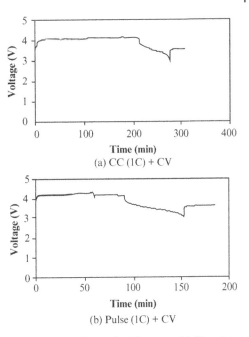

Figure 10.12 Comparison between (a) CC and (b) pulse charging.

10.3.7 Trickle Charging

This method provides a relatively small amount of current for sufficiently long to charge a small capacity battery or to maintain a charge in larger batteries. Trickle charging is typically the last leg of a charging process. In most of the application the trickle charging current is equal to the self-discharge current. The typical trickle current compensates for the self-discharge of a battery that has been idle for a long time. However, this charger is a slow battery charger and may take several hours to complete a charge. During the maintenance, these chargers remain connected to the batteries for an infinite time. Most of the Li-ion technologies cannot handle the over-charging so cannot be connected to trickle charging for long. These batteries either employed with smart charge terminators or avoid using trickle charging.

There are few other specific chargers available. They are either based on technology used or are application specific, such as inductive chargers, intelligent chargers, motion-powered chargers, and solar chargers. Specific chargers are beyond the scope of the book.

10.4 Effect of Charge Control on Battery Performance

Ideally, during charging and discharging cycles of the Li-ion battery there is no active material degradation. The charging and discharging cycles are supported by intercalation and de-intercalation of Li-ions rather than any chemical reaction. In contrast to a lead-acid battery, where lead sulphate is formed on the electrode surface while discharging and is reversed when charging, the charging method has therefore a major influence on the battery performance. The Li-ion battery has a different charging mechanism ideally considered to be unaffected by charging methods. However, this is not true. Charging methods affect the performance of the battery drastically and influence the ability to deliver energy and power.

A charger allows over-charging when the end of the charging is not controlled. Over-charging triggers metal plating on the negative electrode and challenges safety and increases cell impedance. In addition, over-charging gives rise to secondary reactions that are responsible for the loss of Li-ions, thus making the battery life short.

Typical constant current charging raises the concentration polarization, which slows Li-ion diffusion. This reduces the overall power transfer rate of the battery. Instead, pulse charging introduces relaxation intervals that break concentration polarization and improve Li-ion diffusion. As a result, the pulse charging method helps to provide a higher power transfer rate.

In addition, when pulse charging combines with small negative charging, it helps to utilize the active material completely and prevents metal plating on the electrode surface during charging, thus improving battery health.

It is well known that Li-ion batteries do not accept over-charging. This is because, at the end of the threshold voltage, the Li/Li+ voltage becomes close to the over-potential of the negative electrode that allows Li metal plating at the negative electrode. This negative plating increases the risk of a short circuit. The voltage beyond the threshold voltage may raise a secondary reaction and form a passive layer at the surface of the electrode. The passive layer increases the impedance and aging in the battery. The CV step after the CC prevents over-charging and thus ensures safe and prolonged battery health.

Charging with an optimum C_{rate} also improves the performance of the battery, particularly coulombic efficiency. A high C_{rate} charging heats up the battery and aggravates the aging. It also forces higher intercalation, which increases mechanical stress at the graphite electrodes and degrades the electrode quality.

Without a load current, the battery voltage drops from an elevated voltage. After an optimum relaxation time, it achieves an open-circuit voltage that is less than the elevated voltage. The battery, which is charged with complete saturated voltage (with CV steps), keeps the elevated voltage longer compared to one that does not have any saturation.

Some chargers compensate the battery for its self-discharge and for the power consumed by its safety and protections circuits. The compensation is mainly by long-time charging with a very small current and top-up of the charge to make the voltage level ready to use. For example, after removing charging in the LFP battery, the voltage drops to 4.05 V/cell from 4.20 V/cell. The charger tops the voltage back up to 4.20 V/cell. Sometimes a charge starts only when the voltage drops to 4.0 V/cell and recharges the voltage only to 4.05 V/cell instead of 4.20 V/cell. Charging to a smaller voltage removes the voltage stress on the electrodes and increases the battery life.

A fully discharged state of the Li-ion battery is not considered a best choice, neither for storing nor for charging. An Li-ion discharge below a certain voltage is not advisable because of some secondary reactions that possibly create safety concerns.

However, if Li-ion cells are discharged below a certain voltage a chemical reaction occurs that make them dangerous if recharged, which is probably why all such batteries in consumer goods now have an "electronic fuse" that permanently disables them if the voltage falls below a set level. The electronic fuse draws a small amount of current from the battery, which means that if a laptop battery is left for a long time without charging it, and with a very low initial state of charge, the battery may be permanently destroyed.

10.5 Charging Circuits

Charging circuits are improving at a fast pace in terms of high-power capabilities, less losses within the circuits, light weight, high efficiency, and a low charging time. This chapter will cover only a few examples for providing an overview about basic topologies and elements in the circuits.

10.5.1 Half-Bridge and Full-Bridge Circuits

Charges are composed of diode rectifiers or controlled rectifiers, a switching network, boost converters, and filters. Half-bridge and full-bridge topologies are used with a single phase and three phases. Obviously, the half bridge has less components and a low-cost system, but the operation forces high stress on the circuit components. The full-bridge topology has a high cost due to a higher number of components, but it offers high durability by reducing the stress on the components. Figure 10.13 shows the basic bidirectional half-bridge and full-bridge topologies. A few topologies also need a controller and PWM input control to the switch.

10.5.2 On-Board Charger (Level 1 and Level 2 Chargers)

As stated in the introduction section, on-board chargers typically control charging as a part of the power train of the vehicle. This charger can be conductive/non-isolated or inductive/isolated. Isolated chargers make no direct metal connection with the batteries. These chargers can be used in parallel to increase the charging current for reducing the charging time. Non-isolated chargers connect to the battery directly and cannot be used in parallel. Figure 10.14 shows a basic circuit of a unidirectional charger. The circuit has an AC outlet connected to the diode bridge rectifier, followed by a switching network of an interleaved boost converter. The boost converter operates at 180° out of phase. The interleaved boost converter offers a limited power capacity but provides ripple cancellation. Lastly, the capacitive filter provides a controlled DC current at a stable DC voltage at the input of the DC/DC converter. Figure 10.15 is a multilevel charger that controls the charging levels with a series of connected switches S1 and S2. A complete charger off-board and on-board is given in Figure 10.16, together with an electromagnetic interference filter (EMI).

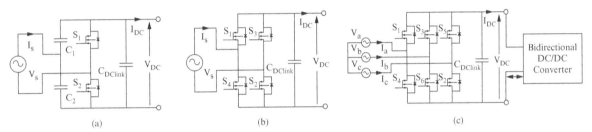

Figure 10.13 Bidirectional chargers: (a) single-phase half-bridge, (b) single-phase full-bridge, and (c) three-phase full-bridge.

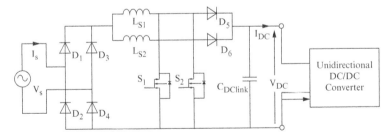

Figure 10.14 Interleaved unidirectional charger topology.

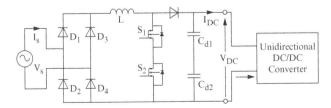

Figure 10.15 Single-phase multilevel unidirectional charger.

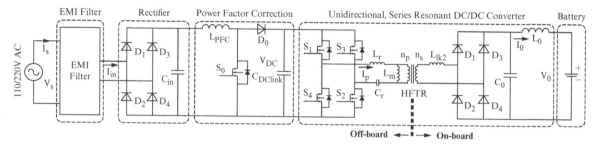

Figure 10.16 Combined on-board and off–board circuit elements of the charging system.

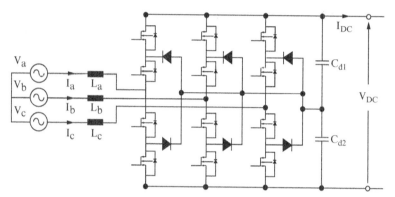

Figure 10.17 Three-phase diode clamp bidirectional charger.

10.5.3 Off-Board Charger (Level 3)

Off-board chargers are high power chargers and have built-in controllers. These are expensive chargers and mainly used as level 3 chargers for EV and HEV automotives. The three-phase level 3 charger diode clamp circuit is shown in Figure 10.17, where each phase of the AC outlet is connected to a switch control network. A diode path provides rectification for each phase. The capacitive filters provide a constant DC at the output.

10.5.4 Fast Charger

Fast chargers offer rapid charging by employing a high C_{rate} charging or combinations of pulse-changing and polarization-curve-based methods. These circuits are often equipped with a controller and cooling system. A controller circuit may be part of the battery or charger or may be split into two. A cooling system helps to control the temperature to a safer limit for the battery while charging at a high C_{rate}.

The CC–CV method is simple in function but has a limitation due to heat generation during a high C_{rate} charging and concentrated polarization. Both factors damage the cells. Thus, pulse charging is the more efficient way to achieve the charging without damaging the cell by heat generation or by concentrated polarization. Most common quick chargers are given by CHAdeMO. Fast chargers are fixed at the location and have effectively no limit of the charging voltage or current. Such high-voltage and high-current chargers are named as a DC fast charger (DCFC) or a DC quick Charger (DCQC).

CHAdeMO is the well-known charging standard for EV quick charging, developed by the CHAdeMo Association. CHAdeMO was formed by its executive members, including the Tokyo Electric Power Company, Nissan, Mitsubishi, and Fuji Heavy Industries (the manufacturer of Subaru vehicles). Toyota later joined as its fifth executive member.

CHAdeMO is an abbreviation standing for "charge de move" or "charge on move." It offers quick charging methods for EV up to 62.5 kW (500 V, 125 A) of direct current with a special connector. The special electric connector combines the charging levels and communication signals through CAN. It is included in the IEC standard IEC62196 (type 4) as a global standard for charging the EV system. CAN communication signals ensure the safety and operating limit control.

Due to cost and thermal issues, AC chargers have limited 240ACV and 75A. In such case, external high-rate DC chargers are appropriate. Table 10.3 provides a most conventional AC charging solution and limits in various parts of the world.

CHAdeMO ensures fail-safe of the charging control at various levels. It offers two communication lines, CAN and Pilot, to avoid any fail during malfunctioning of either communication lines. To control charging, the charger halts DC output immediately with a vehicle coupler drop. Charger runs a "Pre-charge Automatic Safety Check" to examine the circuit insulation and short circuit between the charger and the EV contactor. Figure 10.18 shows

Table 10.3 Conventional AC charging solution and limits in various countries.

Country	Voltage limit	Current limit
USA and Japan	240	30
Canada	240	70
Europe and Australia	230	15
	400-3Ø	32
SAE J1772-2009 Australia	240	80
VDE-AR-E 2623-2-2	400-3Ø	63

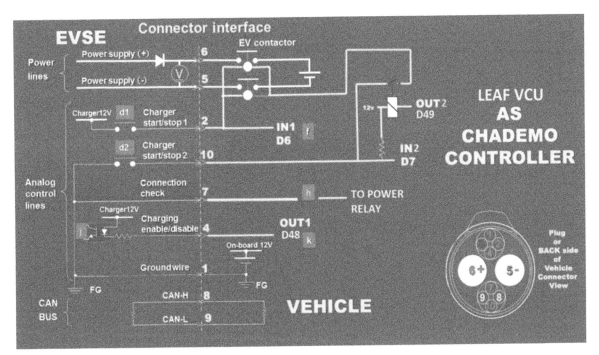

Figure 10.18 CHAdeMO sequence circuit.

the CHAdeMO connector interface and three basic categories of lines. CHAdeMO combines with the DC Power line, Analog control lines for a 12 V supply, and the CAN Bus. A power line provides high DC voltage and current, the Analog line provides 12 V control of various operational and safety switches, and CAN provides the communication link between the vehicle BMS and the charger.

10.5.5 Ultra-Fast Charger

The concept of ultra-fast charging is to provide repetitive high-power charging up to 350 kW/300 A for less than 20 min to extend the driving range. Currently, with the fast charger (500 V, 125 A) it takes 1.5 to 2 hours to fully charge the battery for providing a long-distance driving range up to 500 km. The ultra-fast chargers will not charge the vehicle completely; however, they charge it multiple times to cover the distance. An ultra-fast charger fills the battery according to the need, which is driving speed, required distance to cover, and the initial charged condition.

Big automotive companies like BMW, Ford, Daimler AG, and Volkswagen are teamed up to create an ultra-fast charging infrastructure/network in Europe. The network will be based on a combined charging system (CCS) standard technology. The new infrastructure will extend the DC fast-charging capacity up to 350 kW. These ultra-fast chargers will be used by buses and trucks, as well as, eventually, passenger vehicles.

All Li-ion technologies do not favor ultra-fast charging. Ultra-fast charging reduces the life of the battery and raises safety concerns. In ultra-fast charging, Li-ions are transferred at a fast rate from the cathode to the anode. Graphite anodes cannot accept the fast rate ions and form metallic layer/dendrites. This surface film and dendrites lead to fast capacity loss, increase impedance, and aggravate the risk of a short circuit. Therefore, chargers display a warning of fast degradation in battery health on using ultra-fast charging.

The ultra-fast charging station (UFCS) mechanism and circuit design depend on many factors such as: objective charging time, useful battery capacity at an objective charging time, SoC in the beginning of charging; efficiencies of total energy conversion stages, and UFCS utilization. This mechanism needs to be formulated in terms of two distinctive EV-related terms, i.e. autonomy and the autonomy flow rate.

Autonomy is the average distance an EV is able to cover with the maximally recharged battery in given conditions (capacity, charging current, initial state of charge, etc.).

The autonomy flow rate is the driving distance augmentation in time, expressed in km/min or km/h.

The above two values decide the amount of charge and time of intermediate recharges during the trip. These values also determine an average speed for the total travel time.

The idea of ultra-fast charging is making the charging time (t_{ch}) significantly smaller compared to the actual driving time (t_{dr}):

$$t_{ch} \ll t_{dr} \tag{10.1}$$

where driving time (t_{dr}) is defined as a function of the energy capacity of the battery EEV given in kWh, the difference of the SoC from the starting point to the next stop point, $E_{100(v)}$, which is in kWh/100 km, and the average speed of the vehicle, v:

$$t_{dr} = E_{EV} \frac{SoC_{start(i-1)} - SoC_{stop(i)}}{E_{100(v)} \, v} \tag{10.2}$$

Charging time (t_{ch}) is defined as a function of the desired SoC level of charge, energy capacity of the battery, and available power capacity of the charger R:

$$t_{ch} = E_{EV} \frac{SoC_{start(i)} - SoC_{stop(i)}}{P_{chi}} \tag{10.3}$$

There are infrastructural issues that need to be discussed for accommodating the UFCS load. The short-time peaks due to UFCS on the grid are highly fluctuating and require over-dimensioning of the grid capabilities. These demanding peaks force over-sizing of the infrastructure, including cables, transformers, and switchgears. A possibility of peak mitigation lies in using energy storage buffering between the UFCS and the utility grid. The complete EV charging capacity is delivered by the combination of utility grid and energy storage buffer. This will keep grid balancing. A buffered EV charging station concept is given below in Figure 10.19, where the grid charges the buffer through low power charging (LPC) and later high-power charging (HPC) provides a full charging capacity of EV for ultra-fast charging.

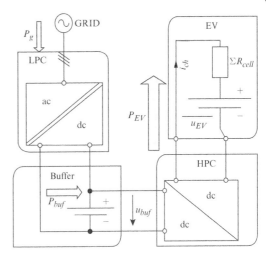

Figure 10.19 Buffered EV charging station R.

10.6 Infrastructure Development and Challenges

It has been claimed that EV is not widely accepted because of two major reasons: (a) due to its cost; (b) due to inadequate charging infrastructure. Charging infrastructure development needs multilevel initiatives from the government and private sectors. The government could give support by making generous and universal policies, standardizing the charging stations capacity and connectors, supporting infrastructural cost, and speeding up the pace. Private sectors, on the other hand, lead in designing and developing compatible systems, providing flexible charging options, planning for utility grid infrastructure, placing metering and billing systems, and reducing the installation and construction cost.

Charging infrastructures are driven by three major factors:

- Penetration rates: which defines how much energy/power is utilized from the grid
- Degree of charging: which defines the levels of charging stations
- Range anxiety: that derives the number of charging stations required within the city and on the highways

The following sections give the available infrastructure scenarios.

10.6.1 Home Charging Station

Home charging stations mostly are level 1, referred to as overnight chargers/garage chargers. These are low-cost systems and mostly customized for the specific car and the user requirements. These systems do not need a separate billing system. However, the systems cannot function as rapid chargers.

10.6.2 Workplace Charging Station

Workplaces can provide level 1 charging with an outlet for employees to plug into their own charging equipment or with the EVSE charging equipment installed at the workplace. Also, an employer can facilitate level 2 chargers. However, proper charging policies and fee structure should be in place for a full utilization of the system. For example, a level 1 outlet may be an easy workplace charging scenario for an employer, but for the employee it may be difficult to bring and use their own cord-sets. Employees may prefer to have a level 1 EVSE installed at the workplace. There should be an appropriate system that caters for the needs of the user and providers.

10.6.3 Community and Highways EV Charging Station

These charging infrastructures are categorized mostly as fast chargers and level 3. An infrastructure development consideration is to satisfy driving range anxiety. Fast chargers and ultra-fast chargers are used to reduce the charging time and make the system at a par with a conventional fuel-filling system. The combination of a rapid charging system with buffering energy storage will improve the level of perpetration into the grid and support the gird balancing. However, these systems are expensive and need government policies in place.

10.6.4 Electrical Infrastructure Upgrades

The existing electrical grid is neither equipped for multiple penetrations from rapid charging stations and nor is it designed to stratify the surge in power due to rapid charging stations in the area. Higher utilization of the localized multiple charging station could easily overload the grid infrastructure. Particularly, a power surge increases distribution transformer losses, voltage deviations, harmonic distortion, and peak demand. This forces the need for additional costs on upgrading the grid distribution system including underground cables, overhead lines, and higher capacity power systems. Typical research shows that the 50% penetration from the uncontrolled EV charging reduces the transformer life by 200–300%. Control charging at the charging station can save fast degradation of the transformers.

10.6.5 Infrastructure Challenges and Issues

- Expensive infrastructure for rapid charging: Although level 3 fast charging provides a method to alleviate range anxiety, the installation and operative cost of the system is high. The cost is further increased with a requirement for grid upgradation or installation of an energy storage buffer.
- Metering and billing policy: Policies and standardization are required for billing and metering systems for the customer and the owner of the charging station. This will involve government utility owners and owners of charging stations. Later a metering and billing data exchange should be built into the vehicle as well.

10.6.6 Commercially Available Charges

Big players of the commercial charger are Siemens, ABB, Zivan, Manzanita Micro, Elcon, Quick Charge, Rossco, Brusa, Delta-Q, Kelly, Lester, Soneil, etc.

Typically, slow chargers vary from a 1 kW to 7.5 kW maximum charge rate. They follow the charging algorithm as charge curves of simply constant voltage or constant current. Public EV charging stations provide 6 kW (host power of 208 to 240 VAC from a 40 amp circuit), where 6 kW will recharge an EV roughly six times faster than 1 kW overnight charging. Fast charging can offer 62.5 kW and ultra-chargers can offer as high as a 350 kW charging system. Table 10.4 summarizes the commercial EV system and available levels of their chargers.

Table 10.4 Summary table for commercial EV chargers.

EV charger	Capacity (kWh)	Range (miles)	Connector type	Level-1		Level-2		Level-3	
				Demand (kW)	Charge time (h)	Demand (kW)	Charge time (h)	Demand (kW)	Charge time (h)
Mitsubishi i-MiEV	16	96	SAEJ1772 JARI/TEPCO	1.5	14	3	7	50	30
Nissan Leaf EV	24	100	SAEJ1772 JARI/TEPCO	1.8	12–16	3.3	6–8	50 plus	15–30
Tesla Roadster	53	245	SAEJ1772	1.8	30 plus	9.6–16.8	4–12	N/A	N/A

10.7 Isolation and Safety Requirement for EC Chargers

EV chargers require safety at various levels due to the sensitivity of the commonly used Li-ion technologies and increasing capacity of high DC charging. Over-voltage protection, over-charge protections, over-current protections, over-heat protections, short-circuit protections, and over-discharge protections are a few of the common safety measures that should be part of any charger. These safety measures are also covered by the standards. Table 10.5 summarizes the common standards for EV chargers.

Safety measures for relatively low-power level 1 charging stations are fairly simple and may be limited to a ground fault sensing device and a circuit interrupting device (CID) or "circuit breaker." However, the charger includes more comprehensive safety measures through communicating with BMS. BMS usually has in-built safety precautions and safety limits.

Uncontrolled over-voltage charging in an Li-ion battery leads to an unsafe direction. On an over-voltage charger, beyond the manufacture specified limit, the cathode material becomes an oxidizing agent, loses stability and produces carbon dioxide (CO_2). The cell pressure rises. If the charge is allowed to continue, the pressure rises further and the safety membrane on some Li-ion batteries burst open. Moreover, the cell might eventually vent with flame. Flame increases the temperature and further raises the risk of thermal runaway. Safety precautions, such as an automatic cut off on charging upon completion or a cut off at a conservative voltage threshold or preventing an over-charge, even after a long-extended charging by controlling the charge current, should be essential. Inexpensive AC powered chargers may cause high ripple voltage and ripple current. This may damage the system. Thus, it should be ensured that ripple should be under the specified limit of the manufacturer.

The charging characteristic is heavily influenced by temperature. Thus, a temperature compensation circuit is needed in the charger to regulate the charging process.

Table 10.5 Safety Standards for EV chargers and connects.

Document name	Document title/section
SAE J-2344	Guidelines for Electric Vehicle Safety
SAE J-2464	EV/HEV Rechargeable Energy Storage System (RESS) Safety and Abuse Testing
SAE J-2910	Design and Test of Hybrid Electric Trucks and Buses for Electrical Safety
SAE J-2929	EV/HEV Propulsion Battery System Safety Standard – Lithium-Based Rechargeable Cells
UL 2202	Safety of EV Charging System Equipment
UL 2231	Safety of Personnel Protection Systems for EV Supply Circuits
UL 225a	Safety of Plugs, Receptacles, and Couplers for EVs
NFPA 70E	Electrical Safety in the Workplace
NFPA 70	National Electrical Code (NEC); Article 220, Branch Circuit, Feeder and Service Calculations; Article 625, Electric Vehicle Charging Systems; Article
DIN V VDE V 0510-11	Safety Requirements for Secondary Batteries and Battery Installations – Part 11
ISO 6469-1:2009 (IEC)	Electrically Propelled Road Vehicles – Safety Specifications – Part 1: On-board rechargeable energy storage system (RESS)
ISO 6469-2:2009 (IEC)	Electrically Propelled Road Vehicles – Safety Specifications – Part 2: Vehicle Operational Safety Means and Protection Against Failures
ISO 6469-3:2001 (IEC)	Electric Vehicles – Safety Specifications – Part 3: Protection of Persons Against Electric Hazards
IEC TC 69	Safety and Charger Infrastructure
IEC TCs 64	Electrical Installations and Protection Electric Shock

Improper design of the battery chargers can produce deleterious harmonic effects on electric utility distribution systems. The active rectifier front end can mitigate this impact. Parasitic load alters the charging cycle and hence it should be avoided while charging. A parasitic load confuses the charger by depressing the battery voltage and preventing the current in the saturation stage from dropping low enough by drawing a leakage current. A battery may be fully charged, but prevailing conditions will prompt a continued charge, causing stress.

References

1 Isaacson, M. J., Hollandsworth, R. P., Giampaoli, P. J., et al., "Advanced lithium ion battery charger," in *Fifteenth Annual Battery Conference on Applications and Advances*, Long Beach, CA, USA: IEEE, 2000, pp. 193–198.

2 Barré, A., Deguilhem, B., Grolleau, S., et al., "A review on lithium-ion battery ageing mechanisms and estimations for automotive applications," *Journal of Power Sources*, 2013, 241: 680–689.

3 Liu, P., Wang, J., Hicks-Garner, J., et al., "Aging mechanisms of LiFePO$_4$ batteries deduced by electrochemical and structural analyses," *Journal of the Electrochemical Society*, 2010, 157(4): A499–A507.

11

Balancing/Balancing Control

11.1 Balancing Control Management and Its Significance

11.1.1 Two Expressions of Battery Capacity and SoC Inconsistency

As discussed in Chapter 9, the battery level is conceptually different from the SoC. Therefore, the battery inconsistency can be described in two different expressions.

The balancing control model discussed in this chapter is applicable to the battery pack with series structure. Supposing the number of the cells in the battery pack is n, and the following two expressions are given.

Expression by battery capacity combined with battery level

For the k cell shown in Figure 11.1, the maximum capacity of the battery is expressed as C_k (Ah) and the current battery level of the cell is expressed as R_k (Ah). The battery level of all cells is respectively expressed by R_1, R_2, ..., and R_n, and the capacity of all cells is expressed by C_1, C_2, ..., and C_n. The above model can be represented graphically, as shown in the Figure 11.2.

Figure 11.1 Battery pack formed by series connection of n cells.

It can be seen in Figure 11.2 that the columns have the same width and their height represents the battery capacity C_1, C_2, ..., and C_n. The dark color parts represent the current battery level R_1, R_2, ..., and R_n of the cell. By analogy, the cells may be considered as a bucket, the bucket capacity corresponds to C_1, C_2, ..., and C_n, and the water stored in the bucket corresponds to R_1, R_2, ..., and R_n.

Expression by battery capacity combined with SoC

The battery pack shown in Figure 11.1 may be expressed by another method. For the k cell, the maximum capacity of the battery is expressed as C_k (Ah) and the SoC is expressed as S_k (%), $k = 1, 2, ..., n$. The battery pack may be expressed as shown in Figure 11.3.

As shown in the figure, the SoC corresponds to the bucket height, which is the same as the maximum SoC of 100%. The battery capacity corresponds to the bucket width (if seen from a three-dimensional perspective, it is the cross section). The water height in the bucket corresponds to the SoC S_k, $k = 1, 2, ..., n$.

Of course, the above two expressions are equivalent, because of the following relationship:

$$R_k = C_k\, S_k \quad \text{(Ah)} \tag{11.1}$$

Battery Management System and its Applications, First Edition. Xiaojun Tan, Andrea Vezzini, Yuqian Fan, Neeta Khare, You Xu, and Liangliang Wei.

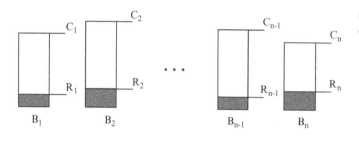

Figure 11.2 Expression by battery capacity combined with battery level.

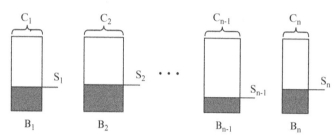

Figure 11.3 Expression by battery capacity combined with SoC.

where $k = 1, 2, ..., n$.

The power battery inconsistency is inevitable for the electric vehicles for the following two causes.

1. Inconsistency resulting from the battery production process

The inconsistency resulting from the battery production process is "innate." As a chemical product, the batteries are delivered with some inconsistency due to the limitation of materials, technology, and other factors. For example, the different batch of the raw materials for the battery may have different chemical characteristics; the same batch of the raw materials has a different granular size and electrical conductivity due to the grinding and storing process; and the battery inconsistency may be caused by the random formation of SEI film around the electrodes during battery formation [1, 2]. These inconsistencies can be manifested in many aspects of battery performance. Here are two examples:

First, the inconsistent battery capacity, namely the difference in C_1, C_2, ..., and C_n shown in Figure 11.2. Second, the inconsistent self-discharge coefficient, that is to say, all cells have a similar SoC when the electric vehicle is parked in the garage and the SoC of the cells may be different due to the self-discharge at a different discharge rate when the electric vehicle is restarted after being idle for one or two weeks.

2. Battery inconsistency resulting from the operating environment

The operating environment had a larger impact on the power battery performance. The factors affecting the battery performance may include temperature, vibration, etc., among which temperature is an important factor causing the power battery inconsistency.

It is not difficult to imagine that the cells in a power battery pack are installed in a relatively closed space according to waterproof, dustproof, compact installation, and other requirements. It is impossible that the cells have fully the same conditions, such as heat conduction, air convection, and heat dissipation in the battery pack, and heat absorption and a release effect is inevitably caused for the chemical reaction and the polarization resistance during operation of the battery. Therefore, it is difficult to achieve an even temperature field distribution in the battery pack, resulting in a battery inconsistency. In the long term, such unevenness may cause different chemical property degradation of the battery, i.e. the different degrees of aging.

11.1.2   Significance of Balancing Control Management

A great deal of practical experience and experimental data show that, due to the objective inconsistency between the cells of the battery pack, if the balancing control management is added in the battery pack, the overall performance of the battery pack can be improved to a certain extent for the following two causes.

1.   The balancing control management is helpful to increase the overall capacity of the battery pack

If the balancing control is not performed for the battery, when the cell of the battery pack is fully charged, the battery management system will stop the charge of the whole series battery by its protection mechanism. Similarly, the discharge of the whole series battery is stopped when the cell with the lower SoC is fully discharged. In other words, the effective capacity of the battery pack meets the cask theory, resulting in failure to effectively use the capacity of the battery pack. Here is a typical example.

As shown in Figure 11.4, a battery pack consisting of four cells (B_1, B_2, B_3, and B_4) has a consistent capacity, namely $C_1 = C_2 = C_3 = C_4 = C_{max}$. The initial capacity of the cells is different before charging due to the different self-discharge rates of the cells, and it is assumed that the initial capacity is highest for the cell B_2 and the lowest for the cell B_3, as shown by the dark color part in Figure 11.4.

The battery pack is charged. The remaining capacity of four cells rise evenly since the battery pack is charged in series. After a period of time, the voltage of the cell B_2 first reaches the protection limit, as shown in Figure 11.5.

Without the balancing control, it is not allowed to charge the battery pack due to the safety management mechanism; otherwise, an undesired safety accident may be caused by overcharging the cell B_2.

At this point, the charging of the battery pack is deemed completed due to failure to continuously charge the cells, although the cells B_1, B_3, and B_4 are not fully charged. After charging, the battery pack starts to work, namely discharging. Since the cells are connected in series, the capacity of the cells is reduced evenly, and the state shown in Figure 11.6 appears after a period of time.

At this point, the cell B_3 has been fully discharged. Although other cells have a dischargeable capacity, the battery pack cannot be continuously discharged since the cells are connected in series. Otherwise, an undesired safety accident may be caused by over-discharging the cell B_3. Therefore, the effective dischargeable capacity of the battery pack is equal to $C_2 - S_2(\text{III}) = C_{max} - S_2(\text{III})$, or $S_3(\text{II})$.

If the balancing control management is performed, all cells can be fully charged on the basis of those shown in Figure 11.5, as shown in Figure 11.7.

During use of the battery pack, the cells connected in series in the battery pack can be fully discharged simultaneously, as shown in Figure 11.8.

Therefore, the effective dischargeable capacity of the battery pack is the maximum capacity C_{max} [instead of $C_{max} - S_2(\text{III})$]. From the practical effect, the overall effective capacity of the battery pack has been improved.

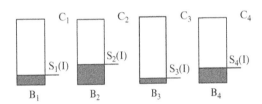

Figure 11.4   Initial capacity of the cells.

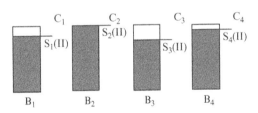

Figure 11.5   After the charge without the balancing control.

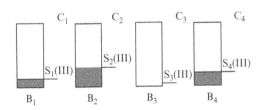

Figure 11.6   After discharge without the balancing control.

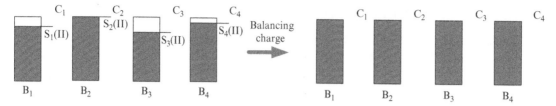

Figure 11.7 After a charge with the balancing control.

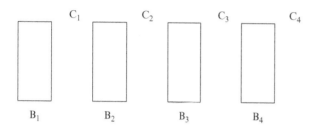

Figure 11.8 After discharge with the balancing control.

2. Balancing control management is helpful in controlling the charge and discharge depth of a power battery

If the SoC is considered changeable from 0 to 100% during the full discharge state to the full charge state, in a practical application, it is best to use each cell at 5%~95%. It is possible to over-charge the battery while causing an irreversible chemical reaction, affecting the battery life, if the SoC is more than 95%. Similarly, it is possible to over-discharge the battery, while causing an irreversible chemical reaction, affecting the battery life, if the SoC is less than 5%.

Theoretical analysis and experimental data show that it is of great significance to implement balancing control management on the battery pack and reduce the charge and discharge depth of the power battery to improve the battery safety, extend the battery life and improve the battery aging (SoH) consistency, specifically discussed below.

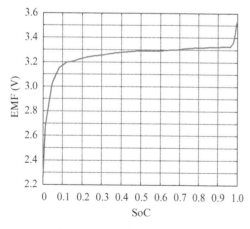

Figure 11.9 Electromotive force curve of a lithium iron phosphate power battery.

First, improvement of battery safety by balancing management.

Figure 11.9 shows the electromotive force curve of a lithium iron phosphate power battery. As shown in the figure, when the SoC is close to 100%, the electromotive force (EMF) of the battery rises steeply. If the charging balancing management is not provided for the battery pack, the following situation may be caused: one (B_2) of four cells is fully charged approximately (that is, the SoC of B_2 is more than 95%), as shown in Figure 11.10, and due to the EMF and the internal resistance, the operating voltage rises quickly, resulting in an over-charge. Similarly, at the end of a battery discharge (SoC is less than 5%), the operating voltage drops sharply, resulting in an over-discharge. If the battery management system provides the balancing management, the balancing control is performed when the SoC is more than 95% or less than 5%, in order to effectively avoid a possible safety accident.

Second, extension of the battery life by balancing management.

As mentioned above, when the SoC is more than 95% or less than 5%, it is possible to cause some irreversible chemical reaction in the battery, thus affecting the battery life. For example, an electric vehicle with several cells connected in series travels about 200 km per day (roughly corresponding to 80% SoC) and four cells of the battery pack are selected for investigation.

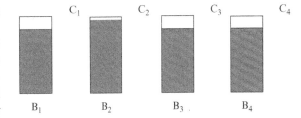

Figure 11.10 SoC of four cells without charging balancing management.

Two conditions (as shown in Figure 11.11):

For the first battery pack without the balancing control (Figure 11.11(a)), the initial SoC is 100% for cell B_2 and 80% for other cells, while the final SoC is 20% for cell B_2 and 0 for other cells. The performance degradation (aging) of cell B_2 accelerated due to frequent work during more than 95% SoC and, similarly, the performance degradation (aging) of other cells is accelerated due to frequent work during 0~5% SoC.

For the second battery pack with the balancing control (Figure 11.11(b)), four cells have an initial SoC of 90% and final SoC of 10%. Compared with the first battery pack, the discharged capacity per day corresponds to 80% SoC, i.e. travelling 200 km. Since the battery does not work at SoC during which it is possible to cause the irreversible chemical reaction, after a period of time, four cells have a lower performance degradation than the first battery pack.

Third, even battery aging can be achieved by the balancing management to improve the SoH consistency.

The SoH inconsistency is detrimental to the battery management system. It will bring greater challenges to the battery level (or SoC) estimation, which is already difficult, and will also affect the implementation of the optimal charging management strategy of the battery pack. In order to improve the SoH consistency of each cell in the battery pack, it is necessary to make the factors affecting battery aging as fair as possible for each cell. The implementation of the balancing control management to ensure the same discharge depth of all cells is an effective way to achieve uniform battery aging [3].

In the example shown in Figure 11.11, the same discharge depth has been achieved in the cells by the balancing control during their working process. As mentioned above, it can not only improve the cycle life of the battery pack but also achieve the uniform aging of the battery pack so that the SoH of the cells is relatively close.

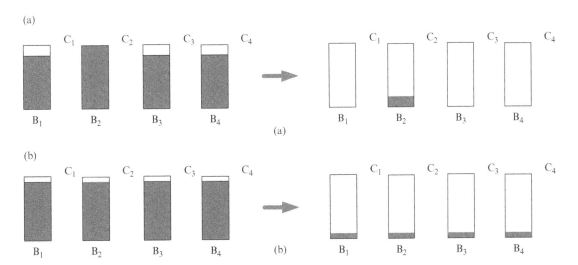

Figure 11.11 Comparison with and without the balancing control.

11.2 Classification of Balancing Control Management

Various methods are available for battery balancing control, and new methods emerge one after another. It is difficult to classify these balancing control management methods, because different classification standards will lead to different classification results. The following attempts to use the form of a table to summarize some common classification criteria and the corresponding types are given in Table 11.1.

It should be noted that there are no absolute criteria for the classification of balancing schemes and nor are there clear boundaries among various methods according to different classification criteria. The same method is classified to "active balancing" according to the criterion of "capacity protection" and can be called "charge balancing" according to the criterion of "action process" [4]. The methods are not absolutely suitable, but the most suitable one should be selected according to the actual demand and cost budget.

11.2.1 Centralized Balancing and Distributed Balancing

Balancing may be classified into centralized balancing and distributed balancing according to the topological structure of the balancing circuit [5].

In centralized balancing, an equalizer is shared by the cells of the battery pack and the energy of the battery pack is distributed by a voltage divider to equalize the capacity transmission among the cells and the battery pack. In distributed balancing, the balancing module is dedicated to individual cells [6].

Figure 11.12 shows a typical centralized balancing topology, in which the same equalizer is used for the balancing management of all the cells in the battery pack (equalizing capacitance).

Table 11.1 Classification of balancing control management.

No.	Classification standard	Main classification
1	Topological structure of balancing circuit	Centralized balancing
		Distributed balancing
2	Balancing action process	Discharge balancing
		Charge balancing
		Bidirectional balancing
3	Protection of battery capacity	Passive balancing
		Active balancing

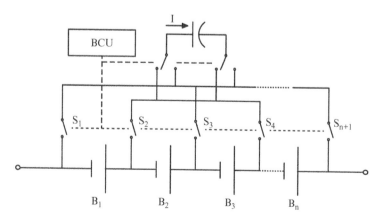

Figure 11.12 Typical centralized balancing topology.

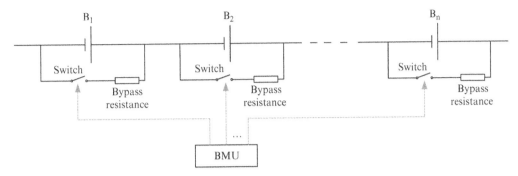

Figure 11.13 Typical distributed battery balancing topology.

Figure 11.13 shows a typical distributed battery balancing topology. As shown in the figure, balancing control is achieved by a bypass resistor connected with each cell in parallel and an electronic switch.

The comparison between these two balancing methods shows that the centralized balancing scheme can quickly collect and transfer the power of the battery pack to the individual cell to be equalized, has a higher balancing speed due to the good performance of its public equalizer, and, moreover, on the whole, the centralized balancing module is smaller than the distributed one [7]. However, in the centralized balancing scheme, the cells are competitive and cannot be equalized in parallel, and a large number of wire harnesses are required for connection between each cell and the equalizer [8]. Therefore, the centralized balancing scheme is not suitable for a battery pack with a large number of cells.

11.2.2 Discharge Balancing, Charge Balancing, and Bidirectional Balancing

According to the different balancing action processes, the balancing control management can be divided into discharge balancing, charge balancing, and bidirectional balancing.

The discharge balancing mode means that the discharge of the cells is equalized to ensure that the discharge of the remaining capacity of each cell in the battery pack is 0, to avoid the situation where some cells have been fully discharged while others still have a dischargeable capacity. After a full discharge, the battery pack is charged at a constant current by the series charge mode until the remaining capacity of each cell in the battery pack is 100%. The whole process is shown in Figure 11.14.

As shown in Figure 11.14, the discharge balancing mode can ensure that the dischargeable capacity of the battery can be fully discharged every time. During the charge, according to the cask theory, the upper cut-off limit is determined by the cell with the minimum capacity. In such a case, the capacity of the battery pack is not fully utilized during charging.

Discharge balancing has disadvantages, such as a high energy loss, and cannot be performed at any time (for example, a high energy loss may be caused by the discharge balancing when the remaining capacity of the battery

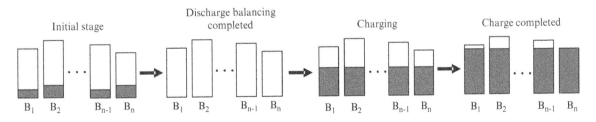

Figure 11.14 Discharge balancing mode.

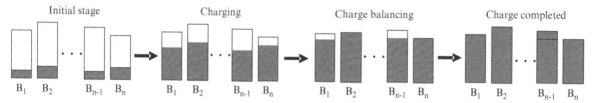

Figure 11.15 Charge balancing mode.

is higher). Furthermore, the discharge balancing requires fully discharge of the cells to improve the discharge depth, and this may affect the cycle life of the battery.

The charge balancing mode refers to balancing the cells by top-aligning the charge mode to ensure that the capacity of each cell in the battery pack reaches 100% during the charge, as shown in Figure 11.15.

The charge balancing mode can ensure that the actual capacity of each cell plays a role during the charge. However, the charge balancing mode does not provide any control for the discharge process, the discharge process meets the cask theory, and the discharge capacity of the battery pack depends on the cell with the minimum capacity. The opposite of discharge balancing is the charge balancing, which is applicable to the battery pack in any state.

The bidirectional balancing mode has the advantages of the discharge balancing mode and the charge balancing mode to provide balancing control during the charge and the discharge. The bidirectional balancing mode can ensure that each cell is discharged until SoC is 0 and is charged until SoC is 100%. Due to addition of the discharge balancing mode, the bidirectional balancing mode still has disadvantages, such as a high energy loss and damage to the battery. However, this mode is conducive to the assessment of the maximum battery capacity (i.e., it is helpful to obtain the maximum capacity C_k of each cell) and can be used to diagnose the SoH of the battery during maintenance of the electric vehicle [9].

11.2.3 Passive Balancing and Active Balancing

The balancing control scheme may be classified into the passive balancing scheme and the active balancing scheme according to the protection of energy in the battery pack during the balancing.

The passive balancing scheme refers to dissipation of the energy in the cells with a higher SoC by parallel resistance and other methods until balancing with other cells in the battery pack. This scheme is implemented as follows: the voltage of each cell is regularly checked. When the voltage of any cell is higher than the average voltage of the battery pack, the parallel resistance of the high-energy cell is connected to dissipate some energy on the parallel resistance until the voltage of the cell is equal to the average voltage of the battery pack. The passive balancing scheme has advantages such as simple control logic and easy hardware implementation, lower cost, and a most common balancing control scheme in the early stage [10]. However, the passive balancing scheme is completed by dissipation of some energy in the battery pack; additionally, the heat may be caused by energy dissipation of the resistance, and a potential safety hazard may be caused for the electric vehicle due to over-heating resulting from poor ventilation.

The active balancing scheme (also called lossless balancing) refers to transfer of the energy from the cell with a high SoC to the cell with a lower SoC by an energy storage element and a series of switch elements to achieve the balancing. The energy storage element used for the lossless balancing scheme includes a capacitor and an inductor. The lossless balancing can remedy the passive balancing, but has disadvantages such as a complicated control logic circuit, and so on.

It should be noted that the active balancing is designed to retain the existing charge and energy of the battery as much as possible during establishment of the balancing control strategy, but the active balancing cannot be truly lossless due to the energy loss of the device. However, in any case, in the same initial state, the active balancing has a lower total energy consumption than the passive balancing and is the mainstream of future development.

11.3 Review and Analysis of Active Balancing Technologies

Existing active balancing technologies are described and compared in this section. The active balancing control is divided into "independent charging type balancing control" and "energy transfer type balancing control."

11.3.1 Independent-Charge Active Balancing Control

It is considered that each cell is charged independently in order to achieve a balancing control of the battery without energy waste. Although it has disadvantages such as a large volume and high cost, this method is quite popular. For example, the independent-charge active balancing control method is widely used in the bus with little sensitivity to volume and weight, and a back-up source for communication. Common schemes are given in Figure 11.16.

Figure 11.16 shows a circuit consisting of one independent charge unit and an $n + 1$ electronic switch. The electronic switches are uniformly controlled by the BCU. The basic idea of this circuit is that the "behindhand" cell is independently charged by the charge unit to achieve the balancing. This method has advantages such as a low energy loss and a high energy efficiency, and disadvantages such as failure to simultaneously complete the balancing control for multiple cells due to competition of the charge unit and poor balancing time efficiency.

Unlike the low charge efficiency of the circuit shown in Figure 11.16, in the scheme shown in Figure 11.17, an independent charge unit is configured for each cell and each charger is controlled by the BCU. Balancing the battery can be rapidly completed using this method with a minimum energy loss, but this method has disadvantages such as large volume, high cost, and more elements of each charge unit resulting in a high failure ratio of the circuit [11]. Additionally, if the battery pack consists of 10 cells, one of which reaches the upper charge limit, it takes plenty of time to alternately charge the other nine cells.

The circuit shown in Figure 11.18 is also an improvement on the scheme shown in Figure 11.16. The "behindhand" cell is charged from the energy of the battery pack by a switching power supply circuit to achieve the balancing effect. Compared with the scheme shown in Figure 11.16, this scheme has advantages such as a high time efficiency, cost reduction to some extent, and wire reduction, with disadvantages such as a lower flexibility and a lower energy efficiency, since the energy is only transferred from the battery pack, rather than from the cell with a higher SoC.

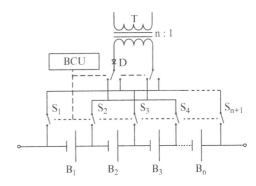

Figure 11.16 Balancing control scheme consisting of an independent charge unit and an electronic switch.

11.3.2 Energy-Transfer Active Balancing Control

The balancing control is designed to equalize the cells. The independent charge balancing control described in the previous

Figure 11.17 Balancing control by installation of an independent charge unit for each cell.

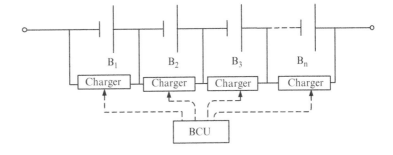

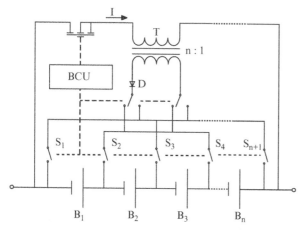

Figure 11.18 Balancing control by an independent charge of the "behindhand" cell from other cells.

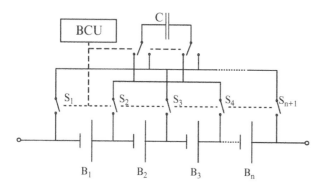

Figure 11.19 Capacitor switching balancing scheme.

section may be understood as respectively charging the cell with a low SoC. The mode described in this section may be understood as a transfer of the energy from the high-energy cell to the low-energy cell, namely an energy-transfer active balancing control, which may be implemented using the following methods.

Capacitor switching balancing scheme

Figure 11.19 shows the circuit structure of the capacitor switching balancing scheme. This method is a typical active balancing strategy, where a balancing capacitor is used as the energy storage element and a switch is used to switch the capacitor between two adjacent cells.

The circuit consists of an $n + 1$ electronic switch and a transfer buffer capacitor C, where the electronic switch is uniformly controlled by the BCU. The basic idea of this method is to transfer the energy from the high-energy cell to the low-energy cell to achieve the balancing. This method has a prominent disadvantage such as an energy transfer only based on battery voltage, a lower energy efficiency, and a high safety risk.

Buck-boost-based adjacent-cell energy transfer scheme

In 2002, Hsieh et al. proposed an adjacent-cell energy transfer method adapted from [12], as shown in Figure 11.20(a), which was improved by Zhao et al. in 2003 [13], as shown in Figure 11.20(b). In this scheme, a buck-boost circuit is provided for each cell. During the balancing, a corresponding switch Q_i is connected to discharge the cell and the discharged energy is stored in the inductor L_i parallel connected with the cell. After the switch is disconnected, the energy stored in the L_i is charged into the cells $i + 1, i + 2, ..., n$ (where n is the total number of cells). The basic idea of this scheme is to transfer the energy from the cell with a high SoC to the adjacent cell with a low SoC to achieve the balancing. This scheme has advantages, such as a relatively simple circuit structure and fewer components. It is noted that, since the current superposition may be caused on the bypass during simultaneous transfer of the charges from the cell i to the cells $i + 2$ and $i + 1$, relevant parameters must be carefully designed to ensure system stability.

PowerPump-based improvement scheme

To control the balancing better, Gary Davison et al. founded the PowerPrecise Company in 2002 and proposed solutions PowerPump and PowerLAN to control the balancing using a special chip.

The TI Company acquired PowerPrecise Company in October 2007 and launched three integrated battery management circuits with a "bq78PL114S12" core on the basis of PowerPrecise's solution, in order to achieve a stronger practicability of this technology. This scheme has a low flexibility; in other words, the TI Company considers that the balancing strategy and the setting of specific balancing parameters are completed by the special chip, and therefore it is difficult for users to modify the strategy and the parameters and expand the information interaction interface.

In order to overcome the deficiency of the TI Company's scheme, the author proposes an improved scheme in the book *Design of EV Battery Management System*, in which the "bq78PL114S12" chip is replaced by any suitable MCU, so that the balancing strategy and the specific balancing parameters are controlled by the BMS in order to improve the flexibility of this scheme [14]. Although this scheme is improved, the energy is still transferred between two adjacent cells and this scheme has a lower time efficiency and energy efficiency if the distance is larger between the high-energy cell and the low-energy cell.

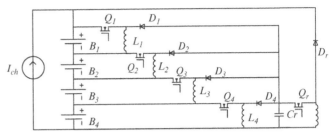

(a) Method proposed by Hsieh *et al.* [12].

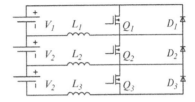

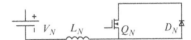

(b) Improvement method proposed by Zhao *et al.* [13].

Figure 11.20 Buck-boost-based adjacent-cell energy transfer scheme.

11.3.3 How to Evaluate the Advantages and Disadvantages of an Active Balancing Control Scheme (an Efficiency Problem of Active Balancing Control)

In fact, for BMS developers, no matter what active balancing scheme is adopted, the following factors should be considered: cost, reliability, and efficiency. It is easy to understand the first two factors. The efficiency can be understood from the following aspects.

Time efficiency

Time efficiency is not a priority for energy-storage battery management systems, but for the electric vehicle, it is hoped to eliminate the battery disequilibrium in a short period of time. Thus, it is hoped to rapidly transfer the capacity from the cell with a higher SoC to the cell with a low SoC in a short period of time, and the balancing control system is required to have a higher charge transfer ability. Generally, for a 100 Ah battery pack, if the difference in the capacity of the cells is 5 Ah, the balancing control system is required for balancing the battery capacity within 1 h to support not less than a 5 A effective transfer current. Therefore, time efficiency is related to the equalizing current. Now the characteristics of the existing active balancing schemes are compared.

1) For the "adjacent-cell energy transfer" scheme, the charges are only transferred between adjacent cells. In addition to the transfer current set for the circuit board, the transfer time efficiency depends on the location of the disequilibrium cells. If the energy is transferred from the first cell to the last cell of the battery pack, a longer balancing time is required.
2) For the "independent charging unit" scheme, each cell is independently controlled, the balancing time is shorter than that of the "adjacent-cell energy transfer" scheme, and the balancing time efficiency depends on the ability of the charge unit.

3) The LTC3300-based control scheme allows the transfer of the charge from a designated cell to the specific cell. Since a high-efficiency charge transfer is achieved in the whole balancing, the balancing time of this scheme is half that of scheme 2.

Charge efficiency

Compared with the passive battery management system, the active battery management system intends to save the energy of the battery to the greatest extent, in order to achieve the purpose of energy saving. Therefore, energy efficiency is a major factor to be considered in battery balancing. However, the definition of the energy efficiency of battery balancing is often vague. For the balancing control efficiency of the battery management system, it may be said that "the efficiency is more than 90%." However, two questions are raised to consider this expression: what does 90% mean – energy efficiency or the charge efficiency? What is the denominator for calculation of the efficiency?

During calculation of the charge efficiency and the energy efficiency, it is possible to cause misunderstandings without having their strict definition, for example,

As shown in Figure 11.21, assume that, before the balancing, the charge level is $Q_0 + Q_1$ for cell B_1, and, respectively, Q_2, Q_3, and Q_4 for B_2, B_3, and B_4; during the balancing, cell B_1 is discharged to charge the cells B_2, B_3, and B_4 (charge transfer); the charge level is $Q_1 + Q_5$ for the cell B_1, and, respectively, $Q_2 + Q_6$, $Q_3 + Q_7$, and $Q_4 + Q$ for B_2, B_3, and B_4 after balancing. Of course, in the actual process, in the case of no external charging, it is not difficult for us to know that there must be

$$(Q_1 + Q_5) + (Q_2 + Q_6) + (Q_3 + Q_7) + (Q_4 + Q_8) \leq Q_0 + Q_1 + Q_2 + Q_3 + Q_4 \tag{11.2}$$

Moreover, since loss is unavoidable, the above equation cannot generally be equated.

Which of the following three definitions is the most appropriate definition of "efficiency?"

1. **This definition is based on "total," which is the total battery level before and after balancing:**

$$\eta_1 = \frac{(Q_1 + Q_5) + (Q_2 + Q_6) + (Q_3 + Q_7) + (Q_4 + Q_8)}{Q_0 + Q_1 + Q_2 + Q_3 + Q_4} \tag{11.3}$$

There is an issue on this definition: the efficiency η_1 relates to the initial and last disequilibrium degrees of the battery and is irrelevant to the performance of the balancing circuit and the balancing scheme.

It can be imagined that, if $(Q_1 + Q_2 + Q_3 + Q_4)$ is very high and the absolute value of Q_0, Q_5, Q_6, Q_7, and Q_8 is low, even if the balancing circuit is poor, the calculated balancing efficiency is very high and even up to 100%. On the contrary, if $(Q_1 + Q_2 + Q_3 + Q_4)$ is very small and the absolute value of Q_0, Q_5, Q_6, Q_7 and Q_8 is high, the efficiency calculated in Equation (11.3) approximates the transfer efficiency of the circuit.

2. **This definition is based on the "ratio of charged and discharged charge:" the efficiency is calculated by dividing the charged charge of the cells by the discharged charge of the battery:**

$$\eta_2 = \frac{Q_6 + Q_7 + Q_8}{Q_0 - Q_5} \tag{11.4}$$

This definition is applicable if the charge is completely different from discharge in the balancing process. In other words, the cell is only in the charging state or the discharging state. This definition is not applicable if the cell is charged and discharged in the balancing process. Why is the cell charged and discharged in the balancing process? For example:

First, as mentioned above, the battery balancing is based on the adjacent-cell energy transfer. For example, as shown in Figure 11.21, if the energy is transferred only between adjacent cells, the energy cannot be directly transferred from the cell B_1 to the cell B_4, but only through the cells B_2 and B_3. The cells B_2 and B_3 are charged and then discharged in the balancing process.

Figure 11.21 Active balancing process based on a charge transfer. (Note: as shown in the figure, B_1, B_2, B_3, and B_4 are four cells, C_i is the capacity, and Q_i is the charge level.)

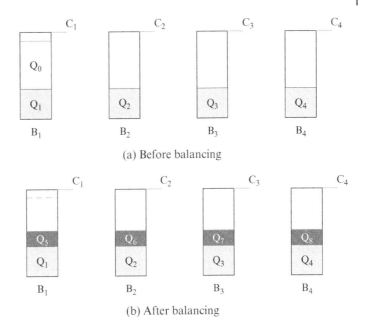

(a) Before balancing

(b) After balancing

Second, since the cells B_2, B_3, and B_4 have a better consistency, the energy is transferred from the cell B_1 with a higher SoC to other cells in the battery pack. Since B_1, B_2, B_3, and B_4 are connected in series, some energy is transferred from cell B_1 and is transferred back to it by a series discharge. The cell B_1 is charged and then discharged in the balancing process.

3. **Definition based on the total transferred energy:**

$$\eta_3 = \frac{Q_5 + Q_6 + Q_7 + Q_8}{Q_0} \tag{11.5}$$

This definition is applicable to judge the performance of the balancing circuit hardware, because it may be considered as the total efficiency of all single transfers. Such a statistical pattern is similar to that of the gross domestic product (GDP): although, at first sight, it seems that sometimes the same data is simultaneously counted in secondary and tertiary industries, but the overall wealth growth ratio of the whole country is not so high, such a statistic can reflect the active degree of the whole national economy operation.

Of course, although this definition is clear, it is not easy to implement it in practice. For the LTC3300-based balancing scheme (to be understood in combination with Figure 11.21), during the balancing, the energy is transferred from the cell B_1 with a higher SoC to the other cells in the battery pack. Since B_1, B_2, B_3, and B_4 are connected in series, some energy transferred from cell B_1 is transferred back to it by a series discharge. During the actual charge measurement, it is easy to measure $(Q_0 - Q_5)$ since it corresponds to the decrease of the charges, while it is difficult to accurately measure Q_5, which is estimated only by the model.

The energy efficiency is not equivalent to the charge efficiency

The above three definitions are based on the charge efficiency. However, the charge efficiency used for the equilibrium statistics is not equivalent to the energy efficiency as the equilibrium potential or open-circuit voltage of the battery varies with the SoC.

The relationship between the charge efficiency and the energy efficiency is understood from three aspects.

1) The charge transfer is accompanied by the energy transfer with the same sign and they are close numerically. Since the increase or decrease in the charge is accompanied by the increase or decrease in the energy, the above defining methods are also applicable to define the energy efficiency. Since the equilibrium potential of the Li-ion battery is similar in the platform area, their difference in numerical value is not large. Generally, the difference is less than 10%.

2) More attention should be paid to the energy efficiency. The charge efficiency is firstly defined since it is easier to understand. However, for a practical balancing circuit, more attention should be paid to the energy efficiency, because, for a real balancing circuit, the index that best reflects its performance is essentially energy efficiency.

3) Attention should be paid to the absolute value of the energy loss while paying attention to the energy efficiency.

In many cases, the "efficiency" is expressed by a percentage. However, for the battery management systems, it is important to consider the absolute value of the energy loss.

First, the following evaluation standard may be designed to judge the merits and demerits of two active balancing schemes: regardless of the single-step transfer strategy, the total energy loss of two balancing schemes is compared when the SoC value is the same before and after the balancing.

Second, the calorific value in the battery pack directly depends on the absolute value of the energy loss, which increases the difficulty of thermal management in the battery management system. This issue is elaborated below.

Thermal management issue caused by time efficiency and energy consumption

In general, it is hoped that the heat generated in the battery pack is as little as possible, so as not to make thermal management more difficult. However, both passive balancing and active balancing will cause additional heat to the battery pack. The "additional" heat described here is relative to the heat produced during the normal charge and discharge process of the battery. The time efficiency and energy consumption discussed above are directly related to the thermal management of the battery management system.

First, it is hoped to perform the balancing control of the battery at a higher current in order to save the balancing time and improve the balancing time efficiency. However, in the case of a given internal resistance of the battery and circuit impedance, the greater the balancing current, the greater is the heat generated in the balancing process, in which case the thermal management of the battery management system is more difficult. Therefore, the "heating" becomes a factor restricting the improvement of the balancing time efficiency.

Second, even in the active balancing control mode, more or less energy consumption may be caused and the consumed energy eventually remains in the battery system in the form of "heat energy." Therefore, the energy consumption should be reduced as much as possible in order to reduce the heat in the balancing process. The battery management system designers must accurately evaluate the energy consumption of the balanced system designed by them in order to determine the corresponding thermal management measures.

Third, unlike the heat produced in the normal charge and discharge process of the battery, the heat produced in the balancing process is not uneven. The uneven heat has an inconsistent impact on the battery degradation and can deteriorate the inconsistencies of the cells in the battery pack if they have accumulated for a long period of time.

11.4 Balancing Strategy Study

How can a balancing control strategy be developed in the case of a given balancing topology? Or, what are the problems to solve during development of the balancing strategy? The "general problems" or "common problems" of the balancing control strategy are discussed in this section. That is to say, these problems must be solved during the development of any balancing control strategy. In other words, the answer to these questions is the design process of the balancing control strategy [15].

11.4.1 Balancing Time

Researchers who are engaged in the research and development of electric vehicles often argue about "whether the balancing should be implemented only at the end of charging," "whether the balancing should be started during the discharge of the battery," and other issues that involve the balancing time. According to the early research on the management strategy of BMS, a consensus has been reached that the balancing control should be implemented at the end of charging. However, with the development of research, there is a need to discuss the necessity and feasibility of starting the balancing control during discharging or charging.

Balancing at the end of charging

Balancing at the end of charging is the most easily understood mode and is widely used at present. This means that, when any cell in the battery pack is fully charged at the end of charging, the balancing control is implemented until all cells are fully charged.

Figure 11.22 shows balancing at the end of charging. As shown in the figure, the cells in the battery pack are charged in series and are expressed as B_1, B_2, ..., B_n. The column height is the same, the column width represents the battery capacity, and the dark color part represents the remaining capacity of the cell. When any cell of the battery pack is fully charged, the balancing control is implemented until all cells are fully charged.

It is generally considered that the balancing at the end of charging can prevent over-charging of the cells while ensuring all cells are fully charged. However, in the case of a small current on the balancing circuit, it may take a long time to complete the balancing at the end of charging.

It is necessary to implement the balancing control in the process of discharging or charging in order to achieve consistency of the cells in the battery pack as soon as possible. The two balancing times are discussed below.

Balancing control in the discharging process

The balancing control in the discharging process means that, if the disequilibrium degree of any cell in the battery pack exceeds a given threshold value during discharging, the balancing is implemented until "the disequilibrium state" is removed from the battery pack.

In fact, if the "passive" balancing mode is used, the discharging balancing is of little significance, because it cannot increase the endurance mileage of electric vehicles. However, if the "active" balancing mode is used, the disequilibrium state may be found in the cells of the battery pack in advance. The energy is transferred from the cells with a high SoC to the cells with a low SoC in order to avoid over-discharging the cells with a low SoC while there is a high remaining capacity in other cells, so as to increase the endurance mileage of the electric vehicle.

Figure 11.23 shows the discharge process without balancing. If series cells are discharged without the balancing measure, the cell with the lowest remaining capacity is the first to be fully discharged and the cells cannot be continuously discharged although they have a high remaining capacity.

Figure 11.24 shows the discharge process with a balancing control. The BMS detects that the SoC of cell B_2 is lower, and the balancing control is implemented in the discharge process in order to avoid discharging cell B_2 to 0 while other cells still have a higher SoC. Therefore, the energy of all of the cells can be used to increase the endurance mileage of the electric vehicle, compared with the discharge process shown in Figure 11.23.

Since the EV battery pack works in series and the cells discharge the same capacity by a series discharge mode, the balancing control in the discharging process depends on a more accurate estimation of the "control in the discharge of each cell," namely, the dark-colored part shown in Figure 11.23. What this part represents is not the capacity of the battery, nor the SoC of the battery, but the "remaining power" of the battery.

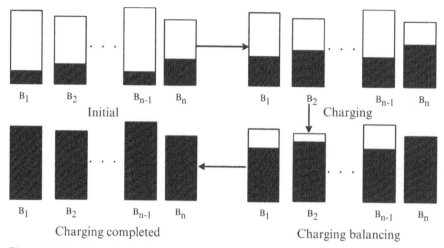

Figure 11.22 Balancing at the end of charging.

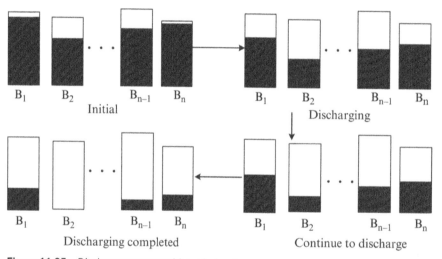

Figure 11.23 Discharge process without balancing.

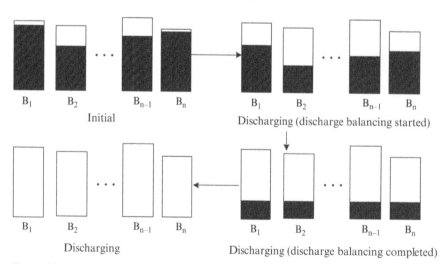

Figure 11.24 Discharge process with balancing control.

Since implementation of the balancing control in the discharging process is beneficial to improving the endurance mileage of the electric vehicles, why has this method not been widely used in practical applications? This is because the SoC estimation error is large in the discharge process.

It is generally believed that the balancing control in the discharging process has the biggest disadvantage of "criterion failure". Suppose that a balancing control algorithm is based on the SoC in the case of a large SoC estimation error. The balancing may be started at the wrong time and the battery's energy will be wasted due to the frequent energy transfer. Furthermore, the transfer of the charge from the cells with a lower SoC to the cell with a higher SoC may do great harm to the endurance mileage of the electric vehicle.

Balancing control in the charging process

The balancing control in the charging process means that, although none of the cells is fully charged in the charging process, the BMS detects that the disequilibrium degree of any cell in the battery pack exceeds a given threshold value and the balancing is implemented until the "disequilibrium" state is removed from the battery pack. The balancing control in the charging process has advantages, such as a short balancing time and removal of cell inconsistency in the battery pack as soon as possible.

Figure 11.25 shows the balancing control in the charging process. The BMS detects the cell inconsistency in advance and decides that it may take a long time to complete the balancing control at the end of charging. The BMS starts the balancing control in the charging process to ensure that those parts other than the dark-colored part are the same and accelerate the balancing of the battery pack.

If the balancing control in the charging process is designed to fully charge all cells simultaneously, the BMS is required to estimate the capacity required to fully charge the cell in the balancing process (i.e. those other than the dark colored part in Figure 11.25). For this, the BMS is required to accurately estimate the capacity and current SoC of each cell. Additionally, it is noted that, if any cell is fully charged in the charging or balancing process, the balancing control is switched to the balancing control at the end of charging.

11.4.2 Variable for Balancing

The balancing control is designed to remove, or at least reduce, the inconsistency of the battery pack. The inconsistency of the battery pack mainly includes capacity inconsistency, inconsistency of the current remaining capacity (Ah), SoC inconsistency (%), inconsistency of the internal resistance, inconsistency of the self-discharge rate, etc.

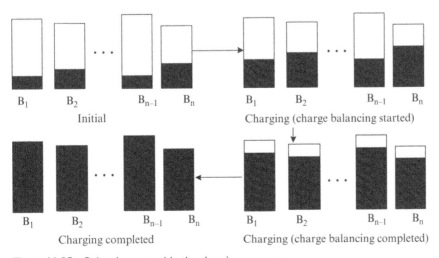

Figure 11.25 Balancing control in the charging process.

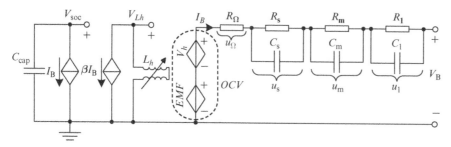

Figure 11.26 Equivalent circuit model of a battery.

When the balancing strategy is developed, which variable should be used as the "battery consistency" criterion? Scholars give different answers, mainly including "terminal voltage," "SoC," and "remaining capacity."

Terminal voltage

The terminal voltage of each cell in the battery pack may be directly measured and many previous balancing strategies were based on the terminal voltage of the cells. However, as shown in Figure 11.26, according to the battery model, due to the effect of the polarization resistance, the EMF hysteresis effect, and other factors, the terminal voltage of the cells cannot reflect the SoC and the remaining capacity of the battery under dynamic conditions. Therefore, it is unreliable to start or stop the balancing control according to the terminal voltage of the cells in the discharging or charging process.

It is assumed that, in the charging process, the cells A and B have different degradation degrees and the same capacity and current SoC, namely

$$\text{EMF}(A) + V_h(A) = \text{EMF}(B) + V_h(B) \tag{11.6}$$

In this case, the balancing is not required. However, two cells have different internal resistances, specifically:

$$R_\Omega(A) + R_s(A) + R_m(A) + R_l(A) \neq R_\Omega(B) + R_s(B) + R_m(B) + R_l(B) \tag{11.7}$$

When the operating current of the battery pack is not 0, the terminal voltage of the cells is not equal, namely:

$$V_B(A) \neq V_B(B) \tag{11.8}$$

In this case, if the balancing is implemented only according to the terminal voltage of the cells, two cells with a consistent SoC and remaining capacity become inconsistent, damaging the battery consistency and causing energy waste.

Of course, the use of the terminal voltage as the balancing implementation criterion is not all bad. If the balancing at the end of charging is selected (i.e. "at the end of balancing mode"), the passive topology is used and the battery pack is charged at a lower charge current and the cells can be fully charged according to the terminal voltage. Specific operations are as follows: the passive balancing topology is used when the voltage of any cell exceeds the pre-set upper limit, the cell is discharged by the bypass resistance to reduce its SoC, the battery pack is charged in series after the discharge for some time, and the above operation is repeated while the charge current and the bypass discharge time are reduced, until the SoC of all cells in the battery pack approximates 100%. In this special case, the balancing circuit topology and corresponding control policy are quite simple, but the balancing time is longer.

SoC

Unlike the passive balancing mode, when the active balancing topology is used, it is very unreliable to start the balancing according to the terminal voltage, but it is more advantageous to start the balancing according to the SoC.

Figure 11.27 Comparison of balancing based on SoC and terminal voltage.

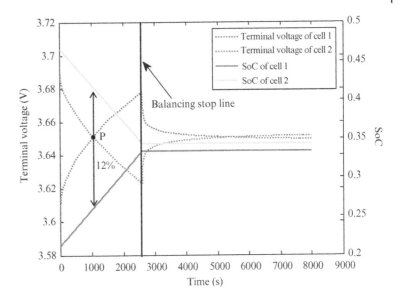

For example, as shown in Figure 11.27, two cells (cells 1 and 2) in the battery pack have poor consistency, a 20% difference in the initial SoC, and a 0.11 V difference in the terminal voltage. The energy is therefore transferred from cell 2 to cell 1 by the active method. In this case, for cell 2, which is being discharged, the terminal voltage is lower than the equilibrium EMF, and for cell 1, which is being charged, the terminal voltage is higher than the equilibrium EMF.

If the terminal voltage is used as the criterion, the balancing control will be stopped since two cells have the same terminal voltage (point P in the figure) at 1100th of a second on the timer shaft. However, the SoC of two cells is not the same at this moment and the difference in the SoC is 12% between two cells according to the vertical coordinate.

If the balancing criterion is the SoC rather than the terminal voltage, the balancing control should be continued after point P. The charge is continuously transferred from cell 2 to cell 1 until about 2500th of a second (considering the SoC estimation error, the SoC of two cells is not fully equal but has about a 2% difference upon stopping the balancing).

Of course, the SoC-based balancing control strategy has a strong dependence on the SoC estimation accuracy and its control effect is also related to the SoC estimation error. For example, the SoC estimation error of the algorithm is ± 3%, and during the balancing control based on the SoC, the balancing stop threshold value can only be greater than 6%. Because the true SoC of the two cells may be 30% when the estimated SoC is 33% for cell 2 and 27% for cell 1, in this case the charge transfer must be stopped; in other words, the active balancing must be stopped.

Remaining capacity

As mentioned in the discussion of the SoC estimation in Chapter 9, if the battery capacity is known, there is a conversion relation between the SoC and the remaining capacity, as shown in the Figure 11.28.

The relation shown in Figure 11.28 can be expressed by

$$\text{SoC} = \frac{\text{Remaining capacity}}{\text{Battery capacity}} \times 100\% \tag{11.9}$$

According to Equation (11.9), the balancing strategy based on the remaining capacity is performed according to the remaining chargeable or dischargeable capacity of the battery after the capacity of the cell is measured and the

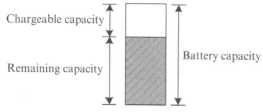

Figure 11.28 Relation between SoC and the remaining capacity.

current capacity of the battery is estimated by the SoC. The balancing is not performed for the cells with the remaining capacity falling within the set range. The discharge balancing is performed when the remaining capacity exceeds the upper limit to reduce the charge of the battery by transfer or dissipation. When the remaining capacity is lower than the lower limit, the cell is charged to obtain the charge from the battery pack so that the remaining capacity of the cells in the battery pack is relatively balanced.

According to different application requirements, the balancing strategy based on the remaining capacity may be classified into the "balancing strategy based on the remaining chargeable capacity" and the "balancing strategy based on the remaining dischargeable capacity." Their difference can be understood according to Figure 11.29, where each capacity state of the battery can be expressed as a dot at a specific location, the horizontal axis represents the battery capacity, and the vertical axis represents the remaining capacity in the current state.

Ideally, when the remaining capacity of all the cells is equal to their capacity, the battery pack is fully charged, and in this case, all dots are connected to produce a ray with a 45° angle from the coordinate axis, as shown by the dotted line in Figure 11.29.

When the remaining chargeable capacity is used as the balancing variable, the balancing target is determined as follows: the cells have the same difference between full capacity and current capacity, as shown by the line parallel to the 45° dotted line in Figure 11.29(b).

Similarly, when the remaining dischargeable capacity is used as the balancing variable, the balancing target is determined as follows: the cells have the same difference between 0 capacity and current capacity, as shown by the line parallel to the *x* axis in Figure 11.29(c).

In order to make the best use of the remaining capacity of each cell, the balancing may be performed simultaneously based on the remaining chargeable capacity and the remaining dischargeable capacity. Balancing the target is determined as follows: the balancing is performed according to the remaining chargeable capacity during charging, in order to fully charge each cell. The balancing is performed according to the remaining dischargeable capacity during charging, in order to fully discharge each cell. As shown in Figure 11.29, this balancing mode may be understood as the dynamic regulation of those shown in Figure 11.29(b) and (c), and can always ensure that the cells can be fully charged simultaneously and fully discharged simultaneously.

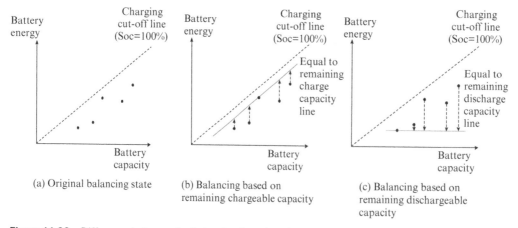

Figure 11.29 Difference between the balancing based on the remaining chargeable capacity and the balancing based on the remaining dischargeable capacity.

The process shown in Figure 11.29 may be understood in combination with the balancing controls described in Section 11.4.1: all cells are fully charged simultaneously by the balancing based on the remaining chargeable capacity, but during discharging, the dischargeable capacity of the battery pack cannot achieve "low alignment" subject to the cell with a minimum capacity. On the contrary, all cells are fully discharged simultaneously by the balancing based on the remaining dischargeable capacity, but during charging, the cells with the minimum capacity are first charged fully. After their combination, the capacity of the cells can be fully used and the maximum dischargeable capacity of the battery pack depends of the average capacity of the cells. According to the above analysis, when using different balancing variables, the dischargeable capacity of the battery pack may be expressed as follows:

$$C_{discharge} = \begin{cases} NC_{min}, \text{based on remaining chargeable capacity} \\ NC_{min}, \text{based on remaining dischargeable capacity} \\ N\bar{C}, \text{based on remaining chargeable capacity and remaining dischargeable capacity} \end{cases} \tag{11.10}$$

where $C_{discharge}$ is the maximum dischargeable capacity of the battery pack, C_{min} is the minimum capacity of the cell, and $\bar{C}$ is the average capacity of the battery pack.

The balancing based on remaining capacity can be performed as follows:

1. Estimation of cell capacity

Under offline conditions, cell capacity can be obtained by fully charging and then discharging the cell. However, a more convenient and rapid online estimation method is required for the battery pack. The common online capacity estimation may be deduced from the equation of the CC method:

$$C_{new} = \frac{\int_{t_0}^{t_1} i(t) dt}{SoC(t_0) - SoC(t_1)} \tag{11.11}$$

The estimation accuracy of the cell capacity depends on the SoC estimation accuracy and the sampling accuracy of the current sensor. The balancing strategy based on the remaining capacity is required to find the remaining capacity of the cells, which is calculated using the following equation:

$$Q(t) = SoC(t) \, C_{new} = SoC(t) \frac{\int_{t_0}^{t_1} i(t) dt}{SoC(t_0) - SoC(t_1)} \tag{11.12}$$

In Equations (11.11) and (11.12), $Q(t)$ is the remaining capacity at time t, C_{new} is the estimated battery capacity, the current i is the function of time t; and $SoC(t)$ is the SoC of the battery at time t.

The remaining capacity of the battery is calculated by estimating the SoC many times. Regardless of the CC method or the open-circuit voltage method, a single SoC estimation may cause more estimation errors of the remaining capacity of the battery than many SoC estimations.

2. Start balancing

If the real-time capacity of the cell is obtained using the CC method during the balancing, the current sampling circuit needs to be set for each balancing channel, increasing the difficulty of developing the balancing circuit board and causing accumulated error during coulomb counting. The balancing target is that during charging, each cell of the battery pack needs to be fully charged and the cells should have the same remaining chargeable capacity, while during discharging, the energy is transferred from the cell with a high SoC to the cell with a low SoC in order to ensure that the capacity of the battery pack is fully used.

For the balancing variable, the balancing starting condition and the variable estimation error should be considered in the balancing strategy.

The balancing strategy achieves balancing of the battery pack by controlling the consistency of the terminal voltage, the SoC, or the remaining capacity. The first problem analyzed in the balancing strategy is selection of a more suitable variable according to different topologies. These variables are analyzed and compared to select the control variables most applicable to the balancing strategy.

11.5 Two Active Balancing Control Strategies

Based on the key indicators and design process of the balancing control strategy discussed in the previous section and the topologies of two representative active balancing schemes, this section introduces the development process of the balancing control strategy by analyzing the variables, the balancing time, and the balancing parameters.

11.5.1 Topologies of Two Active Balancing Schemes

This section proposes two active balancing topologies, including low-level balancing based on small charger, and the lead-acid battery transfer balancing for the EV battery system, and analyzes the development process of the balancing strategy.

Hierarchical balancing based on a small charger

The low-level balancing (LLB) based on a small charger means that the battery pack is divided into multiple groups, the LLB control strategy is implemented in groups while the high-level balancing (HLB) is implemented between the groups. Figure 11.30 shows the specific topology.

As shown in Figure 11.30, the hierarchical balancing consists of LLB and HLB. The battery system is divided into several battery groups, where each group is connected to an independent small charger, the balancing is controlled by the central control unit, the LLB is implemented in the groups, the HLB is implemented between the groups, and the balancing state communication is completed by the control bus between the groups. The balancing topologies of the LLB and HLB are described below.

1. LLB topology

Figure 11.31 shows the LLB topology in the hierarchical balancing based on a small charger. As shown in the figure, the LLB topology is designed according to a bidirectional synchronous fly-back circuit. The balancing circuit consists of a switch circuit and a bidirectional transformer [16].

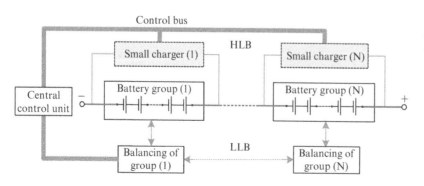

Figure 11.30 Schematic diagram of hierarchical balancing based on a small charger.

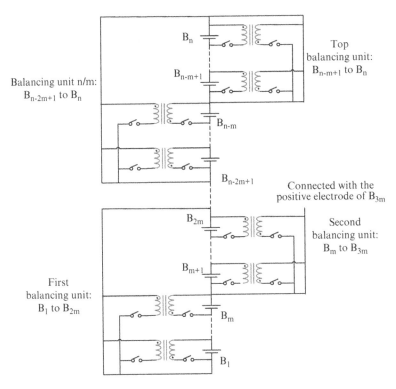

Figure 11.31 LLB topology of hierarchical balancing.

Since the cells in the system are not always evenly distributed, the number of cells in each cell group may not be the same, but the balancing logic is the same. Additionally, the cost and the balancing efficiency are considered when grouping the battery system. In the case of too many battery groups, more chargers are required, causing difficulty in the practical application. Conversely, in the case of too few battery groups, the intra-group balancing efficiency may be reduced, making it impossible to equalize each cell. In a practical application, it is recommended that one group consists of 12 cells.

In the battery group, the cells are connected to the group by a bidirectional transformer. The energy may be transferred between the cells in the group by charging and discharging modes. However, consider that the number of cells controlled by the balancing control chip is different from the number of cells in the battery group; for example, typically LTC6804/LTC6811 series chips of ADI Company are mainly applicable to the control of the battery group with less than six cells. If the battery group consists of 12 cells, at least two chips are required. According to the chip channel number, the battery group would need to further divide the balancing units and consider the balancing relation between the units.

Figure 11.31 shows the battery group consisting of n cells, where $2m$ cells are used as the balancing unit. The energy can be transferred from the cells in the battery group to the balancing unit by the discharging mode, in order to charge other cells in the balancing unit; a cell may be charged from the other cells of the balancing unit by the charging mode. The balancing units are connected and m cells are shared by adjacent balancing units. Charging is completed between different balancing units by using the shared cells as a transmission medium to achieve balancing in the battery group.

Specifically, each cell is connected in parallel with a bidirectional transformer. The primary coil of the transformer is connected in parallel with the battery and the secondary coil of the transformer is connected to the positive and negative electrodes of the balancing unit, where the single cell is located, and is controlled by a switch. Each balancing unit is expanded in the form of a "daisy chain" to accommodate any number of cells. As shown in Figure 11.31, the cell of the first balancing unit consists of $2m$ bidirectional transformers, one end of the secondary coil of the transformer is connected with the negative electrode of the balancing unit, the other end of the secondary coil of the transformer is connected with the positive electrode of the balancing unit, and so on. Since the topmost balancing unit cannot be connected to the next unit in a similar manner, the last m cells can form a balancing unit independently.

Through this connection mode, the problem of matching the balancing chip channel number and the cell number in the battery group can be solved. The balancing chip directly manages m cells and is connected by the balancing unit to cover any number of cells. It has good expansibility. At the same time, each cell can be balanced with the balancing unit and can also achieve energy flow with other cells in the battery group, ensuring the integrity of the battery group.

LLB topology can support bidirectional battery balancing, and the cell can be not only charged through the battery pack but also discharge its energy to other cells in the battery group.

Without considering energy consumption, it is assumed that for the bidirectional transformer in the balancing circuit, the charge current is I_{charge} and the discharge current is $I_{discharge}$. Each cell may discharge or receive the charge to or from other cells. Therefore, the actual balancing current of each cell is affected by the balancing state of other cells, which is the superposition of all currents acting on the cell.

If, in the battery balancing unit of the battery k, k_1 cells are in a charging mode and k_2 cells are in a discharging mode, the balancing current of the balancing unit (excluding the topmost balancing unit) is calculated by

$$I_k = I + k_1 \frac{I_{charge}}{2m} + k_2 \frac{I_{discharge}}{2m} \tag{11.13}$$

where the charge current is defined as positive and the discharge current is defined as negative. If the battery k is in the charging mode, $I = I_{charge}$, if the battery k is in the discharging mode, $I = I_{discharge}$, and if the battery k is charged and discharged, $I = 0$.

Since the cells of the topmost balancing unit are not connected with the next battery balancing unit, its balancing current is calculated by the following equation when the battery k is located in the topmost balancing unit:

$$I_k = I + (k_3 - k_5) \frac{I_{charge}}{2m} + (k_4 - k_6) \frac{I_{discharge}}{2m} + k_5 \frac{I_{charge}}{2m} + k_6 \frac{I_{discharge}}{2m} \tag{11.14}$$

where k_3 and k_4 represent the number of cells in the charging mode and the discharging mode in the second top balancing unit and k_5 and k_6 represent the number of the cells in the charging mode and the discharging mode in the top balancing unit.

The LLB topology structure can reasonably distribute the balancing unit according to the number of cells and the number of specific control chip channels, and has good expansibility. However, according to the topology, when the energy is transferred from the cells of the first balancing unit to the cells of the top balancing unit, the balancing time may be prolonged with the increase in the number of cells. Therefore, for the entire battery pack, especially for a large battery such as the EV battery system, it is necessary to perform hierarchical balancing using the HLB topology. When the battery pack is divided into one group, only level 1 balancing is performed, which is a special case of a hierarchical balancing topology.

2. HLB topology

As shown in Figure 11.30, after grouping the battery packs, a separate small charger is provided for each battery group to charge the battery group and complete the inter-group balancing. The small charger is relative to the large charger of the battery pack, but has a smaller power and volume. The positive and negative electrodes of the charger are connected to the positive and negative electrodes of the battery group respectively, and are connected to the external power supply.

During the hierarchical balancing based on a small charger, the LLB control strategy starts to equalize the cells in the battery group. After LLB of the cells in the battery group, HLB is performed to charge the battery according to "top alignment" to fully charge the battery groups and achieve inter-group balancing. The specific control strategy of hierarchical balancing will be detailed in the following chapters.

Lead-acid battery transfer balancing

Since the starter and other devices, such as the radio, blower, lighting and wiper, of the cars are supplied by a 12 V or 24 V lead-acid battery, the cars must be equipped with a lead-acid battery as the power supply for weak current systems. Because the voltage of the lead-acid battery is similar to that of the cell, charge transfer balancing can be realized by using a relatively simple voltage conversion circuit using the lead-acid battery as an intermediate carrier of the battery balancing.

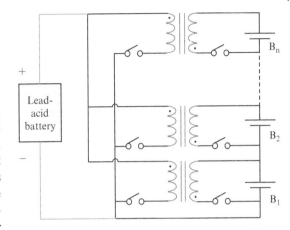

Figure 11.32 Topology of lead-acid battery transfer balancing of the battery pack.

Figure 11.32 shows the topology of the lead-acid battery transfer balancing. As shown in the figure, each cell in the battery pack is connected in parallel with a bidirectional transformer, where the primary coil of the transformer is connected to the cell and the secondary coil of the transformer is connected to the lead-acid battery, with the energy transfer process controlled by a switch. When it is required to charge the cell due to its lower energy, the energy is transferred from the lead-acid battery to the cell through the transformer. When it is required to discharge the cell due to its higher energy, the energy is transferred from the cell to the lead-acid battery through the transformer. The topology can realize the synchronous energy transfer and avoid the complicated circuit design and high cost.

According to the above analysis, the balancing control chip is required for the active balancing schemes: the hierarchical balancing based on the small charger and the lead-acid battery transfer balancing. The LTC3300 balancing chip and the LTC6804/LTC6811 voltage acquisition chip of ADI Company form a feasible scheme in practical applications. Each piece of LTC3300 can be used for bidirectional balancing control of six cells in series and each piece of LTC6804/LTC6811 can simultaneously monitor the voltage of 12 cells. Therefore, the balancing daughter board can be based on 2 × LTC3300 and 1 × LTC6804 in order to equalize the entire high-power battery pack after it has been divided into several battery groups.

Comparison of two active balancing topologies

Due to the simple topology and low cost of passive balancing, unidirectional passive balancing is used at present in most battery management systems to discharge the cells with a high SoC. In the traditional passive balancing topology, the connection of each cell is controlled by a switch array and the excess energy in the cell is dissipated through a resistor to equalize the battery pack.

The topologies of two active balancing schemes are compared by comparison with the traditional passive balancing topology from device complexity, balancing independence, expansibility, balancing type, and vehicle space requirements.

As shown in Table 11.2, two active balancing topologies require the same number of DC/DC conversion circuits as the cells. When the voltage at two ends of the battery group in the hierarchical balancing is more than the voltage of the lead-acid battery, the cost of the DC/DC conversion circuits is more than that of the transfer balancing topology. The LLB of the hierarchical balancing is achieved by the transfer of the energy between the cells. As the cells are mutually affected during the balancing, the control model and the algorithm are relatively complicated. For the lead-acid battery transfer balancing and the traditional passive balancing topologies, the energy is not directly transferred between the cells, the balancing process is independent, and the control strategy is relatively simple.

Table 11.2 Comparison of balancing topologies.

	Hierarchical balancing	Lead-acid battery transfer balancing	Traditional passive balancing
Device complexity	n DC/DC conversion circuits and several small chargers required	n DC/DC conversion circuits required	No DC/DC conversion circuits required
Independence	HLB independent balancing and LLB dependent balancing	Non-direct energy transfer and independent balancing of each cell	No energy transfer of cells and independent balancing
Expansibility	Expansion of battery group generally	Expansion of cell	Expansion of cell
Balancing type	Active	Active	Passive
Space requirement	Location reserved for small charger	Smaller	Smaller

Note: n is the number of the cells in the battery pack.

11.5.2 Hierarchical Balancing Control Strategy

Following the hierarchical balancing topologies described in the previous section, corresponding balancing control strategies are developed in this section and the parameters used in the strategy are discussed.

Selection of balancing variables and balancing time

Different balancing variables may share the same balancing logic. The balancing is described here by the variable SoC because of similarity with the terminal voltage and the remaining capacity. As shown in Equation (11.15) below, the variables used for evaluation of the battery consistency include the variance s^2, the maximum balancing deviation δ_1, and the average deviation of $\delta_{2,I}$ in each cell, where s^2 can reflect the discrete state of the whole battery and δ_1 can reflect the loss of the available capacity of the battery pack due to the "cask effect." The capacity difference between the maximum capacity and the minimum capacity of the cells in the battery pack cannot be effectively utilized and $\delta_{2,I}$ is used to evaluate the deviation of the cell i from the average value of the whole battery pack. During balancing, the balancing judgment can be made according to δ_1 and $\delta_{2,i}$:

$$
\begin{cases}
\overline{SoC} = \sum_{i=1}^{n} SoC_i \\
s^2 = \sum_{i=1}^{n} \left(SoC_i - \overline{SoC} \right)^2 / n \\
\delta_1 = SoC_{max} - SoC_{min} \\
\delta_{2,i} = SoC_i - \overline{SoC}
\end{cases}
\tag{11.15}
$$

where SoC_i represents the SoC of the cell i, $\overline{SoC}$ is the average SoC of the battery pack, and SoC_{max} and SoC_{min} respectively represent the maximum and minimum SoC of the cells in the battery pack.

During development of the specific balancing strategy, the balancing start and stop conditions of the battery pack may be determined by the SoC deviation δ_1 and the threshold value ΔSoC_1 of the battery pack. In the case of $\delta_1 > \Delta SoC_1$, the balancing starts to perform equilibrium charging or discharging of the cells according to their conditions. In the case of $\delta_1 \leq \Delta SoC_1$, all balancing is stopped.

Similarly, the balancing start and stop conditions of each cell may be determined by the SoC deviation $\delta_{2,i}$ and threshold value ΔSoC_2 of the cell. The range of threshold is set to $\left[\overline{SoC} - \Delta SoC_2, \overline{SoC} + \Delta SoC_2 \right]$. When the deviation $\delta_{2,i}$ of the cell is positive and more than " SoC_2, the cell is discharged; when the deviation $\delta_{2,i}$ of the cell is negative and less than ΔSoC_2, the cell is charged; and when SoC is within the range of threshold, no balancing is required.

It can be seen from the above description that for a particular balancing strategy, ΔSoC_1 and ΔSoC_2 are fixed values and do not vary with the balancing process, and SoC, s^2, SoC_{max}, SoC_{min}, δ_1, and $\delta_{2,i}$ variables are changing dynamically with the balancing process.

This process is described by Figure 11.33. The battery balancing is determined for the battery according to the current SoC and the pre-set threshold value of the battery. The battery balancing is divided into three conditions: (1) the equilibrium discharge: the cell with a higher energy is discharged to reduce its energy; (2) the equilibrium charge: the cell with a lower energy is charged from other cells of the battery pack; and (3) no balancing: the cells are in an equilibrium state without a demand for balancing.

Since the terminal voltage of the cell is easily affected by the internal resistance and the balancing current of the cell, and cannot directly reflect the current SoC of the battery, the capacity of the cells during estimation of the remaining capacity should be estimated by considering accuracy and complexity. The SoC is used as the balancing variable for the hierarchical balancing strategy.

While the electric vehicle is being used, the battery pack will not be fully discharged. The discharge process of an electric vehicle battery pack has the following two characteristics. First, during operation of the electric vehicle, it is possible to cause a balancing failure due to relatively complex operating conditions, the unstable discharge current of the battery, and the estimation error of balancing variables. Second, if the balancing is started at the end of a discharge, since the disequilibrium has been formed in the battery, it is impossible to effectively improve the capacity available to the electric vehicle in a short time after starting the balancing. Therefore, it is not suitable in practice to implement the balancing during a discharge. However, under special conditions, such as parking, transient balancing can be carried out, because the current is zero or very small during parking, which is conducive to improving the effect and accuracy of the balancing.

Unlike the discharge process, the EV battery pack is often fully charged. At the same time, the charging time is relatively long and sufficient time is available to carry out the balancing. At the end of an equalizing charge, when a certain cell is fully charged, the battery pack cannot be continuously charged, so the balancing can be started in order to equalize the battery pack and achieve its consistency and "top alignment."

Compared with unidirectional balancing, bidirectional balancing can improve the capacity utilization rate of the battery. Therefore, based on the above considerations, charge balancing can be selected from the perspective of a capacity utilization rate and discharge and parking balancing can be used as a supplement.

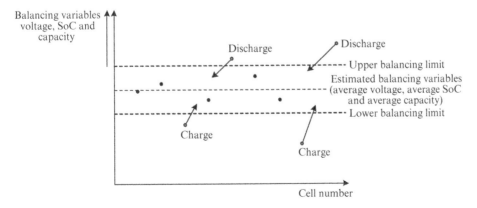

Figure 11.33 Schematic diagram of active balancing.

Hierarchical bidirectional balancing strategy

1. Hierarchical bidirectional balancing control strategy

Figure 11.34 shows the flow chart of the hierarchical bidirectional balancing control strategy, where, when the battery pack is charged and any cell in the battery pack has been fully charged, charge balancing is started and discharge balancing is started during parking.

As shown in Figure 11.34, the hierarchical bidirectional balancing model mainly consists of the following steps:

(a) Battery voltage detection. Any strategy depends on the detection of the battery's basic data. The charging and discharging modes of the battery are determined by the data. When the battery pack is charged, charge balancing is started, and either step (b) or step (d) is performed.

(b) First the battery pack is charged by the main charger and then step (c) is performed after detecting that any cell in the battery pack has been fully charged.

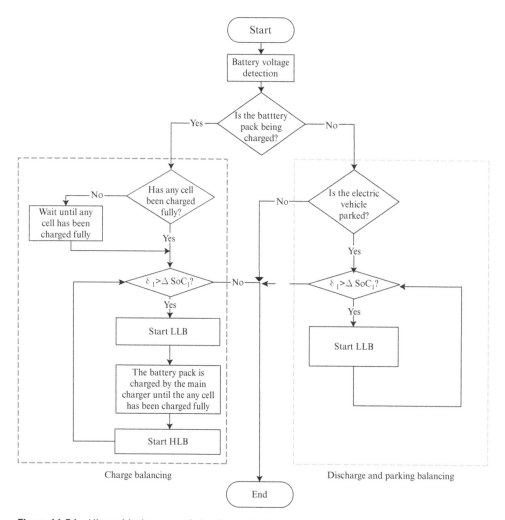

Figure 11.34 Hierarchical two-way balancing control strategy.

(c) The selected balancing variable is determined by whether the balancing has started. When the selected variable exceeds the range of the threshold, namely $\delta_1 > \Delta SoC_1$, LLB is started to equalize the cells in the battery group; otherwise the charge balancing is stopped. Steps (b) and (c) are repeated until the charge balancing is completed. For hierarchical balancing, the battery pack is first charged by the main charger and then the HLB is started, because it is easier to control the fine charging of the battery pack to the SoC = 100% target.

(d) The selected balancing variable determines whether the balancing is to start. When the selected variable exceeds the range of the threshold, namely $\delta_1 > \Delta SoC_1$, LLB is started; otherwise balancing is stopped. Only LLB is started because no condition exists where the small charger can be used during discharging until the cells in the battery group are at an equilibrium.

2. LLB control strategy

For charge balancing and discharge and parking balancing of hierarchical balancing, the LLB can only start after meeting the balancing conditions, in this case the algorithm flow of balancing as shown in Figure 11.35. The main steps, (a) to (d), of the LLB control strategy are given below:

(a) Judge the equilibrium state, monitor the battery voltage, estimate the SoC, judge the equilibrium state of each cell in the battery pack according to the balancing conditions, and then determine the required balancing modes. When completed, start step (b).

(b) Calculate the balancing time and start the balancing. Estimate the SoC for a period of time after the balancing and then perform step (c).

(c) Repeat steps (a) and (b) until $\delta_1 \leq \Delta SoC_1$ and then start step (d); otherwise, repeat steps (a) to (c).

(d) LLB ends.

The battery pack may be equalized by steps (a) to (d). The judgment of the equilibrium state and the balancing operation are discussed below:

As shown in Figure 11.35, the equilibrium state is judged for each cell in the balancing unit and the average $\overline{SoC}$ of the balancing unit is set as the balancing target, the battery balancing scope is set to $\left[\overline{SoC} - \Delta SoC_2, \overline{SoC} + \Delta SoC_2\right]$, and the cells are discharged when the SoC is more than the upper limit and charged when the SoC is less than the range of threshold; otherwise, no balancing is required.

During balancing, if the SoC or the remaining capacity is used as the balancing variable, it is difficult to accurately estimate the SoC due to changes in the equilibrium current, energy consumption, and temperature. Therefore, balancing is started by steps such as "calculate the balancing time → start the balancing → estimate the SoC and judge the equilibrium state,", with the steps continuously repeated (see Figure 11.36). Balancing accuracy can be improved by estimation of the SoC using the voltage one period at a time after balancing. If the SoC is estimated by the CC method or other methods that do not require idling, the idling time is 0. If the SoC is estimated using the open-circuit voltage method, it is necessary to reserve time in order to estimate the open-circuit voltage.

In order to avoid over-balancing and balancing failure in the balancing process of the battery pack, the balancing time is calculated using the SoC with the smallest deviation and capacity of the cell:

$$t = kC\left(\min\left|\frac{SoC_i - \overline{SoC}}{I_i}\right|\right) \tag{11.16}$$

where the minimum difference between the SoC and $\overline{SoC}$ is selected for calculation of the balancing time in order to avoid over-balancing. After the balancing is started, the SoC change in the equalized cell should not be less than 0.1% in order to avoid a balancing failure and not more than 2% in order to avoid over-balancing. In the equation, I_i is the balancing current of the cell i, C is the battery capacity, and k is the correction coefficient. The over-balancing resulting from a capacity or current error is avoided.

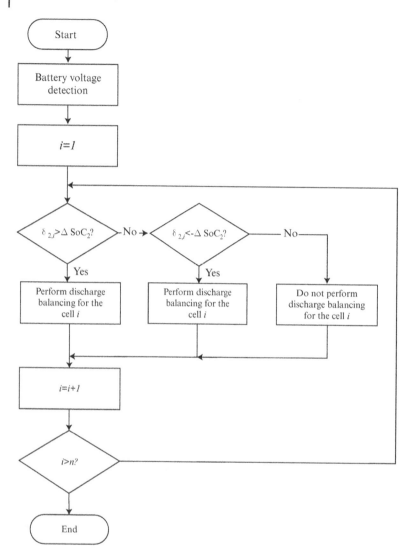

Figure 11.35 LLB control strategy.

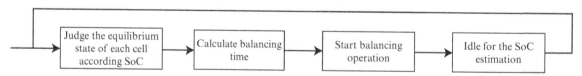

Figure 11.36 Balancing operation.

Assuming that t_0 is the SoC estimation time, k_1 is the SoC estimation number during the balancing, and t_1 is the LLB time (the working current is not 0), the total balancing time is calculated by

$$T = t_1 + k_1 t_0 \tag{11.17}$$

According to Equations (11.16) and (11.17), the LLB time can be optimized by changing the balancing current and the correction coefficient k and optimization of the SoC estimation method.

According to Equation (11.17), the total time of the charge balancing based on a small charger is calculated by

$$T = t_1 + t_{total\ charge} + t_2 + k_1 t_0 \tag{11.18}$$

where, $t_{total\ charge}$ is the total charging time, t_2 is the longest HLB time, and $t_1 + k_1 t_0$ is the longest LLB time of all battery groups.

Similarly, the energy consumption of the whole process is calculated by

$$E = E_{total\ charge} + E_2 + E_{wireloss} + E_{consumption\ of\ battery} + E_{consumption\ of\ board} \tag{11.19}$$

where $E_{total\ charge}$ and E_2 are the energy consumption caused by the charge efficiency of the main charger and the small charger. $E_{wire\ loss}$ and $E_{consumption\ of\ battery}$ are the energy consumption caused on the wire and the internal resistance of the battery. $E_{consumption\ of\ board}$ is the energy consumption caused by boards, including wires of the circuit board, electron component, etc. Because the main charge is an external type, the heating resulting from its charge efficiency does not affect the temperature in the battery box. The following equation is used to calculate the average heat dissipation power in the battery box in the whole balancing process:

$$P = (E_2 + E_{wire\ loss} + E_{consumption\ of\ battery} + E_{consumption\ of\ board}) / T \tag{11.20}$$

11.5.3 Lead-Acid Battery Transfer Balancing Control Strategy

This section discusses the balancing variables and balancing time of the lead-acid battery transfer balancing control strategy and the development of the specific strategy.

Selection of balancing variables and balancing time

The impact of balancing variables on the balancing process is similar. In order to continuously and finely tune the balancing according to the terminal voltage at the end of the balancing, a balancing method based on the remaining capacity is required to accurately estimate the battery capacity, which is quite difficult to realize [17]. Therefore, SoC is selected as the variable to judge the consistency of balancing the lead-acid battery transfer. The balancing is completed for the battery pack in the case of $\delta_1 \leq \Delta SoC_1$ but in the case of $\delta_1 > \Delta SoC_1$, the balancing needs to start. If the conditions are not met, no balancing is required.

According to an analysis of the capacity utilization ratio, the bidirectional balancing can effectively improve the capacity utilization ratio. This balancing can be achieved by discharge balancing of the hierarchical bidirectional balancing only for the battery group, but not for the battery pack. The lead-acid battery transfer balancing strategy is applicable to the battery pack, enabling the balancing of the whole battery pack to be achieved.

Bidirectional lead-acid battery transfer balancing strategy

The lead-acid battery transfer balancing can be divided into bidirectional balancing (target: SoC = 100% for a charging process and 0 for a discharging process), charge balancing (target: SoC = 100%), and discharge balancing (target: SoC = 0) according to the different action processes and balancing targets.

Figure 11.37 shows the specific flow of the bidirectional lead-acid battery transfer balancing, which consists of the following steps:

(a) Battery voltage detection. Determine the charging or discharging mode of the battery pack, start the charge balancing when the battery is being charged, and start step (b); otherwise, start the discharge balancing and start step (d).

(b) Charge the battery pack using the main charger. Start step (c) after detecting any cell in the battery pack that has been fully charged.

(c) Equalize the cells in the battery pack. When the selected variable exceeds the balancing threshold, namely $\delta_1 > \Delta SoC_1$, start the lead-acid battery transfer balancing to equalize the cells in the battery pack (the battery pack is not charged or discharged when the cells in the battery pack have the same SoC); otherwise, stop the balancing. Repeat steps (b) and (c) until the charge balancing is completed.

(d) Start the discharge balancing. When the selected variable exceeds the balancing threshold, namely $\delta_1 > \Delta SoC_1$, start the lead-acid battery transfer balancing to equalize the cells in the battery pack; otherwise, stop the discharge balancing.

From the above process, it can be seen that the steps of the lead-acid battery transfer balancing are similar to LLB of the hierarchical balancing shown in Figure 11.34, except for the balancing objects. The LLB of the hierarchical balancing is applicable to the cells in the battery group, while the lead-acid battery transfer balancing is applicable to the battery group pack. During the balancing by the lead-acid battery, each cell in the battery pack has stronger independence, and the energy is not directly transferred between the cells, but by the lead-acid battery, so as to independently perform the charge and discharge balancing for each cell.

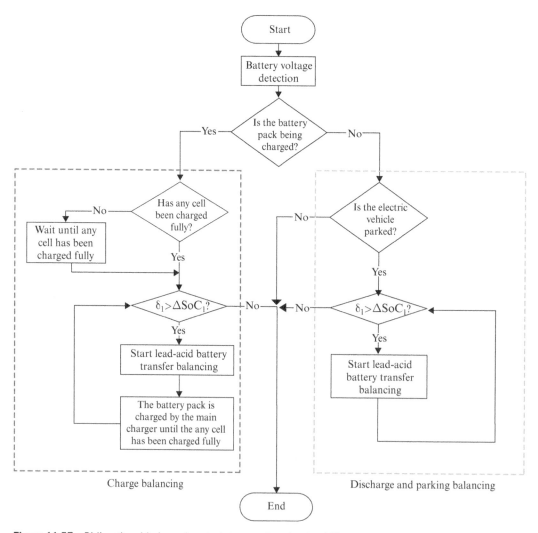

Figure 11.37 Bidirectional balanced control strategy for a lead-acid battery transfer.

It is worth noting that, due to the current limitations of the lead-acid battery, the balancing current cannot exceed the upper limit of the lead-acid battery. When the charging and discharging limitation is reached, the number of corresponding charging and discharging cells should be reduced to reduce the current. In addition, in order to avoid over-charging and over-discharging the lead-acid battery, in specific applications, when the SoC of the lead-acid battery is < 10%, the lead-acid battery is charged by the battery pack. Similarly, when the SoC of the lead-acid battery is > 90%, the energy is transferred to the cell with lower energy in the battery pack to always maintain the reasonable capacity of the lead-acid battery.

11.6 Evaluation and Comparison of Balancing Control Strategies

When balancing control strategies are developed, it is necessary to discuss "which strategy is better." In previous sections, balancing in terms of "required balancing time" and "energy consumption" were evaluated. Based on that, this section will enrich the evaluation indexes of the balancing control strategies.

11.6.1 Evaluation Indexes of Balancing Control Strategies

For the balancing control strategy, the actual balancing effect (whether the equilibrium can converge), time consumption, energy consumption, and impact on the battery life are mainly considered. The balancing evaluation system is established according to the balancing effect, impact on the battery life, time consumption, and energy consumption.

Consumption effect

1. Convergence

The balancing effect is reflected in balancing convergence, that is to say, the balancing strategy converges balancing to a given threshold in any case. For example, if the battery balancing target is set as the average variable $\pm \Delta SoC$, balancing is deemed to normally converge when each cell in the battery group converges to $\left[\overline{SoC} - \Delta SoC, \overline{SoC} + \Delta SoC\right]$. The balancing convergence is an important balancing index. If the balancing cannot be converged and the balancing operation cannot be stopped normally, the energy will be continuously consumed in the balancing process, and this strategy will result in continuous battery energy waste until the battery capacity is 0.

The convergence is the most basic balancing index and discussion of other indicators depends on the normal convergence of the strategy.

2. Capacity utilization ratio η_{cap}

The balancing is designed to eliminate the inconsistency of the battery pack and improve the capacity utilization ratio. The capacity utilization ratio can be defined as the ratio of the maximum discharge capacity $C_{dischage}$ of the battery pack to the mean capacity $\overline{C}$ of all cells:

$$\eta_{cap} = \frac{N\,C_{dischage}}{N\,\overline{C}} = \frac{C_{dischage}}{\overline{C}} \tag{11.21}$$

The most ideal situation is where there is no energy loss in the battery balancing process. After the balancing operation, each cell can be fully charged in the charging process and each cell can be discharged fully in the discharging process. When the capacity of the battery is fully utilized, the capacity utilization ratio is 100%. The capacity utilization ratio of the battery pack is another important index for an evaluation of the balancing effect.

Impact on battery life

During battery balancing, the battery is usually charged or discharged. Research shows that the capacity loss of the battery is related to the accumulated charge and discharge capacity of the battery. The impact of balancing on the battery life is reflected in the accumulated capacity during balancing. When the balancing strategy is evaluated, the impact on the battery life is quantitatively compared to the average accumulated charge–discharge capacity:

$$Q = \frac{1}{n}\sum_{m=1}^{n} \int |i| dt \qquad (11.22)$$

where Q is the accumulated charge–discharge capacity and i is the equalizing current. During the analysis of battery life, Q is used as the evaluation index for the impact of balancing on the battery life. The greater the accumulated charge–discharge capacity in the balancing process, the worse the protection of balancing to battery life becomes. The protection effect of the balancing strategy on the battery life is also one of the strategy evaluation indexes.

Time consumption

Although the time consumption is not a top priority for an energy-storage battery management system, it is hoped that the battery imbalance of the battery management system will eventually be eliminated. Therefore, it is hoped to quickly transfer the energy from cells with a higher SoC to cells with a lower one.

After the initial balancing conditions, such balancing variables, balancing time, balancing parameters, and the initial capacity of the battery pack, have been determined, the balancing time is fixed. However, all possible initial balancing conditions should be comprehensively analyzed for the whole balancing strategy to obtain overall evaluation indexes. The time distribution, maximum time, and average time of balancing are the parameters representing the balancing time. The balancing time distribution reflects the dispersion degree of the final balancing effect under different initial conditions, which reflects the stability of the balancing strategy. The maximum balancing time reflects the balancing time under extreme conditions. In a practical application, the maximum time t_{max} can be defined according to the actual demand, for example, the charging balancing must be completed during charging at night and the total balancing time cannot be more than 12 h. The average balancing time reflects the overall time consumption level of the balancing strategy.

Energy consumption

In the process of charge–discharge balancing, the charge "acquisition" and "loss" are accompanied by energy "acquisition" and "loss." In the process of an energy transfer, the energy consumption may be caused because of the efficiency. The energy consumption can reflect the overall performance and energy protection of the balancing scheme. The absolute value of the energy consumption is directly related to the thermal management of the battery management system. The less energy that is released by the battery pack in the balancing process, the lower the difficulty of the thermal management of the battery management system corresponding to this strategy becomes. The energy consumption distribution, the maximum energy consumption, and the average energy consumption of the balancing are the parameters representing the energy consumption of the balancing. Similar to the time consumption, the average energy consumption of the balancing reflects the overall energy protection level of the balancing strategy.

As shown in Figure 11.38, the energy difference $(E_0 - E_5 - E_6 - E_7 - E_8)$ before and after balancing is the energy consumption of balancing.

In addition, heat dissipation power inside the battery box is also one of the parameters to be considered. The heat dissipation power is defined as the ratio of the energy consumption inside the battery box to time, as shown by

$$P = \frac{E}{T} \qquad (11.23)$$

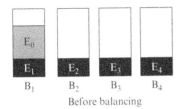 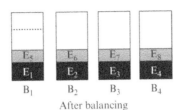

Figure 11.38 Schematic diagram of an energy transfer.

where E is the energy consumption inside the battery (the heat generated outside the battery box is not included in the statistics) and T is the balancing time.

Furthermore, the evaluation of the balancing system also covers safety, feasibility, etc., and the six indexes, including balancing convergence, the capacity utilization ratio η_{cap}, the impact of balancing on the battery life Q, balancing time T, energy consumption of balancing E, and battery heat dissipation power P, are often combined to comprehensively evaluate the strategy. Among them, the balancing convergence is the most important index of the balancing strategy, because, if the balancing strategy cannot converge, the energy will always be consumed by the balancing. The remaining five evaluation indexes are discussed on the basis of convergence.

11.6.2 Comparison of Flows for Balancing Strategies

In this section, the balancing control parameters are analyzed for two balancing control strategies developed in Section 11.5 by the balancing evaluation indexes described in the previous section and the overall performance of two strategies are compared.

The similarities and differences of two balancing strategies developed in Section 11.5 are summarized and analyzed from the balancing target, the balancing time, balancing parameters, and the balancing operation according to the above-mentioned description of the balancing strategies (see Tables 11.3 and 11.4).

In the case of a given balancing variable and balancing time, the equalizing current has a major impact on the balancing time consumption, energy consumption, and heat dissipation power. The simulation results of hierarchical balancing, lead-acid battery transfer balancing and passive balancing are shown in Figures 11.39, 11.40, and 11.41 respectively. With the increase of the equalizing current, the balancing time decreases, but the energy consumption and heat dissipation power increase gradually. All of the time consumption, the energy consumption, the heat dissipation power, and so on are often not optimal. This section compares the strategies in different aspects and obtains the desired balancing strategy by comparing the optimal solution of each strategy.

Table 11.3 Similarities of flows for balancing strategies.

Similarity	Hierarchical balancing strategy, lead-acid battery transfer balancing strategy, passive balancing strategy
Selected balancing variable	SoC
Balancing time	Two-way balancing
Selectable balancing parameter	Equalizing current (1–10 A), number of equalized cells, battery capacity, battery type
Charge balancing starting criterion	Any cell has been charged fully and the battery pack is in a disequilibrium state
Charge balancing stop criterion	The battery pack has been charged fully and the SoC of the battery pack is within the threshold value
Discharge balancing starting criterion	The electric vehicle is parked and the battery pack is in a disequilibrium state

Table 11.4 Difference of flows for balancing strategies.

Difference	Hierarchical balancing	Lead-acid battery transfer balancing	Passive balancing
Charge balancing	LLB combined with HLB	Achieve the consistency of the cells in the battery pack by the energy transfer and then charge the battery pack	Achieve the consistency of the cells in the battery pack by energy dissipation and then charge the battery pack
Discharge balancing	Only LLB	Start lead-acid battery transfer balancing	Start passive balancing
Discharge stop criterion	LLB completed	Maximum difference in SoC of cells is less than ΔSoC_1	Maximum difference in SoC of battery pack is less than ΔSoC_1

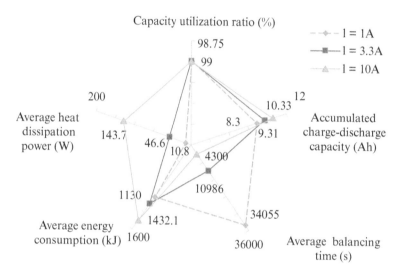

Figure 11.39 Simulation results of hierarchical balancing.

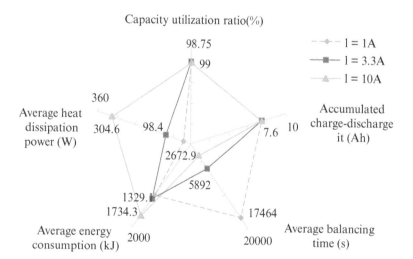

Figure 11.40 Simulation results of lead-acid battery transfer balancing.

Figure 11.41 Simulation results of passive balancing.

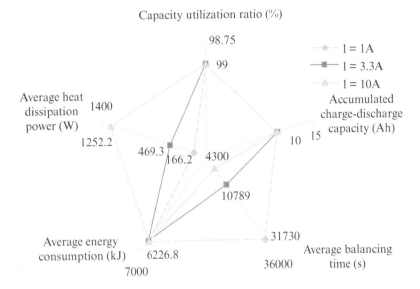

It should be noted that the larger the capacity utilization ratio, the better the balancing effect will be, and the larger the heat dissipation power, energy consumption, time, and cumulative charge and discharge capacity, the worse the balancing effect will be. Therefore, in Figures 11.39, 11.40, and 11.41, the capacity utilization ratio is 100% at the origin of its coordinate axis and 0 at the origin of other axes. The closer each index is to the origin, the better the balancing effect will be.

11.6.3 Comparison of Balancing Time

The balancing time efficiency may be analyzed for the balancing strategy by relative time consumption. The passive control strategy is used here for a comparative study. After the energy consumption and time consumption of the active balancing control strategy are understood, those of the passive balancing strategy are used as the "standard quantity of the passive time consumption" relative quantity. For example, under an initial condition, the balancing time is 1 h for passive balancing, 0.5 h for hierarchical balancing described in Section 11.5, and 0.8 h for lead-acid battery transfer balancing. The "relative time consumption" is 50% for hierarchical balancing and 80% for lead-acid battery transfer balancing. The relative energy consumption may be defined similarly. The relative time consumption and relative energy consumption of balancing are conducive to a better understanding of strategies.

In order to compare and analyze the strategies, the same balancing conditions, equalizing current (3.3 A), and other parameters are used.

Comparison of the relative time consumption

1. **Relative time consumption distribution diagram**

Two balancing strategies are simulated with the passive balancing model to obtain the relative time consumption as shown in Figure 11.42.

(a) Lead-acid battery transfer balancing

At a 3.3 A current, 81.8% of the samples have a time consumption of 50 ~ 60%. Generally, the average balancing time of the lead-acid battery transfer balancing is 55.3% of that of the passive balancing under the same conditions.

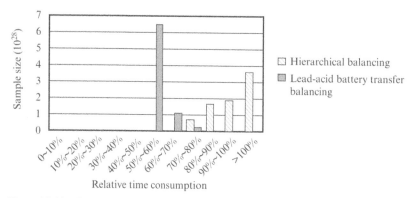

Figure 11.42 Relative time consumption distribution diagram of balancing (I = 3.3 A). Note: 0 ~ 10% represents [0%, 10%], 10 ~ 20% represents [10%, 20%], 20 ~ 30% represents [20%, 30%], and so on.

(b) Hierarchical balancing based on a small charger

At a 3.3 A equalizing current, the relative time consumption of LLB is 53% of the total time consumption of the hierarchical balancing. In the relative time distribution diagram of the hierarchical balancing, 40% of the samples require a longer balancing time than that of the passive balancing. The balancing time is not effectively optimized and the average balancing time is 1.07 times that of the passive balancing, mainly because, under this simulation condition, a longer charging time is required due to the charge current of the small charger used in the HLB, which is 11% of the charge current of the main charger. If the difference in the charge current between the small charger and the main charger is reduced, the time of the hierarchical balancing is gradually close to the time of the lead-acid battery transfer balancing.

Under this balancing condition, the time of the lead-acid battery transfer balancing is shorter.

Comparison of the optimal time

According to the simulation results of the two active strategies and the passive strategy at different equalizing currents shown in Figures 11.39, 11.40, and 11.41, the balancing time of the battery decreases with the increase of the current.

As shown in Table 11.5, the balancing time of two active balancing strategies are shortest at a current of 10 A.

As shown in Table 11.5 and Figure 11.43, in the case of a maximum current, the time consumption of the lead-acid battery transfer balancing and the hierarchical balancing is respectively 62.6% and 87.7% of that of the passive balancing. For the optimal time, the shortest time is required for the lead-acid battery transfer balancing.

In this case, the relative time distribution diagram is shown in Figure 11.44 and is compared with that shown in Figure 11.42, where the relative time distribution is offset towards the left and the absolute balancing time consumption and relative time consumption decrease with the increase in current.

Table 11.5 Comparison of the balancing effect at an equalizing current of 10 A.

Evaluation index	Lead-acid battery transfer balancing	Hierarchical balancing	Passive balancing
Convergence	Converged	Converged	Converged
$\bar{T}$ (s)	2672.9	3770.7	4300.0
Q (Ah)	7.6	10.3	10.0
η_{cap} (%)	99.0	99.0	99.0
$\bar{E}$ (kJ)	1734.3	1432.1	6227.6
$\bar{P}$ (W)	304.6	143.7	1252.2

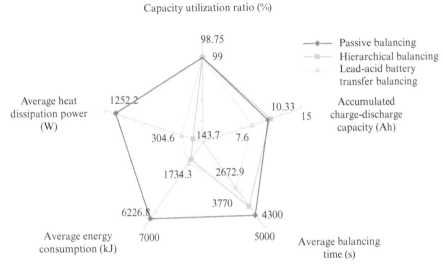

Figure 11.43 Comparison of the balancing strategy evaluation index at optimal time.

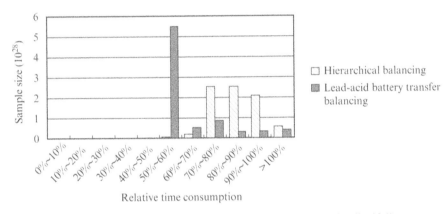

Figure 11.44 Relative time consumption distribution diagram of balancing (*I* = 10 A).

11.6.4 Comparison of Energy Consumption

Comparison of relative energy consumption

Figure 11.45 shows comparisons of relative energy consumption. The relative energy consumption of the hierarchical balancing and the lead-acid battery transfer balancing is less than 30%. Compared with the traditional passive balancing, the energy consumption of the hierarchical balancing and the lead-acid battery transfer balancing is respectively 18.1% and 21.4% of the passive balancing. Compared with the passive balancing, the energy consumption of the active balancing is significantly reduced.

Comparison of optimal energy consumption

Since the balancing energy consumption is related to the specific equalizing current, without the prejudice of generality, the equalizing current of 1 A is selected to analyze and compare the balancing effects of different balancing strategies, as shown in Table 11.6.

As shown in Table 11.6, generally the average energy consumption of the hierarchical balancing and the lead-acid battery transfer balancing is respectively 997.2 kJ and 1257.7 kJ. Figure 11.46 shows the relative energy consumption distribution of the two balancing strategies relative to passive balancing. As shown in the

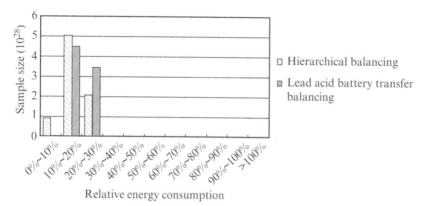

Figure 11.45 Relative energy consumption distribution diagram of balancing (I = 3.3 A).

Table 11.6 Comparison of balancing effect at an equalizing current of 1 A.

Evaluation index	Lead-acid battery transfer balancing	Hierarchical balancing	Passive balancing
Convergence	Converged	Converged	Converged
$\bar{T}(s)$	17464.0	34055.0	31730.0
Q (Ah)	7.6	8.3	10.0
$\eta_{cap}(\%)$	99.0	99.0	99.0
$\bar{E}(kJ)$	1257.7	997.2	6227.6
$\bar{P}(W)$	29.1	10.8	166.2

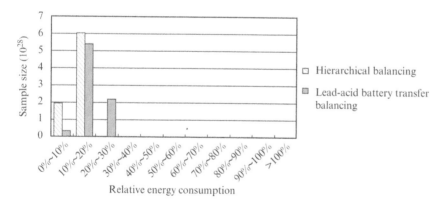

Figure 11.46 Relative energy consumption distribution diagram of balancing (I = 1 A).

above figure and table, the energy consumption of the hierarchical balancing and the lead-acid battery transfer balancing is significantly reduced compared to the passive balancing, respectively by 84.0% and 79.8%. Figure 11.47 shows the comprehensive comparison of all balancing strategies. Compared with the passive balancing, the heat dissipation of the hierarchical balancing and the lead-acid battery transfer balancing has a better optimization effect, and their average heat dissipation power is reduced by 62.8% and 93.5% respectively.

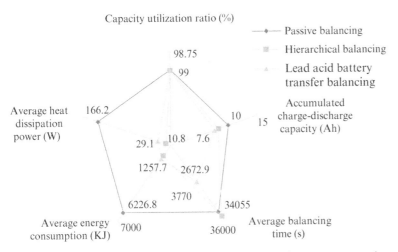

Figure 11.47 Comparison of balancing strategy at optimal energy consumption.

11.6.5 Comparison of the Impact of Balancing on Battery Life

Comparison of an accumulated charge transfer

The accumulated charge and discharge capacity are relatively more stable during balancing because the total charge transfer mainly depends on the inconsistency and the energy transfer efficiency of the cells. The accumulated charge and discharge capacity of the lead-acid battery transfer balancing and the hierarchical balancing are respectively 76% and 83% of the passive balancing. The average charge and discharge capacity of the lead-acid battery transfer balancing is 55 Ah. The lead-acid battery transfer balancing has a major impact on the battery life.

Comparison of an optimal accumulated charge transfer

The comprehensive analysis shows that, at a minimum current, balancing has the smallest impact on battery life. Considering the impact on the life of the lead-acid battery, the hierarchical balancing provides more effective battery protection than the lead-acid battery transfer balancing. Compared with the repeated charge and discharge balancing, the unidirectional charge balancing can provide effective battery protection to avoid continuous charge transfer caused by the balancing during charging and discharging of the battery.

11.6.6 Comparison of the Capacity Utilization Ratio

Considering that the disequilibrium of the battery is caused by the inconsistency of the maximum capacity, the maximum capacity utilization ratio can be realized only by bidirectional balancing. In the process of discharge balancing, the capacity utilization ratio is lower than that of the lead-acid battery transfer balancing, because LLB can only be implemented during hierarchical balancing and cannot guarantee that all cells are fully discharged. At the same time, in the case of inconsistent capacity of the battery pack, the capacity depends on the cell with the minimum capacity during passive balancing. Therefore, the lead-acid battery transfer balancing has the maximum capacity utilization ratio and its comparison with the optimal capacity utilization ratio of the hierarchical balancing strategy is shown in Figure 11.48.

11.6.7 Analysis of the Optimization Case

The balancing strategy is selected in this section according to a specific case. For example, the balancing is required for a 100 Ah lithium iron phosphate battery pack consisting of 96 cells, and specific conditions, such as not more

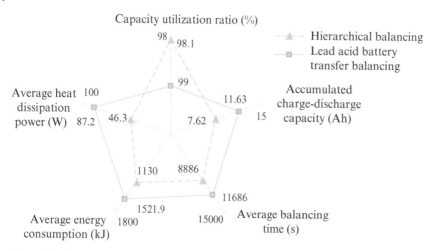

Figure 11.48 Comparison of the optimal capacity utilization ratio.

Table 11.7 List of parameters.

Strategy	Parameters
Hierarchical balancing	For optimal time: 3.6 A equalizing current during LLB, 9 A current of small charger, 12 cells in the battery group. In this case, the maximum balancing time is 4.2 h, the maximum energy consumption is 1380 kJ, and the maximum heat dissipation power is 145 W.
	For optimal energy consumption: 1.2 A equalizing current in the battery group, 9 A current of a small charger, 12 cells in the battery group. In this case, the maximum balancing time is 7.9 h, the maximum energy consumption is 1200 kJ, and the maximum heat dissipation power is 70 W.
Lead-acid battery transfer balancing	For optimal time: 4 A equalizing current. In this case, the maximum balancing time is 2.4 h, the average energy consumption is 1380 kJ, and the maximum heat dissipation power is 130 W.
	Optimal energy consumption: 2 A equalizing current. In this case, the maximum balancing time is 7.9 h, the optimal energy consumption is 1190 kJ, and the maximum heat dissipation power is 50 W.
Passive balancing	Maximum energy consumption does not meet the conditions.

than 150 W maximum heat dissipation power, not more than 1500 kJ average energy consumption, and not more than 8 h maximum balancing time, are met. The parameters of each strategy can be obtained by solving the feasible region, as shown in Table 11.7.

The passive balancing strategy, the hierarchical balancing strategy, and the lead-acid battery transfer balancing strategy are fully compared in this section. The three strategies are compared quantitatively according to different evaluation indexes. Finally, combined with the actual limitations, the parameter selection of the three strategies is analyzed and the following four conclusions can be drawn.

1) When the equalizing current is 3.3 A, the time of lead-acid battery transfer balancing is the shortest. When the equalizing current is 10 A, all three balancing strategies obtain their optimal time and the time consumption of the lead-acid battery transfer balancing and the hierarchical balancing is 62.6% and 87.7% respectively of that of the passive balancing. In comparison, the time of the lead-acid battery transfer balancing is shorter.

2) When the equalizing current is 3.3 A, the energy consumption of the hierarchical balancing is the lowest. When the equalizing current is 1 A, all three balancing strategies obtain their optimal energy consumption. Compared with the passive balancing, the energy consumption of the hierarchical balancing and the lead-acid battery transfer balancing is reduced by 84.0% and 79.8% respectively and the energy consumption of the active

balancing is significantly improved compared with the passive balancing. The energy consumption of the hierarchical balancing is the lowest in the three strategies.

3) When the equalizing current is 3.3 A, the accumulated charge and discharge capacities of the lead-acid battery transfer balancing and the hierarchical balancing are respectively 76% and 83% of that of the passive balancing and the average charge and discharge capacity of the lead-acid battery transfer balancing is 55 Ah. The lead-acid battery transfer balancing has a larger impact on battery life.

4) In bidirectional balancing, all three balancing strategies obtain their optimal capacity utilization ratio, the capacity utilization ratio of the lead-acid battery transfer balancing being the highest, because it achieves active balancing of the battery pack during discharging.

References

1 Speltino, C., Stefanopoulou, A., and Fiengo, G., "Cell equalization in battery stacks through state-of-charge estimation polling," in *American Control Conference (ACC)*, IEEE, 2010.

2 Chun, C. Y., "State-of-charge and remaining charge estimation of series-connected %-ion batteries for cell balancing scheme," in *Telecommunications Energy Conference*, IEEE, 2016.

3 Kötz, R., Ruch, P. W., and Cericola, D., "Aging and failure mode of electrochemical double layer capacitors during accelerated constant load tests," *Journal of Power Sources*, 2010, 195(3): 923–928.

4 Baronti, F., Roncella, R., and Saletti, R., "Performance comparison of active balancing techniques for lithium-ion batteries," *Journal of Power Sources*, 2014, 267: 603–609.

5 Kim, M., Kim, M., Kim, J., et al., "Center-cell concentration structure of a cell-to-cell balancing circuit with a reduced number of switches," *IEEE Transactions on Power Electronics*, 2014, 29(10): 5285–5297.

6 Lindemark, B., "Individual cell voltage equalizers (ICE) for reliable battery performance," in *Telecommunications Energy Conference*, 1991, pp. 196–201.

7 Wei, X. and Zhu, B., "The research of vehicle power Li-ion battery pack balancing method," in *International Conference on Electronic Measurement & Instruments*, Beijing, China: IEEE, 2009, pp. 498–502.

8 Shang, Y., Zhang, C., Cui, N., et al., "A cell-to-cell battery equalizer with zero-current switching and zero-voltage gap based on quasi-resonant LC converter and boost converter," *IEEE Transactions on Power Electronics*, 2015, 30(7): 3731–3747.

9 Haifeng, D., Xuezhe, W., and Zechang, S., "A new SOH prediction concept for the power lithium-ion battery used on HEVs," in *2009 IEEE Vehicle Power and Propulsion Conference*, IEEE, 2009, pp. 1649–1653.

10 Huang, W. and Abu Qahouq, J. A., "Energy sharing control scheme for state-of-charge balancing of distributed battery energy storage system," *IEEE Transactions on Industrial Electronics*, 2015, 62(5): 2764–2776.

11 Li, S., Mi, C. C., and Zhang, M., "A high-efficiency active battery-balancing circuit using multiwinding transformer," *Industry Applications*, 2013, 49(1): 198–207.

12 Hsieh, Y. C., Moo, C. S., and Tsai, I. S., "Balance charging circuit for charge equalization," in *Power Conversion Conference*, 2002, vol. 3, pp. 1138–1143.

13 Zhao, J., Jiang, J., and Niu, L., "A novel charge equalization technique for an electric vehicle battery system," *Power Electronics and Drive Systems*, 2003, 2: 853–857.

14 Tan, X. J., *Design of EV Battery Management System*, Guangzhou: Sun Yat-sen University Press, 2011.

15 Moore, S. W. and Schneider, P. J., "*A review of cell equalization methods for lithium ion and lithium polymer battery systems*," *SAE Publications*, 2001, 01–0959.

16 Lee, Y. S. and Cheng, G. T., "Quasi-resonant zero-current-switching bidirectional converter for battery equalization applications," *Power Electronics*, 2006, 21(5): 1213–1224.

17 Cui, X., Shen, W., Zhang, Y., et al., "Novel active $LiFePO_4$ battery balancing method based on chargeable and dischargeable capacity," *Computers & Chemical Engineering*, 2017, 97: 27–35.

12

State of Health (SoH) Estimation of a Battery

12.1 Definition and Indices/Parameters of SoH

To study a subject, we should first define the basic concept of the studied subject. In this section, such concepts as battery degradation, battery life, and battery SoH will be differentiated and analyzed. The question about what indicators can be used to describe the battery degradation will also be discussed in this section.

12.1.1 Relationship Between Battery Degradation and Battery Life

The battery life can be seen as being conceptually similar to a human lifetime in two ways. First, strictly speaking, the exact duration of the battery is not known until the end-of-life span. Before this, its duration can only be predicted. Second, the mentioned battery life is only suitable for the individual battery. That is to say, the cycle lives of two different batteries with the same brand and the same model are different if they are used in different ways.

A battery life covers a cycle life and a calendar life. This section first defines these two life spans and then makes an analysis on the relationship between battery degradation and these lives.

Cycle life of a battery

Generally speaking, the cycle refers to the process of charging and discharging a battery for one time. The cycle life of the battery refers to the number of cycles during the period calculated from the time when a battery is manufactured to the time when it is no longer in use. The discussion is made as follows:

1) The cycle life of a battery is related not only to the manufacturing process of the battery but also to the operating condition of the battery in the whole life cycle (including the depth of each discharge, charge rate, discharge rate, and operating temperature).
2) The cycle life of the batteries operating under specific conditions is difficult to determine. For example, a battery installed in a hybrid vehicle may be subject to the charging and discharging many times within a minute, but such charging time or discharging time is very short. In this case, it is very difficult to determine the cycle life of the battery.
3) Definition of battery failure criteria. "What are the failure criteria for a battery?" This question is about the definition of battery failure. In a sense, the answer is not unique. In general, the battery failure criteria can be established in three ways: first, the battery has a safety problem, i.e. continuing to use the battery may cause an accident, which generally corresponds to the physical damage of the battery. Second, the actual capacity of the

Battery Management System and its Applications, First Edition. Xiaojun Tan, Andrea Vezzini, Yuqian Fan, Neeta Khare, You Xu, and Liangliang Wei.

battery is below a certain value, such as 60% or 80% of the rated capacity, in which 80% is the index commonly used in the industry. Third, the battery capacity is not low, but the performance is not enough to meet the load requirements. For example, a high current discharge is not available since the internal resistance of the battery is too large.

4) From item 3, the failure criterion of batteries in different applications is different. Therefore, many scholars have proposed that different failure criteria can be established for batteries in different fields, so that batteries can be utilized in a step-by-step manner. For the electric vehicle, the failure criteria can be set as: "the capacity is reduced to 80% of the rated value and the equivalent internal resistance is increased to twice the new battery." For the energy storage system, the failure standard criteria can be set as: "the capacity is reduced to 60% of the rated value and the equivalent internal resistance is increased to five times that of the new battery."

5) To compare the life span of batteries from two manufacturers, a unified evaluation standard is required. Therefore, a standardized test condition is often defined to make the battery life comparable. For example, according to the Chinese vehicle industry standard (QC/T 743-2006), the cycle condition is specified as: "charge the battery by using the constant current and constant voltage two-stage method and discharge the battery at a rate of 0.5 C; then the discharge depth should be 80% of the rated capacity of the battery." Additionally, the battery failure condition is defined as: "after evaluation, it should be considered failed in the case of being less than 80% of the rated capacity." Based on the above-mentioned unified evaluation standard, the horizontal comparison can be made between the life of the batteries from two manufacturers.

Calendar life of the battery

The calendar life of the battery refers to the number of days during the period calculated from the time when a battery is manufactured to the time when it is no longer in use. In the same way as the cycle life of the battery, the discussion about it is also required:

1) The calendar life of the battery is often affected by the cycle life of the battery. That is to say, the calendar life of the battery is related to whether and how the battery is used.

2) Therefore, someone separately studied the "storage life" of a battery while studying the battery. That is to say, after a battery is manufactured, the battery is not charged and discharged (or only charged and discharged for a few times), and then the battery is stored in a certain condition until the battery ages. The battery's calendar life was then evaluated at the moment when the battery was too aged to use.

3) This is similar to the definition of a cycle life. The storage life of a battery is not only related to the manufacturing process of the battery, but is also related to two other aspects: the criteria for battery failure and the storage conditions of the battery.

4) It takes a long time to study the battery's calendar life. For a battery manufactured at current levels, it will take a long time for such a battery to fail by leaving it unused at room temperature. Therefore, many manufacturers select batteries by the method of accelerated aging [1]. The battery is placed at a higher temperature to speed up the aging process. Then the life of the battery at a higher temperature is converted into the calendar life of the battery, but the converted calendar life is not the real calendar life of the battery. It is not accurate since the calculation is not based on the real battery aging process after all.

Relation and difference between battery life and battery degradation

The battery life and battery degradation are related and different. First, battery performance will gradually decline since the date of use, which is an irreversible process, so the more the battery deteriorates, the closer it gets to the end of its life. The remaining life of the battery can be predicted by evaluating the deterioration degree of the battery.

Second, "sudden death" may also occur on the battery, i.e. the battery will suddenly fail while it is being used due to inappropriate use of the battery. Even if the evaluation result of the battery capacity showed that the remaining capacity of the battery was still 90% of the rated capacity yesterday, the battery may fail and become unusable today. In our previous test, the battery was placed in a temperature box at 0°C and charged and discharged using a heavy current. In this test, some particles crystallized at low temperatures and pierced the diaphragm, which caused the capacity of the battery to suddenly decline below 10% of the rated capacity from 95% of the rated capacity. In this case, the battery basically failed.

Example: Would the capacity decline to about 80% if it had been used in this way for another year on the condition that the initial capacity of the battery declined to 90% in

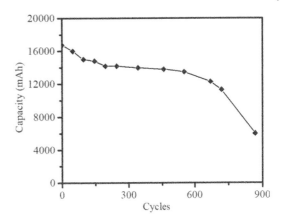

Figure 12.1 Accelerated aging of a battery.

the past year and it was used in this way for another year? This prediction is somewhat reasonable. Many experiments show that this approximate linear rule exists. Additionally, we find that when the battery capacity declines to some degree, heat from the battery is increased significantly due to the increase of internal resistance. Then, accelerated aging of the battery occurs, as shown in Figure 12.1.

We can now understand that "the battery degradation" and "the battery life" are related and different.

12.1.2 Relationship Between Battery Degradation and SoH of the Battery

If "battery management system" is considered as an emerging discipline, there are still many unclear and incomplete definitions for the concepts in this emerging discipline. For example, there is no unified definition for the important concept of "SoC." This results in many vague expressions in the research process and the project acceptance process. The "SoH" discussed in this book is also one of these "important," "common," and "vague" concepts. In this section, different definitions of SoH are briefly reviewed. The author's opinion is given to provide some references for the future theoretical development of this discipline.

Several main ideas about the concept of SoH at present

At present, there is still no unified definition for the concept of SoH. There are different opinions on it in the industry. However, in summary, it is understood mainly based on such aspects as capacity, internal resistance, peak power, and cycle times. The main ideas are shown as follows.

1. SoH is defined by the capacity or the discharging capacity.

Recent literature shows that the numerical definition of SoH, which has been adopted most frequently, corresponds to the attenuation of capacity. Then, the following equation is directly given in many sources:

$$\mathrm{SoH} = \frac{C_{\mathrm{aged}}}{C_{\mathrm{rated}}} \times 100\% \tag{12.1}$$

where C_{aged} represents the current battery capacity and C_{rated} represents the rated capacity of the battery. However, scholars have different ways of understanding the same equation. For example:

First, should the "capacity" in this equation be capacity or energy?

Second, is the "capacity" in the equation the effective capacity or the maximum capacity? The capacity discharged by the battery is different under different conditions (such as temperature and discharge rate). The capacity discharged by the battery is the effective capacity of the battery only under the specific condition. In general, the effective capacity of the battery is less than the maximum capacity.

Third, should the "rated" capacity or "actual" capacity act as the denominator? For a chemical product, the actual capacity of a battery is different from its nominal rated capacity.

Different researchers have given different answers to the above questions. For example, some have shown that SoH is defined by "the capacity discharged:"

$$SoH = \frac{Q_{max}(Aged)}{Q_{max}(New)} \times 100\% \tag{12.2}$$

where Q_{max} (Aged) represents the capacity discharged by the current battery and Q_{max} (New) represents the capacity discharged by the new battery.

However, there is no direct answer to the above questions, even in some of the literature. In the author's opinion, it is inappropriate to discuss this question without a unified definition, and it is also detrimental to the long-term development of this discipline.

2. SoH is defined by internal resistance

Much of the literature shows that the battery degradation is mainly reflected in the increase of ohmic resistance. At the same time, many scholars believe that the life of almost all types of battery is ended due to increased impedance and corresponding power loss. Therefore, "aging of the battery" occurs with the "increase in internal resistance." That is to say, the increase of internal resistance indicates the battery aging and also the reason for further degradation of the battery.

Therefore, some scholars define SoH by the internal resistance:

$$SoH = \frac{R_{EOL} - R}{R_{EOL} - R_{new}} \times 100\% \tag{12.3}$$

where R_{EOL} represents the internal resistance at the end of battery life, R_{new} represents the internal resistance of the new battery, and R represents the current internal resistance.

Of course, the above-mentioned definition is inaccurate since no consideration is given to how the internal resistance is measured, which is the most remarkable problem. As we all know, the internal resistance is a function related to SoC, temperature, and other independent variables. It is inappropriate to discuss the "SoH" without giving the definition of internal resistance.

3. SoH is defined by the number of remaining cycles in a battery.

In addition to capacity and internal resistance, some scholars also define SoH as the ratio of the number of remaining cycles available for the battery to the total life cycle of the battery. However, due to many uncertain factors in the use of the battery, it is impossible to predict the operating environment of the battery. Moreover, it is impossible to predict accurately the number of remaining cycles, so the operability is not good on the basis of such a definition.

The author's opinion and relationship between battery degradation and SoH

The above are the main ideas for the concept of SoH. It seems that many people tend to define SoH by using quantitative indicators, which the present author does not entirely agree with. In this author's opinion, the definition based on the quantitative indicators can be considered as the definition for the narrow SoH, while the definition of the generalized SoH should be based on the literal meaning of the concept itself. Since SoH stands for the state of health, the concept literally means the battery health, which should cover "battery failure" and "battery degradation," as shown in Figure 12.2.

1. SoH should include the warning vector of battery failure

In 2004, Gregory L. Plett proposed that the SoH of the battery is described in the form of a fault vector, i.e. $\bar{s} = [s_1, s_2, s_3, s_4, s_5, s_6, s_7, \ldots]$, where $s_k (k = 1,2,3,4,5,6,7,\ldots)$ represents the component of the vector, which may be equal to 0 or 1 [2]. The meaning of each component is as follows:

- s_1: Is the battery voltage too low or too high? (1 for "Yes," 0 for "No," and similarly thereafter)
- s_2: Is the current in the battery too high?
- s_3: Is the temperature of the battery too low or too high?
- s_4: Is the battery SoC above or below the rated range?
- s_5: Is the self-discharge rate of the battery out of the acceptable range?
- s_6: Is the capacity of the battery decline below the acceptable minimum?
- s_7: Is the internal resistance of the battery out of the upper limit of a certain range?
- ......

The author thinks that the state vector $\bar{s}$ as defined above is close to the literal meaning of the concept of "health" and should be taken into consideration when defining the concept of SoH.

2. SoH also covers the quantitative indicators after attenuation and aging, which are classified as "battery degradation."

After the failure warning vector of the battery has been defined, some quantifiable indicators can be added to describe the "health" state of the battery. These indicators become worse with the use of the battery, so we call them indicators of battery deterioration. In my opinion, "battery degradation" should be included in the concept of generalized SoH and should be included in the indictors used to describe gradual degradation of the battery with use. It should be quantitatively described. See the next section for the question about which indicator is more appropriate to describe the battery deterioration.

From Figure 12.2, the concept of SoH covers the concepts of battery failure and battery degradation, but the above two concepts are easy to express mathematically, which provides good operability.

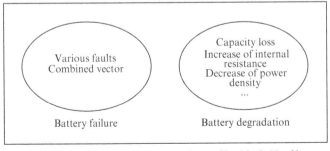

State of health (SoH) of battery

Figure 12.2   Broader SoH covers two parts.

12.1.3 Main Indicators to Describe Battery Degradation

Battery degradation is a gradual and complex process, but we still hope to find some quantifiable indicators to describe the battery degradation. There are two principles for selecting such indicators. First, the indicators are typical and can be used to reflect the degree of battery degradation. In other words, for the same type of battery A and B, x_A represents the degradation degree of battery A and x_B represents the degradation degree of battery B. In the case of $x_A > x_B$, we can judge that the degradation of battery A is more serious than that of battery B. Second, it is operable, that is, it can be obtained by a certain method without much cost (a lot of manpower, material resources, time, etc.). For example, as mentioned earlier, it is not operable to define the degradation by the number of remaining cycles of the battery.

Two indicators used to describe battery degradation

After a long period of experiment and analysis, the author believes that it is appropriate to use "capacity loss" and "DC internal resistance spectrum" as typical indicators to evaluate battery degradation.

1. Capacity loss

As for SoH in a narrow sense, the "current capacity of a battery" is a widely used evaluation indicator and is recognized as the best indicator to reflect the external characteristics caused by battery degradation. In both IEEE 1188-2005 standard and United States Advanced Battery Consortium (USABC), the "battery capacity" is also recommended as the parameter to measure battery degradation. In order to make a capacity loss indicator more comparable, we define the capacity loss C_{loss} as follows:

$$C_{loss} = (1 - C_t / C_{rated}) \times 100\% \tag{12.4}$$

where C_t represents the maximum capacity of the battery at a certain time t and C_{rated} represents the rated capacity of the battery at delivery. Taking the percentage as the final unit will make this indicator more intuitive and comparable.

In practice, there are three problems with the above definition.

First, as the current maximum capacity of the battery, C_t has a certain correlation with temperature, so we may define this capacity as the maximum charged capacity of the battery at 20°C.

Second, $C_t > C_{rated}$ often occurs on new batteries. In this case, the calculated C_{loss} will be negative. This is acceptable since we believe that the greater the number of C_{loss}, the higher the battery degradation will be. Therefore, a negative C_{loss} means a lower degree of battery degradation.

Third, in order to avoid a negative C_{loss}, the denominator of Equation (12.4) can also be set as the initial capacity $C_{initial}$ of the battery. For the high-power battery pack, the operability of such a definition is relatively poor since the C_{rated} of each battery from the same manufacturer and with same model is the same, but their $C_{initial}$ is different. If the denominator of Equation (12.4) is set as $C_{initial}$, this means that the capacity of each battery will be measured once before use. This will reduce the C_{loss} comparability between the two batteries. (In this case, the denominators in the C_{loss} of the two batteries are different.)

2. DC internal resistance spectrum

When the battery operates with load and without load, there is a difference in the voltage between the positive and negative anodes, which shows the characteristic of internal resistance. The formation cause and numerical value determination of internal resistance of the battery are relatively complex. It is difficult to obtain a definite value if the specific condition is not given.

The internal resistance of the battery is mainly related to the temperature, the SoC of the battery, and the battery degradation. A great number of studies have shown that the internal resistance can reflect the battery degradation if a battery is discharged at a constant current under the first two conditions.

As shown in Figure 12.3, the value of internal resistance can be determined by the following equation in practice if the direction of the discharge current is specified to be positive:

$$r(\text{SoC}) = \frac{\Delta U}{I} = \frac{E_B(\text{SoC}) - U_L}{I} \tag{12.5}$$

where I represents the operating current of the battery, EB represents the equilibrium potential of the battery, which is the function of the SoC, and UL represents the operating voltage measured between both the positive and negative anodes on the battery under load.

Figure 12.4 shows the curve of the relationship between the SoC and the charging and discharging internal resistance of a battery under a specific temperature and degradation degree.

According to Chapter 6, the internal resistance of the battery will increase with battery degradation. The equivalent internal resistance of the battery is an important parameter for external characteristics of the battery degradation. Additionally, it is also very operable when evaluating the degradation by acquiring the internal resistance. The internal resistance of the battery is mainly related to the temperature, SoC, and the battery degradation. The battery degradation can be reflected by the internal resistance of the battery under the condition that the first two conditions remain unchanged. In Chapter 6, the degradation rule for the equivalent internal resistance of the battery was analyzed in detail, so will not be repeated here.

Compared with the indicator of "capacity loss," the internal resistance of the battery is related to the SoC. The internal resistances under different SoC values are different. Moreover, according to Chapter 6, the degradation

Figure 12.3 Test the battery by the intermittent discharge at constant current.

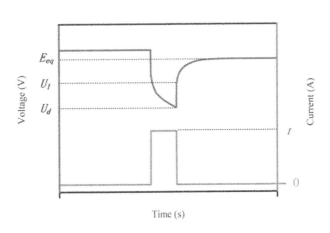

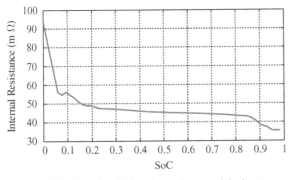

(a) Curve for discharging resistance of the battery

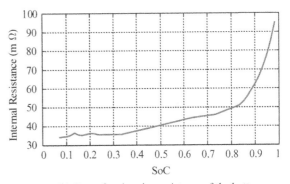

(b) Curve for charging resistance of the battery

Figure 12.4 Characteristic curves for charging and discharging resistance of a battery.

rules for the internal resistance under different SoC values are also different. Taking the charging resistance in Figure 6.22 as an example, the internal resistance in the range from 0.1 to 0.6 of the SoC, which is within the SoC platform area, is relatively close to those in the early and middle stages of the battery life cycle, but the internal resistance increases significantly in the later stage. Therefore, the "internal resistance value" should not be used alone to evaluate during judgment and analysis of battery degradation. The curve of the "internal resistance spectrum" should be used to fully reflect the battery degradation. When the "internal resistance spectrum" indicator is used in practice, it is available to check the tested "internal resistance spectrum" against the standard reference database to judge whether the battery internal resistance is abnormal or not. Then, it is available to qualitatively judge whether the battery is at the end of its life cycle. Based on the above analysis, the "internal resistance spectrum" is used as a qualitative indicator to measure battery degradation.

Inappropriate indicators to reflect battery degradation

When studying the degradation characteristics of batteries, the author also tried using some other indicators, but found that they were not significant enough to describe the battery degradation. Although they are not appropriate to describe the battery degradation, two of them are now shared with readers.

1. AC impedance spectrum

In some of the literature, the AC impedance of the battery is considered to be possibly related to the battery degradation [3]. A possible impedance spectrum is shown in Figure 12.5. However, the author thinks that it is not suitable to characterize the battery degradation by AC impedance spectrum. There are three main reasons.

First, an AC internal resistance spectrum is unstable and very different for different batteries with the same type under different temperatures and different SoC values.

Second, the shape of spectral lines is complex and it is very difficult to differentiate and analyze the parameters, so the fuzzy logic algorithm is recommended.

Third, AC impedance is tested by using EIS equipment, which is difficult to operate. It is impossible for online testing, so it is not very practical and is only suitable for laboratory research.

2. Electromotive force

At a constant temperature, the electromotive force (EMF) is in a fixed relationship with SoC, so SoC is often predicted based on the EMF. Earlier researchers had found a drop in the platform voltage of both lead-acid and

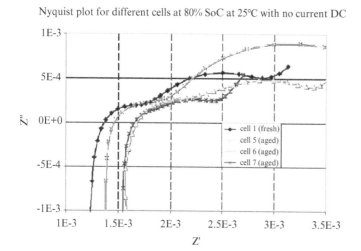

Nyquist plot for different cells at 80% SoC at 25°C with no current DC

Figure 12.5 AC impedance of the battery. (Note: four samples were tested at 25°C and the testing point of the SoC is 80%.)

Li-ion batteries after using for a period of time, so they suspected that the EMF of the battery would decline with the battery degradation [4]. However, for Li-ion batteries with such main components as lithium iron phosphate, the EMF of the battery is only related to the battery's material, production process, charging and discharging temperature, etc., and is irrelevant to the operating current and discharge depth.

Therefore, the EMF of a battery is also not suitable for battery degradation.

Is the available peak power suitable for reflecting the battery degradation?

The available peak power refers to the maximum power provided by the battery under specific conditions, which is equivalent to the maximum discharge (or charge) rate available to the battery. The

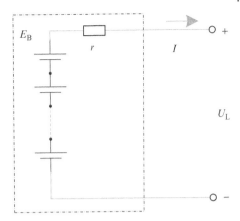

Figure 12.6 Equivalent circuit diagram used to analyze the peak power of the battery.

peak power will gradually decrease with the battery degradation. For example, a battery can discharge at the discharge rate of 5 C at 20°C when the battery has just been delivered from the factory. However, it can only discharge at the discharge rate of 3 C after a period of time due to the battery degradation. Therefore, this indicator can be used to reflect battery degradation.

However, after careful analysis, we know that we can calculate the peak power of the battery by the "capacity loss" and the "DC internal resistance spectrum" that we have known. This has been analyzed in Figure 12.6.

First, if the capacity loss of the battery is known, the exact SoC can be obtained. Second, knowing SoC, we can know the equilibrium potential E_B of the battery at this time. In addition, the internal resistance value r of the battery is obtained through the DC internal resistance spectrum. Third, we generally determine the maximum discharge current available according to the lower limit U_L^{min} of the terminal voltage U_L during discharge of battery, i.e.

$$I_{max} = \frac{E_B - U_L^{min}}{r} \tag{12.6}$$

Based on the known I_{max}, the available peak power can be calculated as follows:

$$P_{max} = (E_B - I_{max}\,r)I_{max} \tag{12.7}$$

Therefore, after taking into account "capacity loss" and "the DC internal resistance spectrum," we do not have to use "available peak power" as an indicator to reflect the battery degradation.

12.2 Modeling of Battery Degradation (Aging) and SoH Estimation

The data-driven method is used to analyze the complex relationship between the capacity loss and the DC equivalent internal resistance of the battery through a machine learning model [5]. Then the capacity loss can be predicted by DC equivalent internal resistance to evaluate the degree of battery degradation further. In this section, the battery degradation model will be established by using the support vector mechanism to build a reasonable training sample and optimize the algorithm based on the actual use condition. Finally, the battery degradation model will be compared with the degradation model, which is based on the artificial neural network.

12.2.1 Support Vector Regression

The support vector regression was originally used to solve the classification problem [6]. If the training set is $\{(x_1, y_1), \ldots, (x_n, y_n)\}$, $\mathbf{x}_i$ represents the N-dimensional input vector $\mathbf{x}_i \in R^n$, y_i represents the input $y_i \in \{1, -1\}$, and the goal of classification is to find a partition hyperplane with the largest interval $f(\mathbf{x})$, where $f(\mathbf{x})$ will be calculated according to the following equation:

$$f(x) = \mathbf{w}^T\mathbf{x} + \mathbf{b} \tag{12.8}$$

The interval refers to the sum of the distances between two heterogeneous support vectors and the hyperplane. From Equation (12.8), the interval can be expressed by $\gamma = 2 / \|\mathbf{w}\|$ and the maximum value of the interval γ can be obtained as follows:

$$\max_{w,b}(2 / \|\mathbf{w}\|) \tag{12.9}$$

In order to maximize the interval, it is necessary to minimize $\|\mathbf{w}\|^2$, so the formula (12.9) can be changed into

$$\min_{w,b}\left(\frac{1}{2}\|\mathbf{w}\|^2\right) \tag{12.10}$$

The constraint condition is $y_i(\mathbf{w}^T\mathbf{x}_i + b) \geq 1$, where $i = 1, 2, \ldots, n$.

The Lagrangian multiplier method is used to add $\alpha_i \geq 0$ into each constraint in Equation (12.10), by which we can solve the dual problem expressed by the Lagrangian function, as shown by

$$L(\mathbf{w}, b, \alpha) = \frac{1}{2}\|\mathbf{w}\|^2 - \sum_{i=1}^{l}\alpha_i\left[y_i(\mathbf{w}^T\mathbf{x} + b) - 1\right] \tag{12.11}$$

where $\alpha = (\alpha_1, \alpha_2, \ldots, \alpha_n)$ and l represents the number of support vectors. As the partial derivatives of w and b in $L(\mathbf{w}, b, \alpha)$ are assumed to be zero, we can solve the dual problem in Equation (12.10) by the obtained partial derivatives and Equation (12.11), as shown by

$$\min\left(\frac{1}{2}\sum_{i=1}^{n}\sum_{i=1}^{n}y_iy_j\alpha_i\alpha_j\mathbf{x}^T\mathbf{x} - \sum_{i=1}^{n}\alpha_i\right) \tag{12.12}$$

The constraint condition is $\sum_{i=1}^{n}\alpha_iy_i = 0$, where $\alpha_i \geq 0$ and $i = 1, 2, \ldots, n$.

Equation (12.13) shows the hyper-plane of the maximum interval finally obtained from Equation (12.12):

$$f(x) = \mathbf{w}^T\mathbf{x} + b = \sum_{i=1}^{n}\alpha_iy_i\mathbf{x}_i^T\mathbf{x} + b \tag{12.13}$$

Generally, the support vector regression machines are only used to solve problems about binary classification. If you want to solve the problem about multiple classifications, you need to improve the support vector regression machines. Generally, there are two methods: (1) modify the objective function, obtain the parameters about the multiple classifications, and combine them to solve the problem about multiple classifications for one time and (2) combine more classifiers to form multiple classifiers.

The capacity loss is obtained based on the DC equivalent internal resistance to further evaluate the degradation of the power battery, which should be included in the regression problems. However, the support vector machines can only be used for classification. Therefore, they need to be improved before they are used to solve the regression problems. The support vector regression machine is the algorithm for solving the regression problems using an optimized support vector machine.

The purpose of support vector regression is to find a regression plane to achieve the shortest distance between all data in the set and this plane. Since it is impossible to make all data on the regression plane, there will be a distance between each data and the regression plane. In order to prevent over-fitting, a tolerance value ε can be given for the distance between all data and the regression plane. If the distance between the data and the regression plane is within the tolerance value, the loss will be 0. If the distance between the data value y and the regression plane $f(x)$ is outside the tolerance value ε, the loss needs to be calculated. Therefore, according to Equation (12.10), we can get

$$\min_{w,b} \frac{1}{2}\|w\|^2 + C\sum_{i=1}^{n} \Phi_e(f(x_i) - y_i) \tag{12.14}$$

where C represents the penalty factor. The loss function Φ_ε is expressed by the following equation:

$$\Phi_\varepsilon(z) = \begin{cases} 0, & |z| \le \varepsilon \\ |z| - \varepsilon, & \text{other} \end{cases} \tag{12.15}$$

where z represents the distance between the data value y and the regression plane $f(x)$.

Due to the possibility of a fitting error, the relaxing factor ξ_i is introduced to obtain

$$\min_{w,b,\xi_i,\xi_i^*} \frac{1}{2}\|w\|^2 + C\sum_{i=1}^{n}(\xi_i + \xi_i^*) \tag{12.16}$$

The constraint conditions are shown as follows:

$$\begin{cases} f(x_i) - y_i \le \varepsilon + \xi_i \\ y_i - f(x_i) \le \varepsilon + \xi_i^* \\ \xi_i \ge 0, \ i = 1,2,3,\ldots,n \\ \xi_i^* \ge 0, \ i = 1,2,3,\ldots,n \end{cases} \tag{12.17}$$

Calculation made by Lagrange multiplication is shown as follows:

$$\begin{aligned} L\left(w,b,\alpha,\alpha^*,\xi,\xi^*,\mu,\mu^*\right) \\ &= \frac{1}{2}\|w\|^2 + C\sum_{i=1}^{n}\left(\xi_i + \xi_i^*\right) - \sum_{i=1}^{n}\mu_i\xi_i - \sum_{i=1}^{n}\mu_i^*\xi_i^* + \sum_{i=1}^{n}a_i\left(f(x_i) - y_i - \varepsilon - \xi_i\right) \\ &\quad + \sum_{i=1}^{n}a_i^*\left(y_i - f(x_i) - \varepsilon - \xi_i^*\right) \end{aligned} \tag{12.18}$$

where $a_i \ge 0, a_i^* \ge 0, \mu_i \ge 0$ and $\mu_i^* \ge 0$ are Lagrangian multipliers and C represents the penalty factor. The greater the penalty factor, the more serious the penalty becomes. As it goes to infinity, it becomes a linearly separable problem.

Take the partial derivative of Equation (12.18), set the partial derivative to be 0, and then substitute the solution of the equation obtained after taking the partial derivative into Equation (12.18) to solve the dual problem of support vector regression:

$$\max_{\alpha,\alpha^*} \left(\sum_{i=1}^{n} y_i(a_i^* - a_i) - \varepsilon(a_i^* + a_i) - \frac{1}{2}\sum_{i=1}^{n}\sum_{j=1}^{n}(a_i^* - a_i)(a_j^* - a_j)x_i^\mathrm{T}x_j \right) \tag{12.19}$$

where

$$\begin{cases} \sum_{i=1}^{n}\left(a_i^* - a_i\right) = 0 \\ 0 \le a_i, a_i^* \le C \end{cases} \tag{12.20}$$

The KKT condition is necessary for the equation

$$\begin{cases} a_i\left(f(\boldsymbol{x}_i) - y_i - \varepsilon - \xi_i\right) = 0 \\ a_i^*\left(y_i - f(\boldsymbol{x}_i) - \varepsilon - \xi_i^*\right) = 0 \\ \left(C - a_i\right)\xi_i = 0, \left(C - a_i^*\right)\xi_i^* = 0 \\ a_i a_i^* = 0 \\ \xi_i \xi_i^* = 0 \end{cases} \tag{12.21}$$

Finally, the obtained solution of the support vector regression machine is shown as follows:

$$f(\boldsymbol{x}) = \sum_{i=1}^{n}(a_i^* - a_i)\boldsymbol{x}_i^\mathrm{T}\boldsymbol{x} + b \tag{12.22}$$

The above equation derivation process shows the mathematical principle of a support vector regression machine.

For non-linear regression, the data can be mapped so that the data are linearly separable in the high-dimensional feature space. However, the mapping from the input space to the high-dimensional feature space will cause an explosive growth of the dimension, resulting in a large amount of computation. In order to avoid the heavy computation, the kernel function $K(\boldsymbol{x}, \boldsymbol{x}_i)$ can be constructed to enable the inner product operation of the feature space within the input space [7]. Then Equation (12.22) changes as follows:

$$f(\boldsymbol{x}) = \sum_{i=1}^{m}(a_i^* - a_i)K(\boldsymbol{x}, \boldsymbol{x}_i) + b \tag{12.23}$$

The kernel function is used to avoid the calculation of the feature space and directly solve the problem in the input space. Then it is not required to solve the non-linear mapping problem.

Two conditions are necessary for the kernel function $K(\boldsymbol{x}, \boldsymbol{x}_i)$.

1) It should be a symmetric function of positive real numbers.
2) The Mercer condition $\iint K(\boldsymbol{x}, \boldsymbol{x}_i)g(\boldsymbol{x})g(\boldsymbol{x}_i)d\boldsymbol{x}d\boldsymbol{x}_i$, $g \in L_2$ is necessary.

The kernel function mainly includes the linear kernel functions, the polynomial kernel functions, the Gaussian kernel functions, and the sigmoid kernel functions.

1. **Linear kernel function**

$$K(\boldsymbol{x},\boldsymbol{x}_i) = \boldsymbol{x} \cdot \boldsymbol{x}_i \tag{12.24}$$

Equation (12.24) is a linear kernel function. It is the simplest kernel function and is mainly used in linear separability. The dimension of the input space is the same as that of the feature space, i.e. it is not required to project into the high-dimensional feature space.

2. **Polynomial kernel function**

$$K(\boldsymbol{x},\boldsymbol{x}_i) = ((\boldsymbol{x} \cdot \boldsymbol{x}_i) + 1)^d \tag{12.25}$$

Equation (12.25) is a polynomial kernel function. It is a kind of non-standard kernel function and is suitable for the data after orthogonal normalization. It is able to map the low-dimensional input space into the high-dimensional feature space, but it has many parameters. If the order number of the function is relatively high, the computation will become very complex.

3. **Gaussian kernel function**

$$K(\boldsymbol{x},\boldsymbol{x}_i) = e^{(-\|x-x_i\|^2 / \delta^2)} \tag{12.26}$$

Equation (12.26) is a Gaussian kernel function. It is a kind of kernel function with strong locality, and is also able to map the low-dimensional input space into the high-dimensional feature space. Its performance is better. It has fewer parameters than the polynomial kernel function, so is widely used.

4. **Sigmoid kernel function**

$$K(\boldsymbol{x},\boldsymbol{x}_i) = \tan h(\varepsilon < \boldsymbol{x}h\boldsymbol{x}_i > +\theta) \tag{12.27}$$

Equation (12.27) is the sigmoid kernel function sourced from the neural network. It is s-shaped and is often used as an "activation function." A multilayer neural network is then achieved.

In machine learning, there is no mature criterion about how to select the kernel function. Some experiments and studies show that the Gaussian kernel function has a better performance and is widely used, since the Gaussian kernel function has a simple form, few parameters, and good smoothness, which can be used for most regression calculations. Therefore, the Gaussian kernel function is selected as the kernel function used in support of vector regression.

12.2.2 Battery Degradation Model Based on a Support Vector Regression Machine

In order to find an appropriate method for evaluating the battery degradation based on DC equivalent internal resistance, it is required to first acquire the training data, i.e. some experiments need to be made in the early stage in order to obtain experimental data.

The laboratory selected three LiFePO$_4$ battery samples, which were respectively numbered A001, A002, and A003 and all were provided with a nominal capacity of 15 Ah. Before the experiment, a capacity evaluation was made to measure their real capacity. Table 12.1 shows the nominal capacity and the actual capacity of A001, A002, and A003. The capacity of the three battery samples is very consistent, so subsequent experiments can be made.

Table 12.1 Sample battery capacity.

Battery No.	A001	A002	A003
Nominal capacity (Ah)	15.00	15.00	15.00
Actual capacity (Ah)	14.84	15.06	14.85

Table 12.2 Capacity loss of sample batteries.

Battery No.	A001	A002	A003
Capacity loss (%)	20.04	20.26	20.14

Next, an accelerated aging experiment will be made on the battery, i.e. the cyclic test will be made. In order to keep the temperature constant during the experiment, an incubator is needed, where the temperature set is 40°C. If the experimental temperature is set too low, the experiment duration will be greatly extended. If the experimental temperature is set too high, the battery degradation speed will be too fast, which will then cause fewer experimental data to be collected. Therefore, after full consideration, 40°C is selected as the experimental temperature.

If the evaluation is required in the cyclic test, including capacity evaluation and DC equivalent internal resistance evaluation, the incubator temperature should be set to 20°C during the evaluation. During a capacity evaluation, the cyclic test can be ended during the capacity evaluation if the measured capacity loss has reached 20% or $\varnothing_R = 0$.

Table 12.2 shows the capacity loss of the three battery samples at the end of the cyclic test experiment.

Establishment of training samples

A cyclic test is made on the three battery samples A001, A002, and A003 until the capacity loss reaches 20% or $\varnothing_R = 0$. The experimental data analysis shows that the three battery samples are effectively evaluated for 37 times, 41 times, and 39 times respectively. Figures 12.7, 12.8, 12.9, and 12.10 show the data of DC equivalent internal resistance and capacity loss, which are obtained from the evaluation. In order to clearly show that the DC equivalent internal resistance changes during degradation of the battery, a DC equivalent internal resistance line of a DC equivalent internal resistance spectrum is drawn for every five test evaluations. This can avoid lines that are too dense.

The DC equivalent internal resistance value corresponding to the range of SoC less than 0.05 and greater than 0.90 should be avoided during selection of the DC equivalent internal resistance value. Then the equivalent DC internal resistance data and capacity loss data of A001 and A002 are used as the training samples for the support vector regression and the experimental data of A003 is used as the prediction samples to discuss the establishment and normalization of the training samples (see Figures 12.7 to 12.12).

The prediction effect is closely related to the input of the support vector regression model. If the input of the support vector regression model is the effective data feature of the

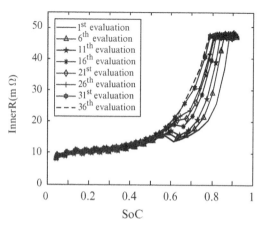

Figure 12.7 DC equivalent internal resistance spectrum of A001.

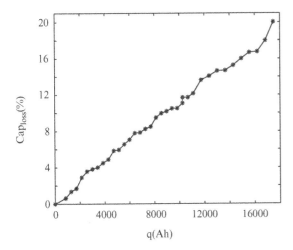

Figure 12.8 Capacity loss of A001.

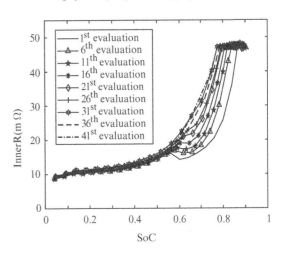

Figure 12.9 DC equivalent internal resistance spectrum of A002.

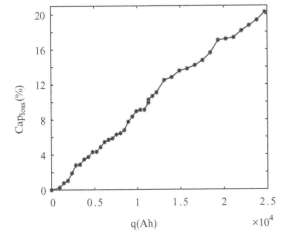

Figure 12.10 Capacity loss of A002.

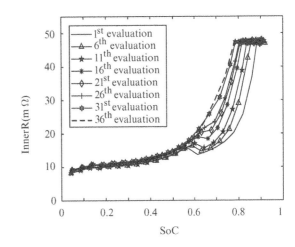

Figure 12.11 DC equivalent internal resistance spectrum of A003.

sample, then the prediction effect will be greatly improved. To discuss how to set up training samples to effectively evaluate battery deterioration, seven schemes for obtaining the training samples are now given.

1) Scheme 1: In the DC equivalent internal resistance spectra of A001 and A002, select all DC equivalent internal resistance data corresponding to SoC = 0.6 as the input of the model and use their capacity loss as the output of the model, i.e. the training set is $\left\{\left(r_{0.6_1}, \mathrm{Cap}_{loss_1}\right), \left(r_{0.6_2}, \mathrm{Cap}_{loss_2}\right), \ldots, \left(r_{0.6_n}, \mathrm{Cap}_{loss_n}\right)\right\}$.

2) Scheme 2: In the DC equivalent internal resistance spectra of A001 and A002, select all DC equivalent internal resistance data corresponding to SoC = 0.7 as the input of the model and use their capacity loss as the output of the model, i.e. the training set is $\left\{\left(r_{0.7_1}, \mathrm{Cap}_{loss_1}\right), \left(r_{0.7_2}, \mathrm{Cap}_{loss_2}\right), \ldots, \left(r_{0.7_n}, \mathrm{Cap}_{loss_n}\right)\right\}$.

3) Scheme 3: In the DC equivalent internal resistance spectra of A001 and A002, select all DC equivalent internal resistance data corresponding to SoC = 0.8 as the input of the model and use their capacity loss as the output of the model, i.e. the training set is $\left\{\left(r_{0.8_1}, \mathrm{Cap}_{loss_1}\right), \left(r_{0.8_2}, \mathrm{Cap}_{loss_2}\right), \ldots, \left(r_{0.8_n}, \mathrm{Cap}_{loss_n}\right)\right\}$.

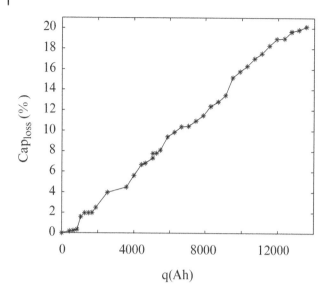

Figure 12.12 Capacity loss of A003.

4) Scheme 4: In the DC equivalent internal resistance spectra of A001 and A002, select all DC equivalent internal resistance data corresponding to SoC = 0.6 and 0.7 as the input of the model and use their capacity loss as the output of the model, i.e. the training set is $\left\{ \left(\overrightarrow{R_1}, \mathrm{Cap}_{loss_1} \right), \left(\overrightarrow{R_2}, \mathrm{Cap}_{loss_2} \right), \ldots, \left(\overrightarrow{R_n}, \mathrm{Cap}_{loss_n} \right) \right\}$, where $\overrightarrow{R_l} = (r_{0.6_l}, r_{0.7_l})$.

5) Scheme 5: In the DC equivalent internal resistance spectra of A001 and A002, select all DC equivalent internal resistance data corresponding to SoC = 0.6 and 0.8 as the input of the model and use their capacity loss as the output of the model, i.e. the training set is $\left\{ \left(\overrightarrow{R_1}, \mathrm{Cap}_{loss_1} \right), \left(\overrightarrow{R_2}, \mathrm{Cap}_{loss_2} \right), \ldots, \left(\overrightarrow{R_n}, \mathrm{Cap}_{loss_n} \right) \right\}$, where $\overrightarrow{R_l} = (r_{0.6_l}, r_{0.8_l})$.

6) Scheme 6: In the DC equivalent internal resistance spectra of A001 and A002, select all DC equivalent internal resistance data corresponding to SoC = 0.7 and 0.8 as the input of the model and use their capacity loss as the output of the model, i.e. the training set is $\left\{ \left(\overrightarrow{R_1}, \mathrm{Cap}_{loss_1} \right), \left(\overrightarrow{R_2}, \mathrm{Cap}_{loss_2} \right), \ldots, \left(\overrightarrow{R_n}, \mathrm{Cap}_{loss_n} \right) \right\}$, where $\overrightarrow{R_l} = (r_{0.7}, r_{0.8_l})$.

7) Scheme 7: In the DC equivalent internal resistance spectra of A001 and A002, select all DC equivalent internal resistance data corresponding to SoC = 0.6, 0.7 and 0.8 as the input of the model, use their capacity loss as the output of the model, i.e. the training set is $\left\{ \left(\overrightarrow{R_1}, \mathrm{Cap}_{loss_1} \right), \left(\overrightarrow{R_2}, \mathrm{Cap}_{loss_2} \right), \ldots, \left(\overrightarrow{R_n}, \mathrm{Cap}_{loss_n} \right) \right\}$, where $\overrightarrow{R_l} = (r_{0.6_l}, r_{0.7_l}, r_{0.8_l})$.

Before training the support vector regression model, we need to normalize the training samples and select the model parameters.

Normalization of training samples and selection of model parameters

The data normalization is mainly based on two considerations. First, the input of training samples is usually not one-dimensional. It may be multidimensional. The evaluation indicators of these different dimensions may have different dimensions and dimensional units. In order to make the indicators comparable, normalization is required to make the normalized indicators in the same order of magnitude. Then the comprehensively comparative evaluation can be made. Second, after normalization, it is possible to speed up solving the optimal solution of the gradient descent.

There are two common normalization methods. One is standard normalization, as shown in the following equation:

$$y = \frac{x - \mu}{\sigma} \tag{12.28}$$

where x represents the data to be normalized, y represents the data after normalization, μ represents the average value of the sample data, and σ represents the mean variance of the sample data. This normalization is mainly suitable for normally distributed samples.

Another method is the normalization of linear functions, as shown in the following equation:

$$y = \frac{x - x_{min}}{x_{max} - x_{min}} \tag{12.29}$$

where x represents the data to be normalized, y represents the data after normalization, X_{max} represents the maximum value in the training data, and X_{min} represents the minimum value in the training data. After such normalization, the data can be normalized to the range of [0, 1] and the data can also be normalized to the range of [–1, 1] by modifying Equation (12.29).

In this experiment, the following equation will be used to normalize the training data within the range of [–1, 1]:

$$y = (y_{max} - y_{min}) \times \frac{x - x_{min}}{x_{max} - x_{min}} + y_{min} \tag{12.30}$$

where y_{max} represents 1 and y_{min} represents –1.

After data pre-processing, the parameters of the support vector regression model should be determined. In this experiment, the method of 50% cross-validation is used to select the parameters.

In the method of 50% cross-validation, first divide the training data into five groups, develop a model by taking turns to use any four groups as a training set, and then use the remaining set of data as a validation set to test the model obtained after training. In this way, the five trained models are finally obtained and the accuracy of the model is measured by calculating the average accuracy of the five trained models.

The e-SVR model is selected as the training model. From the above, the Gaussian kernel function is used as the kernel function. The loss function p of the e-SVR model is set to 0.01. The optimal values of both the remaining penalty factor C and *gamma* in the Gaussian kernel function are selected by the method of 50% cross-validation.

The model is trained after normalizing the sample data and determining the model parameters.

Model training

The seven training sample schemes mentioned in Section 12.2.2 are processed by data normalization respectively and are then input into the support vector regression model for training. The DC equivalent internal resistance and capacity loss data of A003 are used as the prediction data of the model. The influence of the input dimension of the model on the prediction effect is researched, based on the comparison and analysis between the original measurement data of A003 and the predicted one.

The algorithm is optimized to meet the requirements of actual use. First, when the battery is used for the first time, an online evaluation of battery degradation should be made to initialize Cap_{loss} to be 0, which can avoid a negative Cap_{loss}. Second, contrast Cap_{loss_i} and $Cap_{loss_{i-1}}$ while predicting Cap_{loss_i} for $i(i = 2,3,4,...)$. If Cap_{loss_i} is less than $Cap_{loss_{i-1}}$, then $Cap_{loss_i} = Cap_{loss_{i-1}}$, which will avoid the user's confusion about why the capacity loss decreases with the use of the battery. Users easily get confused during frequent degradation evaluation of batteries.

The evaluation results for the seven training sample schemes are shown in Figure 12.13. According to Figure 12.13, the prediction effect is the best in Figure 12.13(d) and (g), while the prediction results obtained by other schemes mostly show an equal capacity loss, as shown in Figure 12.13(a), (c), and (e), or the predicted results are

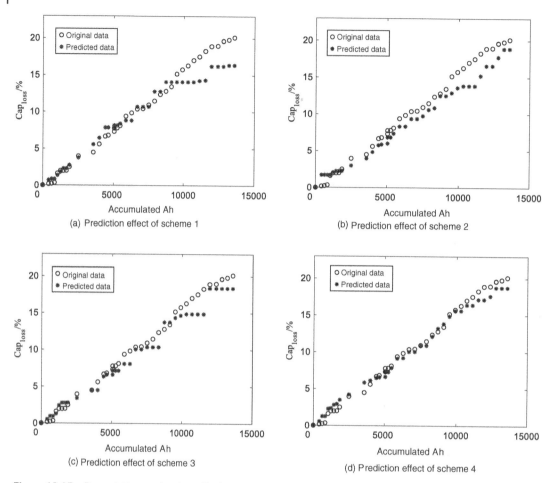

Figure 12.13 Degradation evaluation effect.

Table 12.3 Influence of different input dimensions on the prediction results.

Scheme name / Error type	1/I	2/II	3/III	4/IV	5/V	6/VI	7/VII
The maximum error	3.969	3.695	3.421	1.790	3.318	4.497	1.691
Standard error	1.758	1.562	1.190	0.773	1.105	1.986	0.595

significantly different from the actual measurement results, as shown in Figure 12.13(b) and (f). See Table 12.3 for an analysis of the influence of different input dimensions on the prediction results through the maximum error and standard error RMSE.

The error analysis in Table 12.3 shows that, in general, the prediction effect based on the model input of two dimensions is better than that based on the model input of one dimension, and the prediction effect based on the model input of three dimensions is better than that based on the model input of two dimensions. Additionally, we found that the prediction effect of scheme 7 is better than that of scheme 4, but the prediction effect of scheme 7

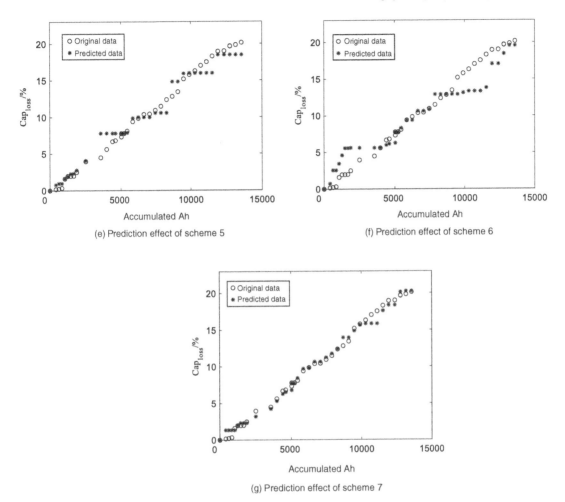

(e) Prediction effect of scheme 5

(f) Prediction effect of scheme 6

(g) Prediction effect of scheme 7

Figure 12.13 Continued

is not much better than that of scheme 4. Therefore, the sample of three dimensions is enough for the model input. In this experiment, the DC equivalent internal resistance data corresponding to SoC = 0.6, SoC = 0.7, and SoC = 0.8 in A001 and A002 samples are selected as the training model to input. In the obtained prediction effect of sample A003, the maximum error is 1.691% and the standard error is 0.595%.

In addition, with respect to the support vector regression model, the training sample and the prediction sample are rebuilt to test the feasibility of evaluating the capacity loss by using the DC equivalent internal resistance.

1) Scheme a: The sample data of A002 and A003 are used for the training model and the sample data of A001 are used as the prediction data.
2) Scheme b: The sample data of A001 and A003 are used for the training model and the sample data of A002 are used as the prediction data.

After rebuilding the training samples of the support vector regression model, their predicted results are shown in Figures 12.14 and 12.15.

Figures 12.14 and 12.15 show that the prediction effect is relatively ideal by training the model through the samples of schemes a and b. See Table 12.4 for the error analysis of the prediction results based on these two schemes.

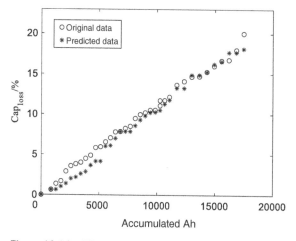

Figure 12.14 Effect of degradation evaluation in scheme a.

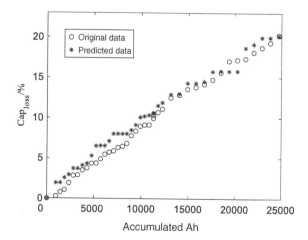

Figure 12.15 Effect of degradation evaluation in scheme b.

Table 12.4 Error analysis of the prediction results.

Scheme name Error type	a	b
The maximum error	1.887	2.143
Standard error RMSE (%)	0.909	1.122

These experimental results show that the predicted results are all relatively accurate through the alternating tests of the three battery groups, A001, A002, and A003. The maximum error in the estimation of capacity loss is not more than 2.500%. This also proves that it is feasible to estimate the capacity loss according to the support vector regression model and the DC equivalent internal resistance.

After the capacity loss is obtained through the DC equivalent internal resistance, the degradation degree of the battery can be estimated according to the DC equivalent internal resistance and the capacity loss. If $Cap_{loss} \geq 20\%$ or $\varnothing_R = 0$, the battery can be replaced.

12.2.3 Steps and Procedures for Evaluating Battery Degradation

The sample selection and analysis and the model parameter selection have been described above. A battery degradation evaluation model based on DC equivalent internal resistance is built through the training of sample data. Sample data should be collected before the samples are processed.

In order to obtain the sample data of the degradation model, the battery first needs to undergo a cyclic test. The capacity loss data and DC equivalent internal resistance data in different use stages are then obtained by means of a capacity evaluation and a DC equivalent internal resistance evaluation. If you find that the capacity loss is more than 20% or $\Phi_R = 0$, you can stop collecting the battery degradation data. It is possible to process the acquired data after stopping the data collection.

From the above section, we know that the prediction effect is best when three-dimensional input is used for the support vector regression model. The specific steps for modeling by using the support vector regression algorithm are as follows:

1) Determine the training set of the model as $\left\{\left(\overrightarrow{\boldsymbol{R}_1}, \mathrm{Cap}_{loss_1}\right), \left(\overrightarrow{\boldsymbol{R}_2}, \mathrm{Cap}_{loss_2}\right), \ldots, \left(\overrightarrow{\boldsymbol{R}_n}, \mathrm{Cap}_{loss_n}\right)\right\}$, where $\overrightarrow{\boldsymbol{R}_l} = (r_{0.6_l}, r_{0.7_l}, r_{0.8_l})$ represents the three-dimensional input vector of the model. Vector $\overrightarrow{\boldsymbol{R}_l} = (r_{0.6_l}, r_{0.7_l}, r_{0.8_l})$ means that the evaluation has been made based on the DC equivalent internal resistance. The SoC represents the DC equivalent resistance points corresponding to 0.7 and 0.8 respectively. Cap_{loss_i} represents the output of the model, which is the capacity loss value obtained by the evaluation made based on the DC equivalent internal resistance for i time.

2) The input and output of the sample set are normalized in order to make the input and the output within the range of [−1, 1].

3) Obtain the optimal hyper-parameter of the model by 50% cross-validation and determine the loss function of the model.

4) Process the training set by the Gaussian kernel function so that the support vector regression machine can optimize the non-linear regression [8].

5) Train the degradation relation model of the DC equivalent internal resistance and capacity loss according to the normalized training set and the support vector algorithm with a set parameter, and store the model.

6) Determine the model testing set, the form of which is the same as that of the training set shown in step 1.

7) Verify the accuracy of the established model by the prediction set.

When the prediction set is used to verify the degradation relation model, the predicted capacity loss value will be obtained. The degradation is evaluated by combining the DC equivalent internal resistance value obtained from the battery in different use stages with the corresponding predicted capacity loss value. The specific flow chart is shown in Figure 12.16.

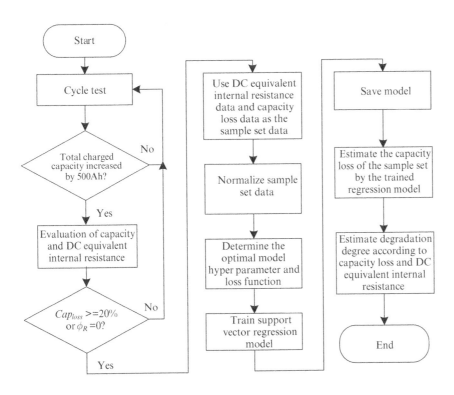

Figure 12.16 Degradation evaluation method for a Li-ion battery based on a support vector machine.

12.3 Battery Degradation Diagnosis for EVs

12.3.1 Offline Degradation Diagnosis of the Power Battery

The most direct diagnosis method for the battery of the electric vehicle means that the diagnostic evaluation is completed for each battery by relevant equipment (e.g. a battery tester) and corresponding test design. Because this test can isolate all instabilities resulting from vehicle operation in an online diagnosis, such as the hysteretic voltage effect, rebound voltage effect, or electromagnetic interference of various devices and temperature effects in the process of vehicle operation, the degradation diagnosis can be more directly and more accurately performed. The off-line degradation diagnosis method is shown in Figure 12.17.

During the offline degradation diagnosis, the EV battery pack is connected to the hardware platform, and the diagnostic evaluation is performed for the battery pack through the hardware platform, including capacity evaluation, charging resistance, and discharge resistance testing. The obtained battery data are stored in the corresponding historical database and then the evaluation data are processed. The data processing results are compared with the corresponding indicators of the reference database. After further analysis, the diagnosis conclusion is fed back to the user, who decides which corresponding battery maintenance strategy to use; for example, if the diagnosis result shows that the capacity of some cells in the battery pack is too low, a battery replacement should be considered. The reference database is a database established after a large number of tests have been carried out on the battery. These tests cover a variety of battery conditions and provide a reference value for the battery performance diagnosis index under each condition.

Two degradation diagnosis indexes are analyzed, mainly during testing data processing.

Offline analysis of the "capacity loss"

The data processing and analysis process of the "capacity loss" indexes are shown in Figure 12.18, including such steps as generating capacity evaluation records, drawing a capacity-accumulated Ah curve and capacity loss-accumulated Ah curve, and comparing the diagnosed "capacity loss" index with relevant standards.

The battery number, the evaluation times, the accumulated Ah, the evaluation capacity, the capacity loss, the evaluation date, and other information are recorded in the capacity evaluation record sheet.

The capacity loss-accumulated Ah curve can be drawn according to the capacity evaluation record table, as shown in Figure 12.19. As shown in the figure, the battery capacity loss curve is obtained for the battery with a 20 Ah rated capacity after six capacity evaluations. The capacity loss obtained by each capacity evaluation is drawn in order to observe the degradation trend of the battery.

The "capacity loss" index is analyzed as follows. According to the data processing results, the current capacity loss C_{loss} of the battery is calculated by Equation (12.4) and the calculated C_{loss} is compared with the "C_{loss} threshold

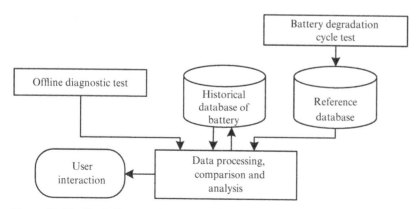

Figure 12.17 Block diagram of the offline degradation diagnosis method.

a." If the C_{loss} is more than *a*, this means that the capacity loss exceeds the pre-warning value and the battery capacity therefore cannot meet the operation requirement of the electric vehicle. Conversely, the capacity loss is within the pre-warning value and the battery capacity meets the operation requirement. According to the professional standard of China QC/T 743-2006, the battery capacity failure condition is "the battery capacity is less than 80% of the rated capacity." According to the definition of C_{loss}, it can be deduced reversely that *a* is 20, that is, when C_{loss} is greater than 20%, the battery capacity failure occurs.

Offline analysis of the "internal resistance spectrum" index

The data processing and analysis process of the "internal resistance spectrum" index is shown in Figure 12.20. The charge and discharge internal resistance curves can be obtained after each charge and discharge internal resistance test. These internal resistance curves can be formed into the charge and discharge "internal resistance spectrum" of the battery. The larger the test number, the more the discharged accumulated Ah is and the more serious the battery aging will be.

The "internal resistance spectrum" is analyzed as follows: the current charge and discharge resistance curve of the battery are obtained according to the data processing results and several points are selected from 20–80% SoC. For example, an internal resistance point is selected every 10% SoC and the internal resistance deviation coefficient γ is calculated at these points, the definition of which is as follows:

$$\gamma = \max\left|\frac{R_{test}(\text{SoC}) - R_{database}(\text{SoC})}{R_{database}(\text{SoC})}\right| \times 100\% \qquad (12.31)$$

where R_{test} (SoC) is the internal resistance of the internal resistance point selected from the tested internal resistance curve and $R_{database}$ (SoC) is the empirical value of the internal resistance provided by the reference database.

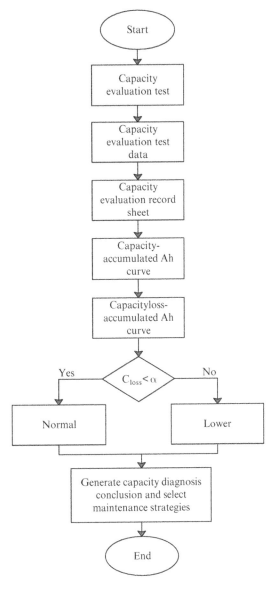

Figure 12.18 Data processing and analysis process of "capacity loss" indexes.

The empirical value selection method is now detailed. First, compare γ with the internal resistance pre-warning threshold β, and when γ is greater than β, this indicates that the current internal resistance of the battery is abnormal. Conversely, the internal resistance of the battery is normal. The internal resistance pre-warning threshold β is the empirical value selected from the battery reference database.

The whole battery life cycle is divided into three stages: an earlier stage, a medium stage, and a later stage. In the platform area (about 20–80% of SoC), the internal resistance of the battery is very close in the earlier and medium stages, but rapidly rises in the later stage. In the historical data of the battery, the empirical value of the internal resistance of the battery during earlier and medium stages is expressed by $R_{database}$ (SoC), which is calculated by

$$R_{database}(\text{SoC}) = \frac{\sum_{i=1}^{N} R_i(\text{Soc})}{N} \qquad (12.32)$$

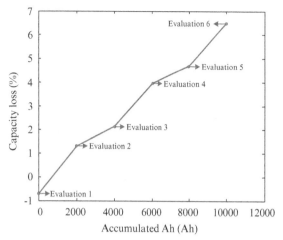

Figure 12.19 Capacity loss-accumulated Ah curve.

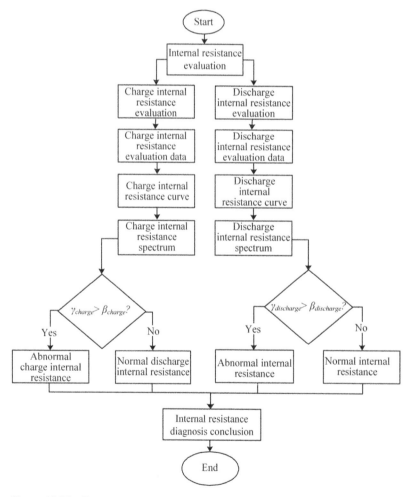

Figure 12.20 Data processing and analysis process of the "internal resistance spectrum" index.

where R_i (SoC) is the interact resistance at an SoC in the internal resistance platform area during evaluation for the i time stored in the historical database and N is the total evaluation times during earlier and medium stages.

The internal resistance pre-warning threshold β is calculated by the following equation:

$$\beta = \frac{\left| \dfrac{\displaystyle\sum_{i=N+1}^{M} R_i(SoC)}{M-N} - R_{database}(SoC) \right|}{R_{database}(SoC)} \times 100\% \tag{12.33}$$

where M is the total evaluation times during the whole battery life cycle stored in the historical database and R_i (SoC) and $R_{database}$ (SoC) have the same definition as in Equations (12.32 and 12.33).

Finally, the off-line diagnosis conclusion of the battery can be obtained by the results of the capacity loss analysis and the internal resistance spectrum analysis. The diagnosis conclusion is fed back to the user, who will decide which battery maintenance strategy is used.

12.3.2 Online Degradation Diagnosis of the Power Battery

Compared with the offline diagnosis method, the online diagnosis method is performed during the use of an electric vehicle and the test time for the degradation diagnosis is saved. During the online diagnosis, additional test equipment is not required and the degradation diagnosis can be completed only by the battery management system installed in the electric vehicle.

Degradation diagnosis based on rebound voltage

The following hardware are required for the online diagnosis: current sensor, voltage sensor, and data storage unit. The function of each hardware is shown in Figure 12.21.

Generally, the battery management system is installed in the electric vehicle. Since the battery management system is equipped with the above hardware functions, in practice additional hardware is not required during online diagnosis. The current sensor can measure the current of the battery in real time and store the current value in the data storage unit. The voltage sensor can measure the voltage value of each cell in real time and store the voltage value in the data storage unit. The historical data of each cell in the battery pack collected by the current sensor and the voltage sensor are stored in the data storage unit. When the online diagnosis time is triggered, the corresponding battery data can be called through the data storage unit for diagnosis and analysis.

This section details the basic principles of online diagnosis by the flow shown in Figure 12.22, including, first, the mathematical expression of the internal impedance module of the battery model; second, the equilibrium potential prediction; third, determination of the diagnostic time; and fourth, an online estimation of the "capacity loss" and "internal resistance spectrum."

Figure 12.21 Online diagnosis hardware and their functions.

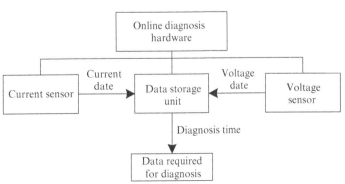

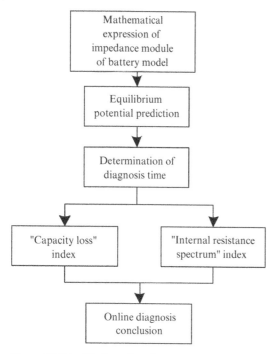

Figure 12.22 Whole online diagnosis flow.

1. Mathematical expression of the internal impedance module of a battery model

The equivalent resistance and voltage rebound of the battery are the over-potential characteristics of the battery. The voltage rebound curve is shown in Figure 12.23. In order to describe this characteristic, the third-order RC network model described in Section 7.4 is used as the circuit model of the battery. The internal impedance module of the battery model is shown in Figure 7.13 and the mathematical expression is shown in Equation (7.6).

2. Equilibrium potential prediction

According to the battery characteristics study provided in Chapter 6, the equilibrium potential refers to the steady-state potential of the battery after charging or discharging. The equilibrium potential is equal to the voltage after enough voltage rebounding time during idling without charging or discharging. This voltage value is expressed as OCV_t in Equation (7.6).

In order to obtain the equilibrium potential, the measured voltage rebound data can be fitted by Equation (7.6) to solve the parameter values in the equation, in order to predict the equilibrium potential through the limited voltage rebound data. As shown in Figure 12.24, the section OA of the curve shows the voltage data of the battery measured t_A seconds before the voltage rebound. The rebound data after t_A (the AB section of the curve) can be predicted by fitting these measured data and Equation (7.6) to obtain the equilibrium potential (the voltage at the point B) of the battery.

3. Termination of diagnosis time

The above equilibrium potential prediction method is based on the measured voltage rebound data. The volume of the measured data undoubtedly affects the prediction accuracy. When the measured voltage rebound data are large enough, the actual voltage rebound time is long enough and the voltage is very close to the equilibrium potential. The more accurate equilibrium potential prediction is achieved by the equilibrium potential prediction method, described in this chapter according to a small number of voltage rebound data.

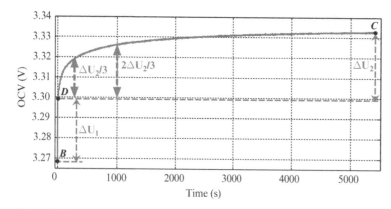

Figure 12.23 Voltage rebound curve.

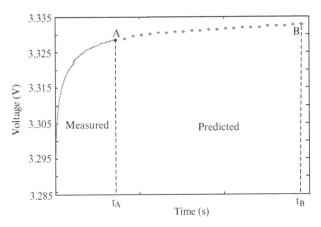

Figure 12.24 Schematic diagram of an equilibrium potential prediction.

In order to investigate thoroughly the data volume required for the above prediction method and the impact of different data volumes on the prediction accuracy, according to the battery charge and discharge resistance test procedure designed in Chapter 6, the batteries of two different brands are selected and respectively tested at different temperatures. The voltage rebound data during different SoCs is collected at a collection frequency of 1s/ time. Tables 12.5 and 12.6 list some prediction results obtained by the above equilibrium potential prediction method.

Tables 12.5 and 12.6 show the equilibrium potential results predicated by the voltage rebound data of the first 30 s and first 60 s. The maximum prediction deviation is 7 mV when the data volume is 30 s and less than 3 mV when the data volume is 60 s. The voltage rebound data of the above two batteries were collected in the laboratory, so the working environment of the batteries was relatively stable with little interference. If data collection is carried out in the electric vehicle, the working environment of the battery is relatively harsh. Therefore, in order to ensure prediction accuracy, when the battery management system is used to collect voltage rebound data in the electric vehicle, the data volume should be at least greater than 60 s.

In conclusion, when the battery is idle for more than 60 s without charge and discharge, a more accurate equilibrium potential prediction can be achieved using the above equilibrium potential prediction method. According to the actual operation condition of the electric vehicle, the diagnosis time may be deemed as the time

Table 12.5 Equilibrium potential prediction result of a brand A battery under different SoCs at 20 °C.

SoC (%)	Data volume for fitting (s)	Predicted value (mV)	Measured value (mV)	Error (mV)
75.62	30	3343	3338	5
	60	3341	3338	3
54.14	30	3309	3303	6
	60	3305	3303	2
36.07	30	3292	3297	−5
	60	3300	3297	3
19.60	30	3268	3261	7
	60	3258	3261	−3

Table 12.6 Equilibrium potential prediction result of a brand B battery under different SoCs at 40 °C.

SoC (%)	Data volume for fitting (s)	Predicted value (mV)	Measured value (mV)	Error (mV)
80.73	30	3326	3333	−7
	60	3334	3333	1
61.45	30	3293	3297	−4
	60	3298	3297	1
42.18	30	3299	3294	5
	60	3294	3294	0
22.90	30	3235	3238	−3
	60	3239	3238	1

when the battery is idle for more than 60 s without charge and discharge. The diagnosis time may appear during waiting for a traffic light, picking up passengers, or is provided by setting a charge timer on the charger when charging the EV battery, which has no major impact on the charging time of the electric vehicle and does not cause damage to the battery. After the battery pack is changed from the charging state or the discharging state to the idle state, the weak current components, such as the air conditioner, display panel, etc., are still operating, while the battery pack still supplies about 2 A of current. The important symbol of this process is the rapid reduction of the current of the battery pack from a larger value to a very small value; for example, reduction of the current from dozens of amperes upon normal driving to about 2 A upon idling. In practical applications, the BMS of the electric vehicle records the current value of the battery pack at every moment in the process of driving. When it detects that the symbol appears and lasts for more than 60 seconds, the BMS determines that it is the online diagnosis time.

4. "Capacity loss" index and "internal resistance spectrum" online estimation

The online diagnosis can be performed for the "capacity loss" index and "internal resistance spectrum," and their online diagnosis principle is detailed hereafter according to the above equilibrium potential prediction method and the diagnosis time respectively.

(a) Online analysis of the "capacity loss" index

In the EV battery pack, all cells are connected in series, and the same current flows through each cell at any moment. Therefore, the capacity change ΔQ of all cells is equal. The battery capacity of any cell can be calculated by the following equation according to the SoC change corresponding to the capacity change ΔQ:

$$C = \frac{|\Delta Q|}{|\Delta SoC|} \qquad (12.34)$$

where C is the battery capacity, ΔQ is the capacity change during the Δt time period, and the ΔSoC is the remaining capacity change during the Δt time period. In Equation (12.34),

$$\Delta Q = \int_{t_1}^{t_2} I dt \qquad (12.35)$$

where I is the discharge current and $dt = t_2 - t_1$.

As shown in Figures 12.25 and 12.26, for the lithium iron phosphate battery, the equilibrium potential has a one-to-one correspondence with the SoC at a certain temperature. This characteristic exists in both the charging and discharging processes of the battery.

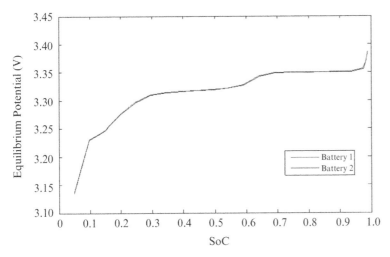

Figure 12.25 Charge equilibrium potential-SoC.

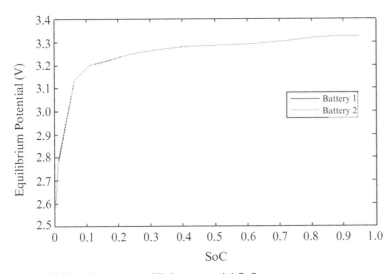

Figure 12.26 Discharge equilibrium potential-SoC.

If two diagnosis times (t_1, t_2) appear during operation of the electric vehicle, the voltage rebound data collected during two diagnosis times is fitted by the equilibrium potential prediction method to obtain the equilibrium potential of the battery. The corresponding SoC_1 and SoC_2 are estimated according to the equilibrium potential-SoC curve to calculate the ΔSoC corresponding to ΔQ and then the current battery capacity. The current "capacity loss" index can be calculated according to the calculated current battery capacity by the capacity loss of Equation (12.4).

(b) Online analysis of the "internal resistance spectrum"

The internal resistance is calculated by

$$r = \frac{\Delta U}{I} = \frac{E_B - U_L}{I} \tag{12.36}$$

where r is the internal resistance, E_B is the equilibrium potential, U_L is the voltage at positive and negative electrodes after loading the battery, and I is the load current.

According to the over-potential characteristic of the battery, when the battery is changed from the operating state to the idle state (without charge and discharge), the battery voltage can rebound until the voltage is stabilized to the EMF of the battery. The rebound voltage is the ΔU in Equation (12.36) and the operating current before the rebound is the I in Equation (12.36).

If the above diagnosis time occurs during the operation of the electric vehicle, the collected voltage rebound data are fit through the equilibrium potential prediction method to obtain the equilibrium potential E_B and the corresponding SoC value of the battery. The equivalent internal resistance r is calculated by the load voltage (U_L) (the battery voltage at the moment before the battery enters the idle state) and the load current (the operating current of the battery at the moment before the battery enters the idle state). In case of multiple diagnosis times, the internal resistance of the battery can be calculated during different SoCs, and the "internal resistance spectrum" index of the battery's internal resistance deviation coefficient γ can be calculated by Equation (12.10) and the reference database of the offline diagnosis method.

So far, the online estimation can be completed for the online diagnosed "capacity loss" and "internal resistance spectrum" indexes in the equilibrium potential prediction method and according to the online diagnosis time and the reference database. Finally, the online battery degradation diagnosis is concluded according to the evaluation criteria for the offline diagnosis method, described in the above section, which are not repeated in this section.

5. Data-driven degradation diagnosis

In addition to the online degradation diagnosis based on the rebound voltage detailed in the previous section, the data-driven degradation diagnosis is available. This section details the online battery degradation diagnosis that is performed by the support vector machine based on the equivalent internal resistance prediction. For the LiFePO$_4$ battery, the capacity loss is generally accompanied by an increase in the internal resistance in the degradation process, that is, there is a certain relationship between the capacity loss and the increase in the internal resistance, as shown by

$$C_{loss} = f(\vec{R}) \tag{12.37}$$

However, it is impossible to explicitly express the relationship between the capacity loss and the increase of internal resistance using this expression. For this, machine learning is required to look for the rule from experimental data. Their relationship is concluded and analyzed by the support vector regression in this section. Compared with other algorithms, the support vector regression is specially used to solve machine learning in the case of limited samples, and has advantages, such as the solution of a high-dimensional problem, lower computation complexity, low generalization error, easy understanding, etc.

In this experiment, according to the data provided in Section 6.2, six 15 Ah lithium iron phosphate batteries are selected for test and verification, among which three batteries were test samples at 40 °C and the other three from 20 °C, to evaluate the prediction effect of the proposed method under different conditions. The capacity and internal resistance rules of the battery have been detailed in Section 6.2 and are not repeated in this section. To facilitate modeling and analysis, the original relative capacity C_n / C_0 is changed to the relative capacity loss C_{loss}, shown in Figure 12.27, and $C_{loss} = 1 - C_n / C_0$.

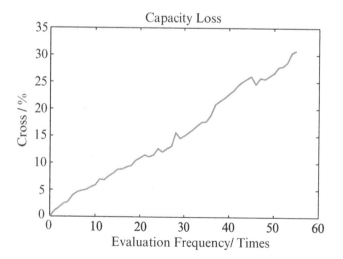

Figure 12.27 Relative capacity loss growth curve of a battery.

At each temperature, two batteries are selected as training samples for the support vector regression model and one battery is used as a prediction sample. First, the experimental data at 40 °C are used to verify the feasibility of establishing the relationship between the capacity loss and the equivalent internal resistance by the support vector regression, and the experimental data at 20 °C are used to verify the expandability of the model with temperature.

The resistance values corresponding to 0.6, 0.7, and 0.8 SoC are selected from each internal resistance curve and are denoted as R_{1n}, R_{2n}, and R_{3n}. The capacity loss corresponding to the evaluation is selected from Figure 12.27, R_{1n}, R_{2n}, and R_{3n} are used as the input of the training model, and C_{loss} is used as the corresponding output to obtain the regression result and the prediction result, as shown in Figure 12.28.

For the online-used power battery, the battery degradation estimation error may be caused by the measurement error of the sensor. If the voltage measurement error of the sensor is ±5 mV and the current measurement error is ±20 mA, then

$$U'_{ocvcn} = U_{ocvcn} + e_{1n} \tag{12.38}$$

$$U'_{cn} = U_{cn} + e_{2n} \tag{12.39}$$

$$I'_{cn} = I_{cn} + e_{3n} \tag{12.40}$$

$$R_{icn} = \left(\frac{U'_{ocvcn} - U'_{cn}}{I_c} \right) \tag{12.41}$$

$$Q_{cn} = \frac{I'_{cn} \times t_{cn}}{3600} \tag{12.42}$$

where U'_{ocvcn} ($n = 1, 2, 3,...$) is the charge open-circuit voltage measured every time, e_{1n} is the measurement error, U'_{cn} ($n = 1, 2, 3,...$) is the charge operating voltage measured every time, I'_{cn} is the current measured every time, and e_{3n} is the measurement error. After the data with the measurement error is trained, the prediction results are obtained as shown in Figure 12.29.

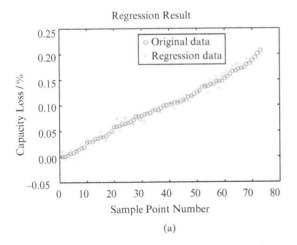

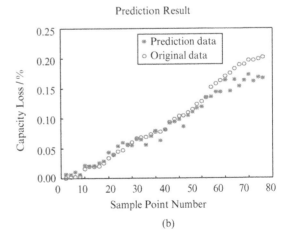

(a)　　　　　　　　　　　　　　　　(b)

Figure 12.28 Regression result and prediction result.

The original data are compared with the prediction results to perform the error analysis as shown in Figure 12.30, where the absolute value of the maximum error percentage is not more than 4%.

The temperature is the most important factor affecting battery degradation. The battery experimental data used for training and prediction are obtained at an experimental temperature of 40 °C. In order to study the extensibility of the temperature during the prediction of battery degradation by the support vector regression, the battery data obtained at the experimental temperature of 20 °C are here used as the training sample and the prediction sample respectively in order to obtain the training result and the prediction result, as shown in Figure 12.31.

Through data processing, the maximum error is 0.0113 and the RMSE is 0.0041 in regression training and the maximum error is 0.0113 and the RMSE is 0.0044 in the prediction, that is, the relative error and RMSE are small, indicating that the method is also accurate in predicting the degradation degree of the battery at 20 °C.

Therefore, the support vector regression model can predict the degradation degree of the battery by the internal resistance data of the battery at different temperatures and has good expansibility.

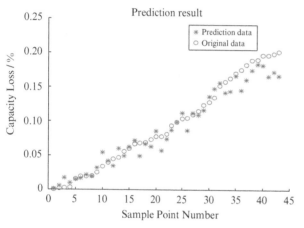

Figure 12.29 Prediction results and data.

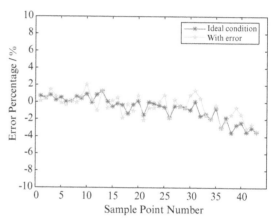

Figure 12.30 Error analysis.

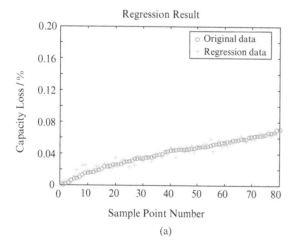

(a)

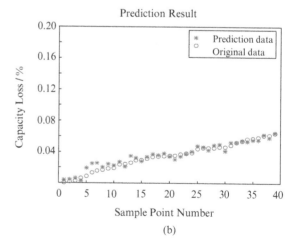

(b)

Figure 12.31 Training result and prediction result.

The battery degradation can be predicted online by the equivalent internal resistance, so that users can timely know the state of the EV battery, in order to decide whether to replace the battery, ensuring the traveled distance of electric vehicles and improving safety.

References

1 Ecker, D., Gerschler, J., Vogel, J., et al., "Development of a lifetime prediction model for lithium-ion batteries based on extended accelerated aging test data," *Journal of Power Sources*, 2012, 215: 248–257.

2 Plett, G. L., "Extended Kalman filtering for battery management systems of LiPB-based HEV battery packs: Part 2. Modeling and identification," *Journal of Power Sources*, 2004, 134: 262–276.

3 Sakamoto, T., "Evaluation of Li-ion battery in early degradation stage – AC impedance spectrum analysis for marine energy storage," *Marine Engineering*, 2021, 56(2): 273–283.

4 Doerffel, D. and Sharkh, S. A., "A critical review of using the Peukert equation for determining the remaining capacity of lead-acid and lithium-ion batteries," *Journal of Power Sources*, 2006, 155(2): 395–400.

5 Lyu, Z. and Gao, R., "A model-based and data-driven joint method for state-of-health estimation of a lithium-ion battery in electric vehicles," *International Journal of Energy Research*, 2019, 43(14): 7956–7969.

6 Hertz, T., Hillel, A. B., and Weinshall, D., "Learning a kernel function for classification with small training samples," in *Proceedings of the 23rd International Conference on Machine Learning*, 2006, pp. 401–408.

7 Yang, D., Wang, Y. j., Pan, R., et al., "State-of-health estimation for the lithium-ion battery based on support vector regression," *Applied Energy*, 2018, 227: 273–283.

8 Li, X., Yuan, C., and Wang, Z., "State of health estimation for Li-ion battery via partial incremental capacity analysis based on support vector regression," *Energy*, 2020, 203: 117852.

13

Communication Interface for BMS

BMS performs internal communication between its master and slave modules and external communication with other system devices like the charge controller, power controller, energy management system, etc. Both communications, internal and external, require a high speed and robust data transfer. Interfaces and protocols should be compatible to accommodate multidevice communications. In addition, interfaces should have bidirectional capabilities and buffer capacity.

An overview of the BMS system in Figure 13.1 describes a master–slave BMS configuration. The master is a central control of BMS that communicates internally with cell/pack level hardware and externally with high level hardware like a laptop or other controllers. Each slave is attached to a battery pack and transfers the essential pack information to the master. The pack essential information consists of the pack and cell voltages, temperatures, pack current, SoC, SoH, and other information related to safety and control. The master BMS is responsible for processing the received information and then communicates with external devices for charge and safety controls [1]. In Figure 13.1, slaves talk to the master with a serial protocol RS-485 (Modbus) and the master connects with other external devices through protocols RS-485 and CAN. A few common protocols used for data exchange between the BMS master and slave are the Controller Area Network (CAN)/FlexRay, Modbus, etc.

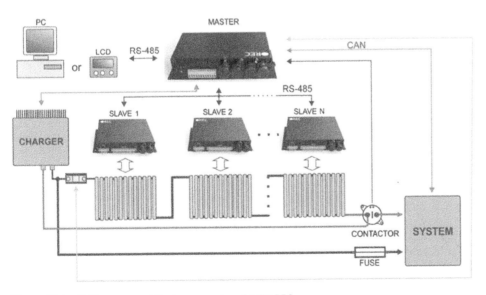

Figure 13.1 BMS system architecture overview. Credit: REC.

Battery Management System and its Applications, First Edition. Xiaojun Tan, Andrea Vezzini, Yuqian Fan, Neeta Khare, You Xu, and Liangliang Wei.

The communication through sensors is another inter-system communication that any BMS must have. Sensor communication is by an analog/digital I/O or pulse width modulation signals. Most importantly, the communication with sensors needs isolation between the high voltage of the pack and the low voltage of communication signals.

Compared with other systems, the external communication between various controllers is simpler. It uses serial communication, DC bus-serial communication over a power line, and wireless communication. The scope of this chapter is limited to the wired communication.

With the advancement of the technology, every battery-powered system requires communications between BMS and other system devices. The communication between BMS and external devices optimizes power usage of the system and improves battery performance. Such battery systems are often called intelligent batteries. The communication may simply be a data link used for performance monitoring, data logging, diagnostics, setting system parameters, or system control signals. The communication protocol selection is based on the application where it has been used rather than on battery technology.

Figure 13.2 is an example of communication between BMS and an external power inverter and depicts an intermediate power control software (PCS) interfacing layer to facilitate communication between BMS and a

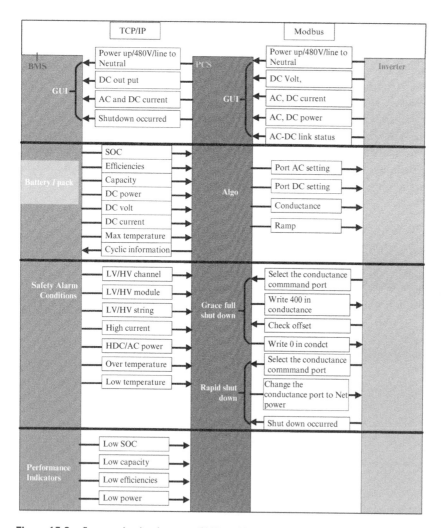

Figure 13.2 Communication between BMS and PCS.

power inverter. PCS can be hosted in the external system or it can be a part of an inverter controller. In the example, the BMS interface is divided into the following four sections:

a) The GUI interface that BMS receives from a power inverter. This information is not necessarily available to BMS. However, it is helpful to monitor the power controller through BMS when the system is connected to a less intelligent power controller.
b) Battery/pack information that BMS supplies to PCS to control charging and discharging modes.
c) Safety limits or safety alarm conditions that BMS shares with PCS to initiate errors, warnings, and shutdown.
d) Performance indicators that BMS supplies to PCS are optional information from BMS to alert the system and to generate pre-warning conditions.

Communication protocols used in Figure 13.2 are TCP/IP between high-level controllers and Modbus from PCS to the inverter hardware.

Ideally, a communications protocol includes a method of addressing the devices connected to the bus, data word format or the message, a priority setting, transmission sequence of the bits, control signals, error detection, handling or correction, and transmission speed. In addition, the BMS system designer should follow some basic limitations for safety concerns when setting up the internal and external BMS communications. The battery pack generally offers a high DC voltage. Special care must be taken to guarantee the separation of the high voltage (HV) present in the battery system from the low voltage (LV) of the communication channels to ensure the safety of the user and the system. In the battery pack, cells are connected in series. The voltage shift between series-connected cells is huge. The ground signal at the first cell may be much higher than the last cell ground signal. Apart from software protocols to achieve isolation, there are other hardware solutions, such as

- Isolated serial communications,
- Optical isolation,
- Wireless serial communications.

In addition, the BMS design should include onboard diagnostics of the communication channels to ensure the reliable data transfer between the controllers and the BMS modules [2].

13.1 BMS Communication Bus and Protocols

There are specific and general-purpose communication buses and protocols often used for internal and external communication in the BMS system.

The most commonly used BMS communication buses and the protocols are given in the following list:

Communication Bus

- SMBus
- Control Area Network (CAN) Vehicle Bus
- Ethernet
- Fibre Channel (high speed, for connecting computers to mass storage devices)
- I^2C
- RS-232
- RS-423
- RS-485
- SPI

- Universal Serial Bus (moderate speed, for connecting peripherals to computers)
- EnergyBus

Communication Protocols

- Modbus
- CANopen

13.1.1 System Management Bus (SMBus)

The SMBus is a joint contribution from Intel and Duracell. The bus is designed to set up a communication between low-power smart battery systems (SBSs) and the battery chargers [3]. The smart battery provides information about battery specification, current conditions, and its usage history. The charger uses the information to determine the optimum charging profile in order to manage the real-time load. The primary objective of the SBS and its combination with the charger is to prevent the battery from over-charging or deep discharging and to extend the life of the battery by protecting it from damage.

The SMBus is a serial, single-ended, two-wire bus. It was originally derived from I^2C for communication with low-bandwidth devices such as a laptop's rechargeable battery subsystem. The SMBus is neither accessible to the user nor configurable. The bus is usually not smart enough to identify its functionality. However, the power management bus (PMBus) is an improved version of the SMBus with advanced features. The SMBus clock frequency range is 10–100 kHz, which is extended in the PMBus up to 400 kHz.

However, the SMBus is introduced for a specific communication between the smart batteries and the charger. All the other general-purpose communications interfaces and protocols mentioned below are widely used with the BMS.

13.1.2 BMS: Internal Data Communication

BMS internal communication can be either on-board or between the board communications (master and slave architecture). Integrated Circuit (IC) packages use serial buses for transferring data to reduce the number of pins in the package. Serial communication limits the speed of the data transfer compared to parallel communication, but saves on cost. A few examples of such low-cost serial buses often used for BMS internal communication follow:

- SPI,
- I^2C,
- UNI/O,
- 1-Wire.

The Universal Serial Communication Interface (USCI) modules support multiple serial communication modes. Different USCI modules support different modes and are categorized as USCI_A and USCI_B. USCI_B is more commonly used in BMS. USCI Bx modules support the I^2C mode and the SPI mode, as shown in Figure 13.3, which shows three clock signals and two mode selections for pins for I^2C and SPI.

Serial peripheral interface (SPI) bus

The SPI bus is a transmission protocol that regulates the synchronous serial data (bits) using a clock. As shown in Figure 13.3, SPI includes a clock line and separate transmission and receiver pins for data transfer. The chip select pin allows the bus to enable or disable the operation. In the master–slave mode, the SPI provides serial data transmission by multiple devices using a shared clock provided by the master. The chip select pin is always controlled by the master. SPI offers bidirectional communication faster than I²C. There is no defined limit to any maximum clock speed, so SPI is capable of attaining a potentially high speed. Unfortunately, SPI does not offer any error detection mechanism for the data transfer. The receiver does not provide any confirmation on successful transmission via handshaking and there is absolutely no option for flow control via the hardware.

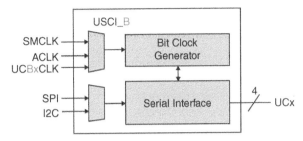

Figure 13.3 Universal serial communication interface B.

The SPI mode features include a 7–8 bit word length with 3 or 4 pins for SPI operation. It works in the master–slave mode and uses independent transmit and receive shift registers and buffer registers. SPI offers a continuous transmit and receive operation with an independent interrupt capability for receive and transmit messages.

Inter-integrated circuit (I²C) bus

The I²C is a two-wire synchronous bus but is a low-speed bus. It offers bidirectional communication and runs with data rates up to 3.4 Mbits/s. It is a half-duplex bus and was originally designed for communication within a system between internal modules rather than for any external communications. Like SPI, it is suitable for master–slave applications. Master initiates and regulates communication with multiple slaves.

The I²C uses two lines to transmit data and clock signals, with an additional line for the reference. The I²C bus uses a lower number of pins compared to SPI. Typically, I²C is used for internal communications within BMS embedded systems.

13.1.3 BMS: External Data Communication

BMS communication with external devices is important for any automated system. Application from consumer electronics like a mobile phone, laptop, or for bigger systems like e-mobility and grid-connected energy storage, BMS communication is essential. BMS needs to talk to external devices, for example charge controllers, power inverters, converters, and energy management systems, in order to regulate the battery power according to the usage and to ensure the real-time safety of the entire system [4]. Typically, inter-device communications protect the battery from damage by controlling charging and environmental conditions and extending the life of the battery. On the other hand, information about the battery charge and health conditions allows the system to utilize the power up to the maximum limits and reduces the overall cost. Table 13.1 summarizes the common features of

Table 13.1 Summary table for BMS external data communication.

Serial Transmission Standard	Cable length/ feet	Speed/bps	Mode	Possible number of devices
RS 232	50	20 k	Full duplex	1 channel
EIA-485/ RS485	4000	100 k–10 M	Half duplex	32 multiple
USB	16	3 G	Half duplex	127 multiple
CAN	3280–131	50 k–1 M	Full duplex	–
LIN	131	19.2 k	Half duplex	16
FlexRay	–	10 M	–	2

the data communication interface standards. The lengths of the cable and the data speed affect the quality of the data. The data speed depends on the cable length.

13.1.3.1   RS232 and RS485 with Modbus

RS232 and RS485 are actual physical electrical specifications for communication networks used with Modbus, which is the communication protocol widely used today. Most of the vendors support either Modbus or CAN or both. There are several forms of Modbus. Modbus/RTU is used on serial systems (e.g. RS-232 and RS-485).

RS232 is a standard for serial transmission of the data between two external devices. Separate transmit and receive lines provide full duplex communication between the devices. The allowed data rate with RS 232 is 20,000 bits per second with cable lengths up to 50 feet. RS-232 is not widely used in BMS.

RS485 (EIA-485) is a standard similar to RS232 for serial data transmissions between multiple devices. RS485 is limited to 32 channels for multiple connections. Normally, RS485 offers half-duplex communication. It uses a differential balanced line over a twisted pair for noise immunity. The speed limit in RS485 depends on the cable length and ranges for data rates up to 100 K bits per second or 10 M bits per second. Maximum allowed cable lengths are up to 4000 feet.

13.1.3.2   USB Universal Serial Bus

The USB is the most popular plug and play standard for connections between computers and peripheral devices using a star topology. It is a half-duplex master–slave bus, with a 4-pin standard connector, incorporating two pins connected to a twisted pair for carrying a differential data signal, a ground (earth) line, and a 5 V power rail.

The data rate has significantly increased with subsequent versions up to 3 gigabits/s. With star topology, up to 127 devices including hubs can be connected to the bus. The cable lengths are limited to 16 feet with USB.

The host or the master controls the communication through a unique address of each device so that only one device can actually transmit or receive data to or from the host at one time. In battery applications, the USB connection is used for monitoring the battery state or for setting control limits.

Figure 13.4 is an example of communication buses used in BMS, where a system block of a Li-ion battery uses an MSP430 microcontroller and bq76PL536 battery modules by Texas Instruments. A bq76PL536 is a battery monitoring IC with secondary protection. Multiple bq76PL536s are connected to monitor batteries without additional isolation between ICs. A high-speed SPI is used for internal communications between a microcontroller and bq76PL536 to ensure reliable communication through a high-voltage battery stack. The BMS or microcontroller communicates with an external host or battery charger using USB communication or asynchronous serial communication such as RS232 or RS485.

13.1.3.3   CAN Bus

The controller area network (CAN) bus is an ISO standard for a two-wire serial communication and is used in a real-time distributed system. Initially, it was designed for integrating intelligent sensors in a centralized multiplexing system. It simplifies the wiring harness and reduces the size and weight of the wiring loom by multiplexing the signals on to a single shared broadcast data bus.

The CAN bus allows 12 V, or other potential power, to be distributed around the system. The system controller manages all sensors and actuators through the bus. It is a robust communication bus and is designed to tolerate a high level of noise in an automotive system and still to maintain secure communications. Scaling-up of the system with a CAN bus is easy as new nodes can be added without reconfiguring the rest of the nodes.

The CAN bus provides various transmission rates corresponding to different bus lengths. A high speed is allowed up to 1 Mbps with a 131 feet bus length, whereas low speed is 50 kbps with a 3280 feet bus length.

Standard CAN microcontrollers are available from a variety of semiconductor manufacturers. This provides an off-the-shelf cost-effective solution to the automotive industry for a BMS controller with CAN as its main communications channel.

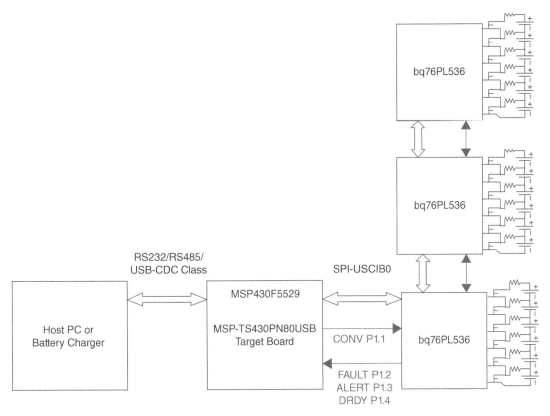

Figure 13.4 RS232–RS485 Communication Standards in BMS.

13.1.3.4 LIN Bus

The local internet network (LIN) bus is another automotive communications standard, similar to the CAN bus. However, unlike CAN, it is a single wire and operates at 20 kbps. LIN uses distributed multiplexers and standardized smart connectors IC hardware that allow simple, low-cost IC solutions. It offers more flexibility and uses less wiring but more electronics than the CAN Bus.

Mainly, the LIN bus was conceived for automotive applications. However, it is used in many other applications such as dishwasher and washing machines.

13.1.3.5 FlexRay Bus

The FlexRay bus is a step advancement over the CAN bus. It has recently been specifically developed for more complex control applications, like sophisticated engine management systems planned for future automotive use. It operates on a data rate of 10 bp and offers both synchronous and asynchronous data transfers.

FlexRay is based on a time-triggered architecture known as time division multiple access (TDMA) in contrast to CAN, which is event-driven. In TDMA, communication is organized into pre-defined time slots. High priority signals are guaranteed access to the synchronous pre-determined time slot channels. In addition, low priority signals get the bus only when bus is free to transmit data asynchronously.

The FlexRay bus can therefore support fast-responding dynamic control systems rather than just the simpler sensors and actuators permitted with the CAN bus.

13.2 Higher-Layer Communication Protocols

In order to find a generic solution to provide overall compatible communication protocols, it is necessary to address the application layer interface along with lower layers. The set of protocols, from data to application layers, offers a higher flexibility and configuration capabilities. CANopen is among them. CANopen is the embedded network protocols that have been designed to develop compatibility between different manufactured devices.

CANopen is a communication standard developed for embedded systems used in a wide range of industries, automation, and motion applications. It combines higher layer communication protocols and profile specifications. Mainly, CANopen covers the following:

• A network programming framework,
• Device descriptions,
• Interface definitions, and
• Application profiles.

CANopen is the set of protocols that provides a standard communication between devices and applications from different manufacturers. It is based on the CAN lower layer protocol. In terms of the OSI model, as shown in Figure 13.5, CANopen employs the layers from a network to an application layer. The two bottom layers, data and physical, are usually defined by CAN, although a CANopen device profile supports devices using other lower layer protocols such as Ethernet Powerlink and EtherCAT. The physical layer defines the lines used, voltages, high-speed nature, etc., while the data link specifies the CAN frame-based (messages) protocol.

At the higher layer, the CANopen standard consists of an addressing scheme, several small communication protocols, and an application layer defined by a device profile. Basic CANopen covers the top five layers: network (addressing, routing), transport (end-to-end reliability), session (synchronization), presentation (data encoded in standard way, such as data representation), and application. The application layer describes how to configure, transfer, and synchronize CANopen devices.

The CANopen internal device architecture consists of three logical parts, as shown in Figure 13.6:

• The stack of protocols handles the communication via the CAN network.
• The application software provides the internal control functionality and processes the hardware interfaces.
• One of the central themes of CANopen is the object dictionary (OD). The CANopen object dictionary interfaces the protocol as well as the application software. OD is essentially a table that stores configuration and process data. It contains references (indices) for all used data types and stores all communication and application parameters. Obviously, the OD is important for CANopen device configuration and diagnostics. Within the network, CANopen master can read the OD index of CANopen slaves, which is a unique identifier used to access each slave for communication.

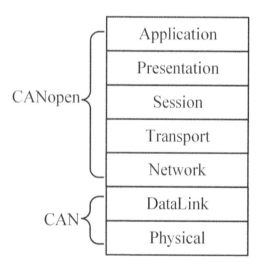

Figure 13.5 CAN and CANopen in the OSI layer model.

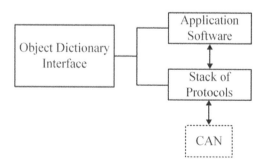

Figure 13.6 CANopen internal device architecture.

The CANopen provides many advantages over the other network protocols. As an example, it offers a standard communication object (COB) for time-critical processes and network management data. The device structure and device behavior can be configured through its application layer. The CANopen is an essential "plug and play" communication solution that eliminates the complexities in communication and allows manufacturer-specific functionality implementation. CANopen provides a sophisticated error detection and correction mechanism.

Industries are embracing the CANopen technology. CANopen is already used in various applications, such as medical equipment, off-road vehicles, maritime electronics, railway applications, or building automation. Soon BMS also intends to adopt CANopen for its future generation.

Readers are encouraged to review the CiA DS 301 (CAN in Automation) specification for more information and recent updates such as CiA 401 for I/O modules and CiA 402 for motion control.

13.3 A Case Study: Universal CiA EnergyBus for a Low-Emission Vehicle (LEV)

EnergyBus is a combination of a cutting-edge connector with communication interface protocols. It is a single standard for an LEV worldwide that provides energy and service data transmission safely through the same connector (see Figure 13.7). At the same time, it is a universal standard for all electrical components of LEVs that are connected through the connector. These electrical components communicate with each other using the CANopen protocol. EnergyBus with LEV-BMS ensures improved safety of the LEV.

EnergyBus consists of a standardized set of connectors that connects electric components of LEV, such as batteries, chargers, motors, sensors, and human interface. Performance of the components has been enhanced due to a uniform safe exchange of the information over the same connectors.

Figure 13.8 describes the EnergyBus framework, which is an application interface for EnergyBus devices used in e-bikes. The software is written in C language and follows the CANopen/EnergyBus standard CiA 454. The flexible

Figure 13.7 Energy Bus connector.

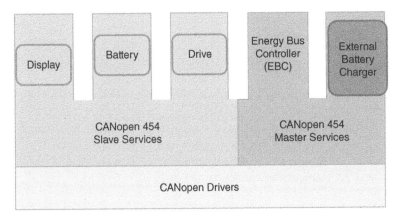

Figure 13.8 Energy Bus framework.

Figure 13.9 E-bike ST2.

and configurable user interface easily integrates with the CANopen communication services. The framework is designed for master/slave modes. The modular structure allows an easy integration of different devices like display, drive, battery, charges, and EBC (EnergyBus controller). The other single device can also be combined with the EBC. The CANopen communication protocols run automatically in the background.

Standardization is the key for LEVs to appeal to the mass market. Moreover, EnergyBus has made it possible for LEV to develop. Objectives of EnergyBus are as follows:

- It provides the compatibility in connections of electric components and in communication protocols.
- Compatible components make product development easy and cost-efficient. Overall, it contributes to the fast development of high-quality vehicles for the market.
- Development of EnergyBus has also promoted public LEV infrastructure such as rental stations or swapping machines without the monopoly of a single manufacturer.

EnergyBus mainly provides a universal connector and diagnostic tool. Servicing and maintenance of small electric vehicles are challenging without universal diagnostic tools and compatible spare parts. EnergyBus connectors and bus systems provide one charger for every battery, independent of the size and the chemistry. This helps vendors to offer services for the products from different manufacturers at an affordable cost. On the other hand, diagnostic tools can quickly identify the issues in the LEV efficiently. This can help in easy replacement of the faulty part. A diagnostic tool provides battery internal information such as SoC and SoH, together with a history of the battery. Information about battery history determines the wrongly diagnosed cases and separates them from the warranty coverage.

The universal diagnostic tool with advanced features makes the diagnostics more efficient and reduces the overall cost compared to specific diagnostic tools supplied by the individual manufacturers. The E-bike is one of the important applications of EnergyBus. The Impulse, the first E-bike based on EnergyBus (see Figure 13.9), was introduced by TourDeSuisse (TDS) in Switzerland in 2012 and is followed by many other companies like Stromer.

Apart from the E-bike, EnergyBus has been widely employed for other applications such as Public Charge Stations, Island Power Systems, repeater stations in mobile communication networks, off-grid housing, village DC grids, and off-grid water pump stations.

References

1 Sakamoto, T., "Evaluation of Li-ion battery in early degradation stage – AC impedance spectrum analysis for marine energy storage," *Marine Engineering*, 2021, 56(2): 273–283.

2 Alzieu, J., Smimite, H., and Glaize, C., "Development of an onboard charge and discharge management system for electric-vehicle batteries," *Power Sources*, 1995, 53(2): 327–333.

3 Liu, V.-T. and Yu, M.-T., "Design of power management for electric vehicles," *Advanced Science Letters*, 2012, 9(1): 92–98.

4 Wu, Y. J. and Chung, J. G., "Efficient controller area network data compression for automobile applications," *Frontiers of Information Technology & Electronic Engineering*, 2015, 16: 70–78.

14

Battery Lifecycle Information Management

When the electric vehicle is driven, the battery management system collects the dynamic information such as current and voltage every moment and updates the SoC, SoH, and other real-time status of the battery. In engineering, generally, these data generated online will be saved into the local memory of the BMS, regularly taken out, and classified into a unified data management center. The data generated in the whole using a process from the factory to elimination is saved in order to realize the information of the battery in the whole life cycle, which can facilitate checking the work history of the battery at a later period and then providing data support for judging the battery degradation or finding out the cause of battery failure.

The informatization of the full battery lifecycle is actually about three questions that people are most concerned about, i.e.

1) What type of information should be saved?
2) How can the information be saved and managed?
3) How often is the information saved?

Before these three questions are answered, it is first necessary to answer the questions about the real-time battery data and the data size. The corresponding specific information scheme can be designed only after calculating the total data size. For this purpose, this section will systematize the possible data generated by the battery during its use in order to analyze the processing process of the battery from collection to final output and introduce the specific method for achieving the database.

14.1 Data Type of Power Battery

The basic information data of the battery that is involved in the information on the full life cycle covers the battery voltage value, the current value, the temperature value, and the system time. These data are obtained in different ways and in different lengths. In order to save and manage the relevant information reasonably, all kinds of data should be fully analyzed in combination with data operation and saving requirements to establish a complete battery information database.

In order to be able to analyze and process the data easily at a later time, it is also necessary to record the relevant information while establishing the system information database in addition to such original information that can be directly collected, such as current and voltage. Therefore, a relatively complete battery information database should cover at least:

1) Basic data information of cells, such as the ID number;
2) Voltage value of cells;

Battery Management System and its Applications, First Edition. Xiaojun Tan, Andrea Vezzini, Yuqian Fan, Neeta Khare, You Xu, and Liangliang Wei.

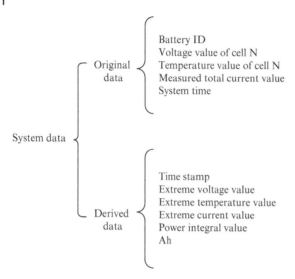

Figure 14.1 Data type of power battery.

3) Temperature value of each temperature sensor;
4) Total voltage of the battery system;
5) Total current of the battery system;
6) Current system time;
7) Minimum and maximum voltage values and their corresponding battery IDs;
8) Minimum and maximum temperature values and their corresponding battery IDs;
9) Maximum current value in a given time period and its corresponding moment;
10) Maximum power integration and AH within a given time period.

According to the above information, the battery data can be classified into two categories, as shown in the classification map of Figure 14.1.

14.2 Vehicle Instrument Data Display

Displaying the battery information has always been one of the important functions of the battery management system. Through the instrument, the battery management system is able to inform the relevant personnel of the current state of the battery. An old-fashioned battery instrument, shown in Figure 14.2, is able to effectively show the operator the remaining capacity of the battery to help the operator to determine whether to recharge or to stop charging the battery. On the panel, there are several different areas to assist people in determining whether a battery has reached the upper limit of its cycle life through the charging voltage rebound.

Of course, this instrument is not designed especially for electric vehicles. In this section, the question will be discussed about how to display the information about the

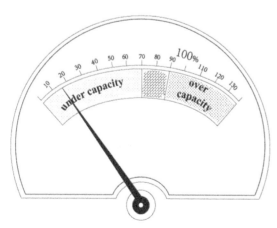

Figure 14.2 An old-fashioned instrument displaying battery information.

battery pack through the instrument on an electric vehicle. First, what information needs to be displayed on the vehicle instrument is clarified. Then, two ways of displaying information on the instrument will be discussed. On the one hand, the relevant battery information can be displayed by upgrading the traditional instrument. On the other hand, according to the characteristics of electric vehicles, it is possible to use the current mature liquid crystal display technology to develop an instrument for the battery management system.

14.2.1 Battery Information Displayed on the Vehicle Instrument

The battery information to be displayed on the vehicle instrument can be basically classified as follows.

1. Information to be displayed during normal driving

In the process of driving an electric vehicle, the information to be provided to the driver mainly includes: battery temperature information, total voltage and current information of the battery pack, the remaining capacity information, and information on the estimated remaining mileage. The refresh rate of such information does not need to be too high. For example, voltage and current information needs to be refreshed every second and the remaining capacity information needs to be refreshed every 10 seconds. This is enough information for a driver and will not cause a large burden on the BCU in the battery management system.

2. Information on charging during normal parking

The electric vehicle stops and is plugged in, which indicates that the battery pack starts to be in a charging state. At this time, relevant charging information needs to be conveyed to the driver through the instrument, such as the total voltage, temperature information, charging current of the battery pack, remaining capacity information, and information on the estimated end time of charging. For the advanced battery management system, it is also a requirement to display the information on the charging mode, such as the fast-charging mode, the slow-charging mode, and whether to add the balancing control. With respect to displaying the above information, the refresh rate should also not be too high. Any refresh rate in the range from 1 to 10 seconds is acceptable.

3. Danger warning information

During driving or charging, any abnormal condition monitored by the battery management system should be reported to the driver or operator through the instrument in a timely manner. These warning messages cover over-voltage alarm, over-current alarm, over-temperature alarm, low remaining capacity alarm, etc. Some failure information is also included, such as a communication network failure and self-check failure. At this time, in addition to the alarm given in a display manner, this may need to be combined with a sound alarm.

14.2.2 Upgrade Based on a Traditional Instrument Panel

Since some information displayed on the traditional automobile instrument panel no longer exists or is not very important on an electric vehicle, modifications can be made on the basis of the original traditional automobile instrument to allow information of the battery to be displayed by the modified instrument, as shown by the following example.

Figure 14.3(a) shows a traditional automobile instrument. The "fuel gauge" is meaningless for an electric vehicle. Moreover, the engine speed and water temperature gauge are not as important as those in a traditional automobile. Therefore, these three instruments can be respectively used to display SoC of the battery, operating current, and operating voltage, as shown in Figure 14.3(b). The speedometer and odometer in the middle of the instrument on the electric vehicle are the same as those on the traditional automobile, so they can be kept.

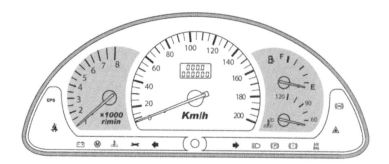

(a) Traditional automobile instrument

Figure 14.3 Electric vehicle instrument modified based on a traditional automobile instrument.

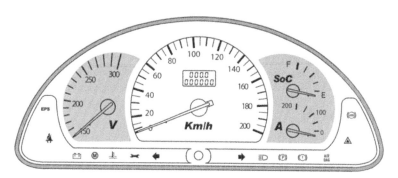

(b) Modified electric vehicle instrument

The electric vehicle instrument modified but based on the traditional instrument has advantages in that the modified electric vehicle instrument has the same style as that of the traditional automobile, which is in line with a driver's usual habit. Moreover, only a slight modification is required, so it is not necessary to make many changes to the design.

14.2.3 Design of the New Instrument Panel

With the development of technology, it is possible to use liquid crystal display screens as an automobile instrument in terms of cost and reliability. Therefore, currently, LCD screens are mounted on many new models of automobiles in order to display the relevant information on the vehicle, which has the following advantages. First, it features a friendly interface and large amounts of information. The instrument designed by using the LCD screen has various styles. It can support the arrangement and combination of various characters and graphics, so it is available to flexibly transmit information to the driver. Second, it is designed flexibly and can easily be upgraded. Through the LCD screen, the instrument can be designed by programming customization.

Furthermore, the instrument software and interface can be upgraded as required. It is in line with the current trend of personalized use of the automobile. Third, many LCD screens support touch screen input, which has good interactivity. The interactivity of traditional instruments is poor, while the new liquid crystal instruments can facilitate the interaction between the users and the vehicle's control system.

Now, based on previous work experience, some suggestions will be given on the new instrument panel design.

1. **The other functions of the vehicle instruments should be combined to comprehensively display the relevant information on a running vehicle and then replace the traditional instrument.**

The instrument of an electric vehicle should also display other relevant automobile information in addition to the information on the battery pack, such as vehicle speed information, mileage information, light control information, and gear information. Figure 14.4 shows a typical electric vehicle instrument displaying the basic information on the vehicle and the remaining capacity information on the battery pack. In this way, the information can be displayed completely through the LCD screen, which overcomes the disadvantages of the traditional instrument.

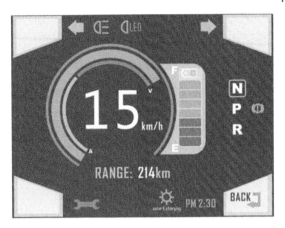

Figure 14.4 Electric vehicle instrument displaying all kinds of information.

2. **Different types of information on the battery pack should be displayed through the split screen.**

Due to the flexibility of the electric vehicle instrument, different interfaces can be used to transmit information according to different states of the vehicle. For example, if the electric vehicle is being charged, the speed information and gear position information are no longer important. Figure 14.5(a) for the interface shows the charging information of the battery pack.

Figure 14.5(a) shows that the information related to battery charging can be effectively transmitted to the driver. The information includes the temperature value at the highest temperature point in the battery pack, the minimum temperature value, the maximum voltage value of a cell, the minimum voltage value of a cell, the charging current, the charging mode, and the estimated charging time. In addition, when any communication error occurs between the battery management system and the vehicle control unit, the driver can also be notified through another interface, as shown in Figure 14.5(b).

3. **The touch screen should be considered to facilitate users to enquire about specific information.**

Currently, many LCD screens are available to input information through the touch screen, which facilitates the interaction with the driver. Therefore, the touch screen function can be considered on the automobile instrument. In the interface of Figures 14.4 and 14.5, four gray icons are designed in the four corners to facilitate the drivers

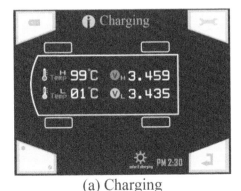

(a) Charging   (b) Error alarm

Figure 14.5 Different ways to display an instrument corresponding to different states.

when checking the status of the electric vehicle and the battery at any time. By touching these icons, the driver can get the corresponding system information.

14.3 Battery Data Transmission Mode

To achieve the management of battery information in the full lifecycle, the real-time data generated during running of electric vehicles need to be saved locally in the battery management system and to be transmitted to the data center through the data communication system in order to facilitate unified management and maintenance in a later period.

At present, the remote transmission of data is rarely considered in the conventional battery management system, where there is no effective way of data communication. To meet the requirements on the full lifecycle management of battery data, the communication modules should be added to the traditional design scheme to ensure the mutual communication between the local information and the information of the remote data center. In this section, according to the characteristics of battery data transmission, a data transmission module oriented to the battery management system is introduced. The traditional data transmission channel is expanded so that the local information of the battery management system is connected to the data center in a variety of network forms so that the real-time online transmission of the battery data is realized.

14.3.1 Hardware Implementation of Data Transmission

The data of the power battery can be collected and saved through BMS and reported to the data center in real time through the data transmission module under specific conditions in order to realize data transmission. The hardware structure of the data transmission module is shown in Figure 14.6.

The data transmission module consists of an embedded processor, a power management module, a communication module, a data storage module, a data transmission channel switch module, a mobile communication module, a wireless network, and a wired network module, in which the power management module is connected with a

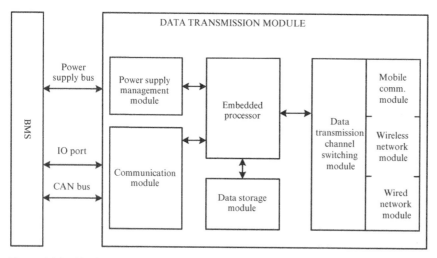

Figure 14.6 Hardware structure of the data transmission module.

battery management system through the power bus [1]. The data communication between the communication module and the battery management system is made by an IO handshake port and a CAN bus. Through the mobile communication module, the wireless module, and the wired network module, the data are connected to the remote data center in different forms.

The functions of each module in the data transmission module are described as follows.

The battery management system and the data transmission module are connected through three types of signal lines, which are respectively the power supply bus, the IO handshake port, and the communication bus. Since most BMSs currently on the market are powered by 12 V low-voltage batteries, the power bus of the data transmission module can be mounted on the power bus of a 12 V lead-acid battery. The IO handshake port is used for handshake detection while connecting the BMS with the data transmission module. In the IO handshake port, it is possible to have a signal line triggered by the electrical level. At the BMS end, the IO port is grounded by a 10 kΩ pull-down resistor. At the data transmission module side, a high level of 12 V is at the IO port. When the BMS detects the high level, it needs to connect the data transmission module to the BMS. After the BMS successfully "shakes hands" with the data transmission module, it will transmit the collected real-time information to and interact with the data transmission module by means of a CAN communication bus.

The embedded processor is a core computing unit of the data transmission module, and is powered by the power management module. It is able to extract and save data from the data storage module and communicates with the data center through the communication module, in which the data storage module may be composed of two SD cards with SDIO interfaces. These two SD cards can work alternately. When one SD card is in the read state, the received data can be saved in the other SD card. The SD card mentioned here is equivalent to an FIFO queue. When the SD card is full, the original data will be overwritten at the original storage location of the SD card. When the network signal is detected and the saved battery data is completely uploaded, the SD card will be emptied and the new battery data information will be written again.

The mobile communication module, the wireless network module, and the wired network module are connected in parallel. The modules are switched by an embedded processor through the data transmission channel in order to select the connection channel with the data center. The mobile communication module may be a short message transmission module or a 3G/4G/5G data transmission module. The wireless network module may be a WIFI module or a Bluetooth module. The wired network modules may be all kinds of communication modules supporting wired LAN, with an RJ45 interface outside. These three communication modules can be switched freely according to the actual condition. Therefore, the data center can be connected in a variety of ways to ensure the stability and reliability of the data transmission.

14.3.2 Control Flow of Data Transmission

The battery management system is available to collect the real-time information of the battery pack and transmit it to the data center in different network modes through the internal data transmission module corresponding to each battery pack and the intelligent data transmission module of the entire battery system. See Figure 14.7 for the overall flow of data transmission.

After collecting the information of the battery pack, the battery management system will send a handshake signal to the data transmission module through the IO handshake port, which indicates that the battery management system is online. After successful handshaking, the data transmission module is connected to the battery management system through the communication bus, and has been in a listening state for data acquisition. After it obtains the communication data of the battery management system, it will start to acquire the network connection method and identify the current network environment. When any communication network is available, the relevant data will be sent to the data center through such a network.

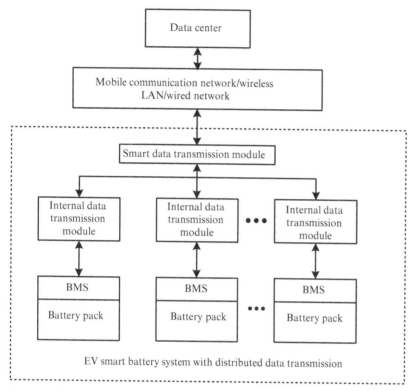

Figure 14.7 Overall flow of data transmission.

According to the different network states, the data transmission module is available to transmit data using the mobile communication network, the wireless network, or the wired network. The specific switching process is as follows:

1) The embedded processor reads the battery data information saved in the data storage module in turn. Then start step 2.
2) The data transmission channel switch module connects and detects the wired network module. If the wired network module is connected, start step 3; if the wired network is offline, start step 4.
3) The wired network module transmits the battery data information to the data center. Then start step 8.
4) The data transmission channel switch module connects and detects the wireless network module and detects the wireless network module. If the wireless network module is connected, start step 5. If the wireless network module is offline, start step 6.
5) The wireless network module transmits battery data information to the data center. Then start step 8.
6) The data transmission channel switch module connects and detects the mobile communication network module. If the mobile communication module is connected, start step 7. If the mobile communication module is offline, start step 10.
7) The mobile communication module transmits the battery data information to the data center. Then start step 8.
8) After the battery data information is sent, delete the battery data information that has been sent from the data storage module. If the battery data information in the data storage module is empty, start step 9; otherwise, start step 1.
9) The embedded processor transmits the successful signal to the battery management system through the communication bus and the communication module.
10) The embedded processor writes the battery data information into the data storage module and sends the failure signal to the battery management system through the communication bus and the communication module.

The traditional battery management system is combined with a data transmission module. Then each electric vehicle can be used as a terminal of the Internet to transmit the information collected by the vehicle battery management system to the data center through the mobile communication network, wireless, or wired network of the data transmission module in real time, which can ensure the safety of the battery in use. Additionally, whenever a wireless or wired network is connected, the battery operation data will be transmitted synchronously to the data center. Then, it can be ensured that the battery of each electric vehicle is monitored and traceable throughout its lifecycle.

14.3.3 Hierarchical Management of Power Battery Data

In previous chapters, the data types, the data volumes, and the methods for realizing the data transmission that are generated during running of the electric vehicles are technically analyzed. From a macroperspective, the storage logic and the feasibility of the hierarchical management of the battery data will be discussed and analyzed, as well as the shortcomings of the current data management models.

The hierarchical management of power battery data covers two levels: the internal data management of the vehicle and the big data operation management. The internal data management of the vehicle is made for the logical relationship between the battery information and other information of the vehicle, the data prioritization, the storage path allocation, and determination of the proper time for data storage and transmission, etc. The operation of big data is made for the data interaction between the electric vehicle terminals and the data centers of automobile manufacturers, automobile operators, and national and local governments, in order to solve the problems such as who sends, who keeps, and who supervises the data.

The internal data hierarchical management of electric vehicles is shown in Figure 14.8, in which the vehicle data covers in-vehicle battery data and other vehicle data. The in-vehicle battery data are collected in real time by BMS and saved in BMS memory. The other vehicle data are collected by the vehicle controller from the in-vehicle motor controller, power-assisted steering, throttle gear, and other signals, and are saved in the vehicle controller memory. The figure shows that the two types of data are independent from each other and are collected and saved by the vehicle controller and the battery management system respectively, which reflects the modular and hierarchical management of the data.

If the network environment permits, the relevant data can be obtained from the vehicle controller memory and BMS memory through the data transmission module mentioned above and sent to the target data center. Considering the bandwidth, traffic, and other problems during the data transmission as well as the possible difference in the requirements of different target data centers, it is not certain that all data may be sent during the data transmission. The key information can be extracted for the transmission through the specific screening logic. Questions, such as which data should be chosen, who the data will be sent to, and who will be responsible for the reliability of the data, should be analyzed from a more macroperspective at the level of hierarchical management of the big data operation.

The hierarchical management model of the big data operation of the current electric vehicle is shown in Figure 14.9. The figure shows that the collected data are sent by the electric vehicle terminal to the automobile manufacturers and vehicle operators. In order to achieve centralized and unified management of the operational data, the automobile manufacturers report the data to the local government data centers and national data centers. In addition, the automobile operators also report their own data to the local government data centers. Moreover, some information channels are also

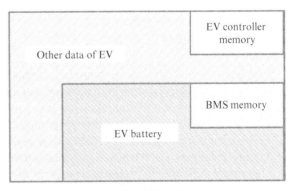

Figure 14.8 Data hierarchical management inside an electric vehicle.

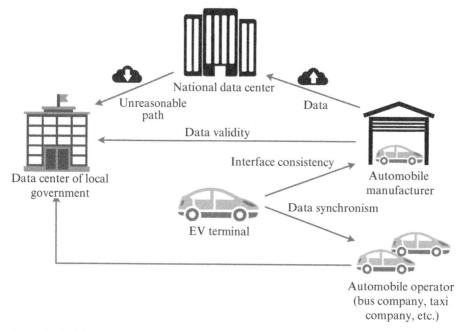

Figure 14.9 Hierarchical management of the big data operation of the current electric vehicle.

reserved between the national data centers and the local government data centers for the data interaction and data verification. However, in practical use, this model still has the following problems:

1) Interface consistency issue. The EV terminal sends real-time data to the automobile manufacturers. However, at present a uniform data interface is not available for different vehicle models of the same manufacturer or the vehicle models of different manufacturers, so at present, data type, data format, communication frequency, and so on may not be consistent. Therefore, it is possible to cause disordered management, which would prejudice the follow-up data traceability and maintenance.

2) Data synchronism issue. Since the target needs vary with the organizations that need different data from the EV terminal, the data obtained by the automobile manufacturer and the automobile operators cannot be mutually supplemented and verified, resulting in disordered management.

3) Data validity issue. Since the data center of local government obtains the data from the automobile manufacturer and the automobile operator, rather than the EV terminal, a validity and reliability issue is caused. The relative data are uploaded by the enterprises and it is difficult for the government to effectively supervise the data of the EV terminal, prejudicing unified management and accountability.

4) Data real time issue. The automobile manufacturers upload the data to the national data center, but a uniform time requirement has not been established at present, which causes a timeliness issue. The data center cannot directly control the real-time information of the EV terminal and it is difficult to grasp the relevant dynamic data in time when an abnormal situation arises, which reduces the supervision ability of the data center.

5) Path irrationality issue. The data of the EV terminal are transmitted and saved through multiple chains. The local government and the national data center respectively collect a set of data through different routes, with duplication and redundancy, which may lead to data conflict and the decentralization of supervision responsibilities. This could cause difficulties in implementation of the data supervision responsibilities of different organizations.

Therefore, in the full lifecycle management of the EV terminal and its battery data, it is necessary to unify the data transmission type, define the data transmission chain, and implement the data supervision responsibility, in order to ensure data effectiveness and reliability.

14.4 Information Concerning a Full-Power Battery Lifecycle

14.4.1 Database Structure of a Power Battery

In order to realize the informatization of a power battery, the battery management system is required to collect and store the original data of each cell and the corresponding derived information at every moment. With the increase in use time, the amount of stored data increases gradually.

Therefore, in order to effectively manage the battery data and improve data enquiry and processing efficiency, in the development of the battery management system, in addition to the front-end battery data collection, we should also consider the back-end data storage form, determine the internal data relationship, and sort out the structure of the battery database.

1. Conceptual structure of the battery database

The conceptual structure of the battery database is a conceptual model of different battery information established according to the hierarchical relationship of battery data types. Specifically, an entity relationship tool diagram (E-R diagram) is used to depict the composition of the system data and its relationship.

The conceptual structure design of the database can be divided into two stages: a local view design and a global view design.

The local view of the conceptual structure of the battery database is shown in Figure 14.10. Two problems should be considered during the local view design. First, when used as an attribute, there can be no more

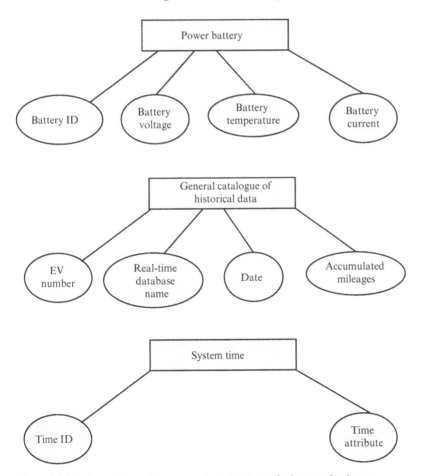

Figure 14.10 Local view of the conceptual structure of a battery database.

information expression in the specified environment, that is, it is the smallest information unit in the system. Second, if an attribute has more information expressions, such an attribute should be considered as an entity. As shown in Figure 14.10, the battery database can be divided into three entities: battery, general directory of historical data, and system time. Each entity has several attributes which basically coves all kinds of data information described in the previous section.

The local E-R diagrams are integrated into an overall E-R diagram of the system in order to establish the global view of the EV battery database, as shown in Figure 14.11. Because redundant data can destroy the integrity and consistency of the database, repeated and conflicting information should be eliminated as much as possible in the global view.

2. Logical structure of the power battery database

The logical structure design of the battery database refers to the transformation of the global and local E-R conceptual models obtained from the conceptual structure into a specific data relationship model. The data information generated by the power battery can be further divided into two categories: real-time data and historical data, according to the generation time and function of the formation. Among them, the real-time information refers to the current operation state information of the EV, while the historical data is formed by importing the real-time information into the historical database. Through the historical data, the complete information of the full lifecycle process of the battery can be mastered. When it is required to backtrack and diagnose the battery, the required evaluation results can be obtained through the extraction and processing of the historical data.

To this end, the basic data types are further divided into two parts: real-time information database and historical information data, and the logical structure of the database is introduced.

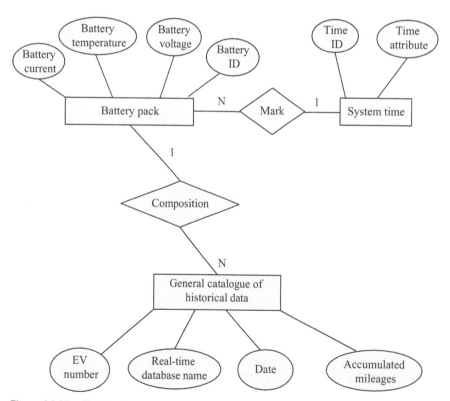

Figure 14.11 Global view of the conceptual structure of the battery database.

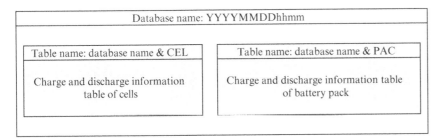

Figure 14.12 Real-time information database.

The real-time information database is named according to the starting date (year, month, and day) and time (hour and minute) of a certain travel; the format is: YYYYMMDDhhmm. Two data tables are designed in the database to store all the basic data of the travel in real time. The structure of the real-time information database is shown in Figure 14.12.

The information of the cells in the battery pack is named by the database name + "CEL." It is designed as shown in Table 14.1.

The information of the battery pack is named by the database name + "PAC" and is designed as shown in Table 14.2.

The design of the database and data sheet is described below:

Sheet name
CEL is the abbreviation of the cell and is used to indicate the cell information.
PAC is the abbreviation of the package and is used to indicate the battery pack information.

Data width
CellID refers to the cell ID number, and the data width is expressed by 1 byte (0~255) for the battery pack consisting of 100 cells.

Table 14.1 Information of cells in the battery pack.

Field name	Description	Key assignment?	Data type	Data width
CellID	Cell ID number	Y	Unsigned integer type	1 byte
TimeID	Time ID number	Y	Unsigned integer type	3 bytes
Temper	Battery temperature	N	Single precision floating point type	4 bytes
Volt	Battery voltage	N	Single precision floating point type	4 bytes
State	Equilibrium state	N	Boolean type	1 byte

Table 14.2 Information of the battery pack.

Field name	Description	Key assignment?	Data type	Data width
TimeID	Time ID number	Y	Unsigned integer type	3 bytes
Time	System time	N	Unsigned integer type	3 bytes
Current	Total current	N	Single precision floating point type	4 bytes

TimeID refers to the time ID number, 24 h a day corresponding to 86,400 s, which is expressed by 3 bytes since 2 bytes (0~65,535) is not enough.

State is the equilibrium state, including two state values: equilibrium and disequilibrium, Boolean type, expressed by 1 byte.

Time refers to the system time. If minutes and seconds are used, the time is expressed by character type and 6 bytes. If the second is used, the time is 24 h a day corresponding to 86,400 s, which is expressed by an unsigned integer type and 3 bytes.

Tempe, Volt and Current refer to the temperature, voltage, and current of the battery respectively. The single precision floating point type is used, the data width of which is 4 bytes in a general programming environment (–3.4 × 10–38 ~ 3.4 × 10–38), representing 6~7 valid digits, and meets the data scope and accuracy requirement.

The historical information database is named by the battery pack name; for example "information database of battery pack N." Two sheet classes are designed in the database: the historical data retrieval sheet, which is used to store the battery pack information of each travel, such as vehicle number, real-time information base name, date, accumulated mileage, etc., and the battery pack information sheet imported by the real-time database, i.e. the information sheet of the cells and the battery pack. The structure of the historical information database is shown in Figure 14.13.

The general catalogue of the historical data is designed as shown in Table 14.3.

The design of the database and the data sheet are described below:

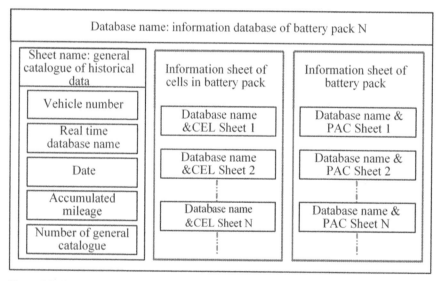

Figure 14.13 Historical information database.

Table 14.3 General catalogue of the historical data.

Field name	Description	Key assignment?	Data type	Data width
TabNum	Number of general catalogues	Y	Unsigned integer type	2 bytes
NowNam	Real time database name	N	Unsigned integer type	5 bytes
Tester	Tester	N	Character type	8 bytes
Kmiter	Tested mileage	N	Single precision floating point type	4 bytes

a) Field name and number

The field name consists of English words or their abbreviation described in the Description, and the data width is more than 8 bytes.

This sheet is used to record the relevant information of a travel and the fields may be added as required, such as current weather, travel location, etc.

b) Data width

TabNum is the number of general catalogues and automatically increases by 1 when new travel data are entered. The data generation times are not more than 20 according to a specific requirement, 2 bytes (0~65,536) of data width are used, and the travel records of 3276 days (about 9 years) can be expressed.

NowNam is the real-time database name, including information of travel starting and ending times, and is expressed by character type and 12 bytes, or the unsigned integer type ranging from 200801010000 to 999912312459 and 5 bytes (0~1099511627775) are required.

Tester is the vehicle number information in Chinese, where a Chinese character has 2 bytes width, the location of four charters is reserved, and 8 bytes are required.

Kmiter is the accumulated mileage. The single precision floating point type is used, the data width of which is 4 bytes in a general programming environment ($-3.4 \times 10^{-38} \sim 3.4 \times 10^{-38}$), representing 6~7 valid digits, and meets the data scope and accuracy requirement.

14.4.2 Power Battery Data Volume Estimation

The database design rationality criterion: on the premise of ensuring the correct logical structure, the information and storage capacity are compressed as much as possible to achieve data content storage without redundancy and compact logical structure, in order to meet the storage capacity requirements of the battery management system. This section intends to estimate the information volume for the real-time and historical data of the battery and evaluate the basic scale of the database to provide a reference for the design of the battery management system.

1. Data volume estimation of real-time information

The storage space and processing rate of the data are estimated. Suppose that the battery pack consists of 100 cells and each battery management board reports data once per second, according to the above design, 100 records are increased per second in the cell information sheet and one record is increased per second in the battery pack information sheet. According to the data width specified in Tables 14.1 and 14.2, the data volume generated per second is

$$(1+3+4+4+1)\,\mathrm{B} \times 100 + (3+3+4)\,\mathrm{B} = 1310\,\mathrm{B} \tag{14.1}$$

The data volume generated per hour is

$$1310\,\mathrm{B} \times 3600 = 4{,}716{,}000\,\mathrm{B} \approx 5\,\mathrm{MB} \tag{14.2}$$

The existing embedded hardware platform or PC can store and process 5 MB of data volume according to its storage capacity and processing rate.

2. Data volume estimation of historical information

Suppose that the battery pack consists of 100 cells, each battery management board reports data once per second, and the test time is 3 h, according to the above design, the data volume per test is

$$\left[(1+3+4+4+1)\,\mathrm{B} \times 100 + (3+3+4)\mathrm{B}\right] \times 3600 \times 3 + (2+5+8+4)\,\mathrm{B} = 14{,}148{,}019\,\mathrm{B} \tag{14.3}$$

The data volume per test is about 15 MB. According to the actual requirement, suppose there are three tests per day and four test days per week. The data volume per year is

$$14,148,019 \text{ B} \times 3 \times 4 \times 52 = 8,842,511,875 \text{ B} \approx 8 \text{ GB} \tag{14.4}$$

The existing PC can store 8 GB of data volume per year according to the storage capacity.

The basic data volume of the battery is estimated above and the relevant results show that the database of the battery can be managed in a common software platform. As mentioned above, two data tables are established for each charge and discharge process of the battery. According to the current demand for electric vehicles, the data generation times are not more than 20 times every day, and 7200 per year, and the corresponding data table is not more than 14,000. Therefore, the data can be managed in the existing commercial database management software, such as Microsoft Access™, SQLserver™, and other platforms.

14.5 Storage and Analysis of Historical Information of a Battery

The previous battery management systems did not value the storage of the historical information of the battery. With the development of research in this field, more attention has been paid to the storage and analysis of the historical information. Two issues are mainly solved in this section: first, how to store the historical information of the battery and, second, how to analyze the historical information of the battery after the event.

14.5.1 Necessity for Storage of Historical Information

In the battery management system, it is necessary to store historical information for the following four reasons.

First, in the research and development stage of electric vehicles, it is of great significance to grasp the historical information of the battery for the debugging of electric vehicles. For example, for automobile safety, through the analysis of historical data, we can understand the change rules of voltage and temperature in the battery operation process and analyze whether there are safety risks. Again, in terms of vehicle performance, through the analysis of historical data, the battery energy consumption can be calculated in order to analyze the endurance mileage of the vehicle. At the same time, the battery consistency in the actual driving process can be evaluated and analyzed.

Second, for electric vehicles, the historical information of the battery is the basis for fault diagnosis and battery maintenance. The overhaul and maintenance of the battery pack will become an important part of the regular maintenance of electric vehicles in the future. The repair shops of electric vehicles should be equipped with appropriate computers and software to collect and analyze the historical data stored in the battery management system, in order to judge the SoH of the battery and determine the maintenance measures to be adopted. For example, if the performance of the cell is particularly degraded, the battery should be replaced. The consistency of the battery can also be judged based on the historical information to decide whether to conduct a comprehensive diagnosis or overall balancing of the battery pack.

Third, the historical information of the battery is helpful to accurately assess the SoC of the battery. As discussed in Chapter 7, the SoC of the battery at a given moment is related to historical information over a long period of time. Storage of the historical information of the battery is of great significance for a battery state assessment. For example, the open-circuit voltage is often used to estimate the SoC of the battery, provided that the battery has been idling for the past few minutes. In the case where the driver immediately restarts the vehicle after a stop, without the historical information of the battery the open-circuit voltage upon restarting is used as the initial value for the SoC estimation and a larger SoC estimation error may be caused since the battery voltage is rebounding.

Fourth, the historical information of the battery can be used as the basis for the battery balancing management. After the battery is delivered, its performance and capacity will be degraded to a certain degree. The EV batteries are used in a more complex environment than other batteries, so the degradation degree of each cell may not be the same after using the battery for a period. By analyzing the historical data of the battery, the actual capacity of each cell at the current moment can be calculated, in order to provide the basis for the balancing control of the battery.

14.5.2 Achievement of Historical Information Storage

The historical information can usually be stored as shown in Figure 14.14. In other words, BMC transmits the collected information of each cell to BCU and then BCU stores the data after filtering as needed. In the figure, there are two data storage routes, one expressed in the solid arrow and the other expressed in the dotted arrow, and the two dotted line virtual boxes respectively represent the embedded vehicle system and the PC computer.

Route 1 is applicable to the R&D of the electric vehicle and the PC architecture-based test device is installed in the electric vehicle. The BCU transmits the data collected from the BMC to the PC by the communication port, while the real-time battery database is physically established on the PC and is stored in the mass hard disk in the form of the database file.

Route 2 does not depend on the vehicle-mounted PC, corresponding to the transmission mode 2 in the figure. The real-time database is built into the FLASH memory of the upper computer of the on-board battery management and the data are stored in the most compact way (RAW files). The data can be imported into a PC computer for storage or analysis after the electric vehicle has been running for a period of time.

There are some minor issues with battery history storage that are worth noting. Attention is paid to minor issues during the historical information storage of the battery.

First, the above practices are the general implementation method. Recently, a number of engineers have proposed that FLASH memory should be added to each BMC to back up an additional set of data in a redundant manner. Such redundancy has two advantages. On the one hand, redundant backup is beneficial to prevent data loss, because the communication network of the electric vehicle is often unreliable due to interference, and BCU may be required to deal with many urgent tasks, which may result in data loss if it is not able to process BMC data temporarily. On the other hand, in the battery pack of the electric vehicle, it is allowed to replace the cells. If the data of the cells is stored by the BMC and the data of the battery pack is stored by the BCU, the information mismatch caused by cell replacement can be avoided.

Second, in the early test phase, the author tried to use the memory card commonly used by common digital products (such as CF card, SD card, etc.) to

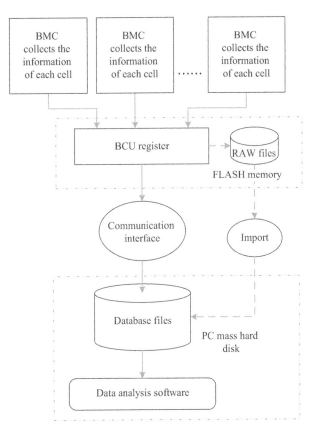

Figure 14.14 Implementation scheme for historical information storage of the battery.

store on-board data. Later, it was found that such a memory card was not reliable when used in an electric vehicle for a long time, and it was easy to cause data loss due to the damage of the memory card. Therefore, as the core part of the electric vehicle, a more reliable FLASH chip should be used in the battery management system, which needs a high reliability design.

Tables 14.4 and 14.5 show examples of the storage formats for cell information and battery pack information.

The information storage feasibility is estimated by the following specific example. Suppose the battery pack consists of 100 cells and each BMC reports the data once per second, according to the above design, 100 records are increased per second in the cell information sheet and one record is increased per second in the battery pack information sheet. According to the data width specified in Tables 14.4 and 14.5, the data volume generated per second is

$$(1+3+2+2+1) \times 100 + (3+3+2) = 918 \text{ B} \tag{14.5}$$

The data volume generated per hour is

$$908 \times 3600 = 3,268,800 \approx 3.3 \text{ MB} \tag{14.6}$$

The 3.3 MB information volume per hour can be stored and processed by the embedded hardware platform or the PC according to their storage capacity and processing rate.

14.5.3 Analysis and Processing of Historical Information

The previous section discusses how to achieve the historical information storage of a battery. This section discusses the post-processing analysis of the historical information.

Figure 14.15 shows the analysis flow diagram after storage and processing of the historical data. In the figure, the real-time database obtains the unprocessed original data with compact data format directly from

Table 14.4 Storage format of cell information.

Field name	Description	Key assignment?	Data width
CellID	Cell ID number	Y	1 B
TimeID	Time ID number	Y	3 B
Temper	Battery temperature	N	2 B
Volt	Battery voltage	N	2 B
State	Equilibrium state	N	1 B

Table 14.5 Storage format of battery pack information.

Field name	Description	Key assignment?	Data width
TimeID	Time ID number	Y	3 B
Time	System time	N	3 B
Current	Total current	N	2 B

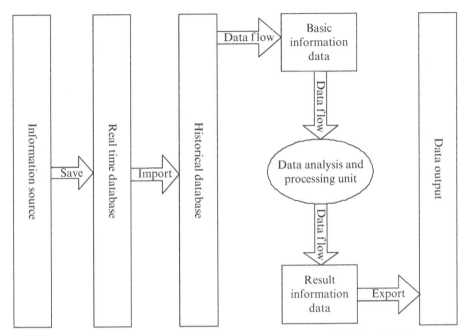

Figure 14.15 Analysis flow diagram after storage and processing of the historical data.

the BMS through the communication interface, and the data may be stored in the form of RAW files in the FLAH chip. In order to facilitate the retrieval of the historical information, before an analysis is made of the historical information, the index information of the real-time database should be extracted and placed in the general information table in order to constitute the historical database, which is stored on the PC in the format of database files.

Many basic data of the battery can be obtained directly from the historical database. The data analysis and processing unit is mainly designed to conduct data mining in these basic data in order to obtain relatively rich derived data.

In fact, the information derived from the original information can be very rich, and according to the specific research and analysis needs, analysis software is compiled.

Some examples of the most basic derived information are given below:

1) Display basic data information of the designated travel.
2) Search basic data information of a cell during a trial-run experiment.
3) Calculate 10 minimum voltage values and cell ID during a trial-run experiment.
4) Calculate 10 maximum voltage values and cell ID during a trial-run experiment.
5) Calculate 10 maximum temperature values and cell ID during a trial-run experiment.
6) Calculate 10 maximum current values and time during a trial-run experiment.
7) Calculate 10 maximum current values and time during a trial-run experiment.
8) Calculate the maximum power integral and Ah of a period of time during a trial-run experiment.

After analysis and processing by the software, the battery data can be output in various forms, such as reports and graphs, as shown in Figure 14.16.

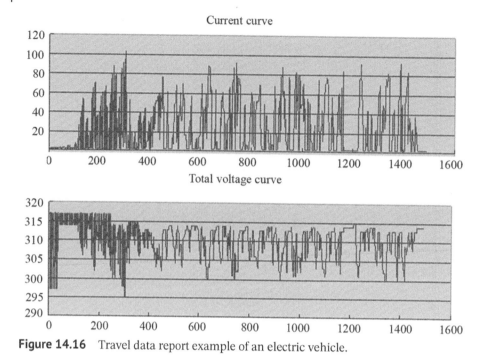

Figure 14.16 Travel data report example of an electric vehicle.

Statistical table of extreme value during a single charge and discharge of an EV battery pack

Experiment time: June 4, 2008 16:47 Experiment state: normal Experimented by Li Si

Number of 10 cells with minimum voltage and voltage value			Number of 10 cells with maximum temperature and temperature value			Maximum current and occurrence time	
Cell No.	Voltage	Occurrence time	Cell No.	Temperature	Occurrence time	Current	Occurrence time
184	2.674	17:47:11	106	38.7	17:48:51	104.0	17:47:10
192	2.674	17:47:11	184	38.7	17:44:09	102.0	17:47:08
142	2.694	17:47:11	186	37.9	17:48:09	94.0	17:47:06
186	2.791	17:47:11	158	37.5	17:44:09	92.0	17:21:12
62	2.830	17:47:11	132	37.4	17:44:09	91.0	17:46:59
130	2.923	17:47:11	160	37.2	17:44:09	91.0	17:04:19
176	2.943	18:41:14	104	37.0	17:44:09	91.0	17:17:07
24	2.948	17:47:11	108	37.0	16:50:30	89.0	17:47:04
106	2.967	18:41:14	82	36.9	17:47:58	89.0	16:59:51
160	2.972	17:47:11	110	36.9	17:44:09	88.0	17:46:28

14.6 Battery Detection System Based on a Mobile Terminal

This section describes how to monitor the battery in real time by the mobile client and to store and analyze the battery data by the server, in order to realize the remote monitoring and fault diagnosis of the battery.

Figure 14.17 shows the system function diagram of the remote battery monitoring and fault diagnosis system designed according to the requirements of the intelligent battery system and the actual requirements of the application scenarios on electric vehicles.

The battery detection system based on the mobile terminal mainly consists of a server and a mobile client. The server is mainly responsible for battery data storage, battery data diagnosis and analysis, and the response to the request from the mobile terminal and data supply. The mobile client is mainly responsible for battery data acquisition, the display of the data in different forms and alarming, and other functions. To sum up, the overall architecture of the detection system is shown in Figure 14.18.

In Figure 14.18, the whole system is divided into left and right parts. The cloud-based server is arranged on the left and is mainly responsible for battery data storage and processing. The BMS communication module communicates with the server according to the formulated coding rules. After the data are uploaded to the server via the

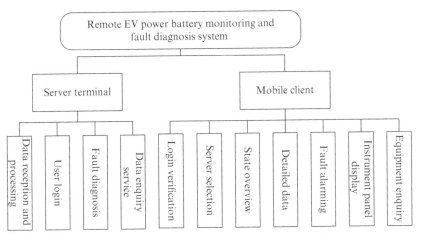

Figure 14.17 Remote power battery monitoring and fault diagnosis system.

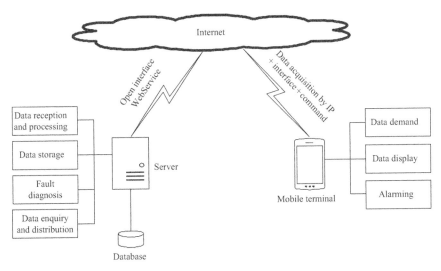

Figure 14.18 Overall architecture of the battery detection system based on the mobile terminal.

Internet, the server decodes, classifies, and writes the data to the data table. The fault diagnosis based on the battery data is completed by the independent software module of the server and the diagnosis results are written on the alarm report in the database for storage. The server opens different request instructions through the established Web Service and the mobile client sends instructions to obtain data. After receiving the instructions, the server queries the results in the database and returns the data through the Internet. The Android-based mobile terminal is arranged on the right and mainly displays the battery data and the alarming.

14.6.1 Server Program Design and Implementation

The server program includes battery data reception and processing, battery diagnostic program, and battery data enquiry services. Figure 14.19 shows the realization process of the battery data reception and processing program.

The battery degradation diagnosis and fault analysis are also implemented in the server program, as shown in Figure 14.20.

The battery data enquiry service implementation process includes: database operation and interface call, as well as WebService release and server IIS deployment, as shown in Figure 14.21.

14.6.2 Design and Implementation of the Mobile Terminal

In order to meet the functional requirements of the mobile terminal and provide convenience to users and simplify operations as much as possible, the designed client interface and interactive operations are shown in

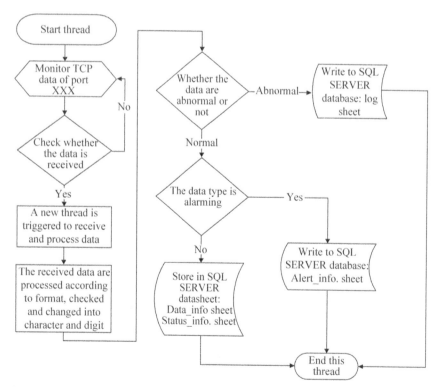

Figure 14.19 Data reception and processing flow charge.

Figure 14.20 Diagnostic program implementation process.

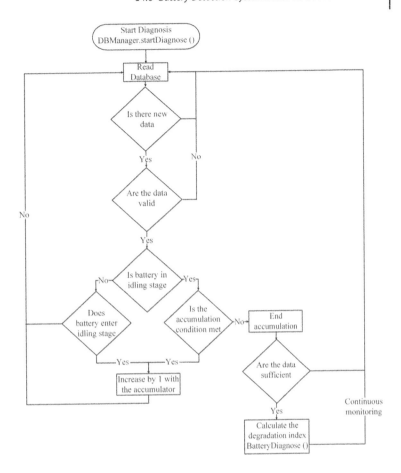

Figure 14.21 Enquiry service implementation process.

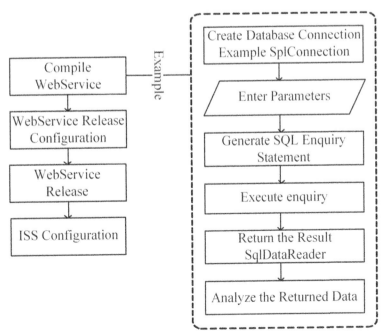

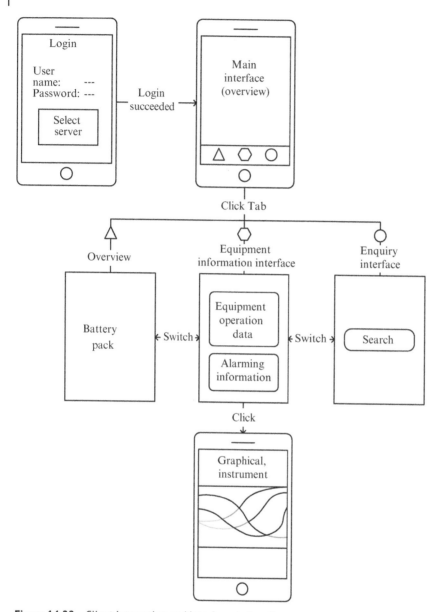

Figure 14.22 Client interaction and interface design diagram.

Figure 14.22, which are mainly composed of three hierarchical progressive interfaces and diversified data display functions, which are provided in the main interface.

Figure 14.23 shows the screenshot of some tested interfaces: (a) client login interface, (b) vehicle overview interface, at which all states of the vehicle can be browsed, (c) equipment details interface, which displays details of the EV battery, (d) equipment enquiry interface, and (e) graphical and instrument interface, by which the change of battery voltage and current and speed can be viewed.

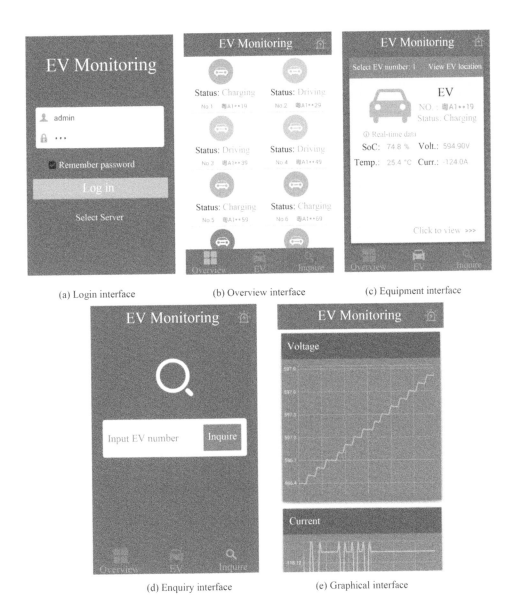

(a) Login interface   (b) Overview interface   (c) Equipment interface

(d) Enquiry interface   (e) Graphical interface

Figure 14.23 Schematic diagram of the client.

Reference

1 Barsukov, Y. and Qian, J., *Battery Power Management for Portable Devices*, Norwood, MA, USA: Artech House, 2013, pp. 111–138.

Part IV

Case Studies

15

BMS for an E-Bike

In practice, all manufacturers of E-bikes today rely on a storage solution with lithium-ion (Li-ion) technology. E-bikes reliability and effective distance covered with support of the battery strongly depends on the health and charging state of the battery. A common bike battery voltage today has a range between 24 V and 52 V nominal. The battery pack consists of individual cells connected in series. In general, the series-connected cells never behave in the same way, even if they belong to the same batch from the same manufacturer. They always partially differ in SoC, capacity, internal resistance, self-discharge rate, and temperature behavior. These differences will intensify with the age of the battery, usually due to a thermal imbalance in the pack. Due to the different rates of aging, the cells run further and further apart towards the end of their lives. Thus, to provide a dependable and high-performance solution for E-bike users, the battery pack should be in perfect balance. In other words, all cells should perform more or less the same. The requirement creates scope for a balancing system to be in place.

In a battery, the weakest cell determines the total capacity of the pack while the stronger cells still have some remaining capacity. Therefore, the load could never utilize the full capacity of the pack. In order to counteract this effect, passive voltage/charge balancing in Li-ion batteries has been frequently used in the past. However, today, sophisticated active voltage/charge balancing systems are often used to extract the remaining capacity from a battery. Active charge balancing, also known as active balancing, allows the entire average battery capacity to be used, as the stronger cells support the weaker ones.

This chapter covers the effect of active and passive balancing and its use for an E-bike battery pack, testing methodology, testing results of active and passive balancing, and an active hardware solution. We particularly describe each section with reference to a case study on the Stromer E-bike.

15.1 Balancing

The battery management system (BMS) precisely monitors every cell of a battery. Each cell is accessed individually and the cell voltages, current, and temperature are measured. The battery management system maximizes the driving range and provides reliability and safety to the battery. Without BMS, it is hard to determine a strong or a weak cell within the pack. In a battery pack, the rate of aging in each cell is different and the cells deviate significantly from one another towards the end of their service life. This phenomenon is also referred to as "cell drift." The weakest cell in the battery pack determines the total capacity of the pack. During charging, the stronger cells remain under-charged until the weakest cell reaches its complete charged state. In other words, the weakest cell has the highest voltage drop across its higher internal resistance and reaches the cut-off voltage much faster than a stronger cell. In this case, the charging process must be terminated according to the weakest cell so that the weak cell does not over-charge. Over-charging the cell with a high internal resistance can damage the cell or whole pack or, in the worst-case scenario, it leads to the risk of fire in the pack. When unloading, the behavior is

Battery Management System and its Applications, First Edition. Xiaojun Tan, Andrea Vezzini, Yuqian Fan, Neeta Khare, You Xu, and Liangliang Wei.

similar. The weakest cell reaches its discharge cut-off voltage earlier than the others and limits the discharge duration, while other cells are still left with energy or capacity.

In order to prevent the cells of the battery from running apart, today almost all battery management systems have either passive or active balancing. The task of the balancing is to bring all the cells (voltage or charge) to the same level within the tolerance band while charging or whenever they drift apart more than a tolerance band. The process helps to maintain all cells at the same level of voltage continuously and achieves a complete charging state in all cells at the same time. By balancing, the entire capacity of the pack can be fully utilized.

15.1.1 Passive Balancing

The challenge of a battery management system is to balance the number of series-connected battery cells while cycling (charging or discharging). Balancing is all the more difficult when series-connected cells have different residual charges in the beginning of the charging process. This initial charge imbalance makes some cells reach their maximum voltage (charge state) earlier than others and terminates the charging process. Because fully charged cells cannot be over-charged, the other cells, those that have not yet reached the maximum voltage, remain under-charged. Passive balancing is developed to maintain the cell voltages in such a way that each cell reaches the maximum voltage together and attains a maximum input capacity. In passive balancing, the charging energy of the cells, which have a higher voltage potential, is converted into heat via a bypass resistor connected in parallel with the cell. As shown in Figure 15.1, when the upper voltage threshold is reached, cell A is partially discharged via a resistor in order to bring cell B to a higher charge level. This method is used especially during the charging process, when the energy supply is available from the charger.

The balancing method is employed with the Stromer charger. At the end of the charging process, the balancer turns on and off a few times. The balancer unloads the weaker cell briefly and brings the weaker cell voltage down to the same level as that of stronger cells.

When discharging, the weaker cell voltage is lower than that of a strong cell and the same balancing process would not balance the cells. Thus, discharging must stop the process as soon as the weakest cell has reached the lower voltage threshold. As a result, the residual energy remains unused in the stronger battery cells. Since passive balancing methods are not effective during discharging and cannot improve the energy utilization, another solution must be found.

15.1.2 Active Charge Compensation

With active balancing, excess charge is transferred from one cell to another. Instead of the energy stored in the cells being converted into heat losses, the cells support each other in this method. The charge transfer between the cells can take place during charging and discharging, as well as during the state of rest. Therefore, the cells remain balanced at all times. During charging, the active compensation can carry a charge from a weaker cell to a stronger one. When unloaded, the stronger cells can support the weaker ones. In contrast to passive balancing, active balancing permits high-compensation currents. The active charge balancing can be accomplished by various methods, for example with switched capacitances or inductances, as well as with the aid of a highly efficient fly-back converter. For more information about the selected compensation procedure, refer to Section 15.4 on active balancing solution.

Figure 15.2 illustrates the advantages of active balancing with cells of different capacities. The energy quantity between the maximum and minimum cell capacity is not converted into heat during charging, as is the case with passive charge balancing, but it is transferred between the cells. As the weaker cells are charged more rapidly, they release their

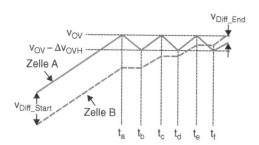

Figure 15.1 Passive charge compensation (*Source:* www.elektronikpraxis.vogel.de).

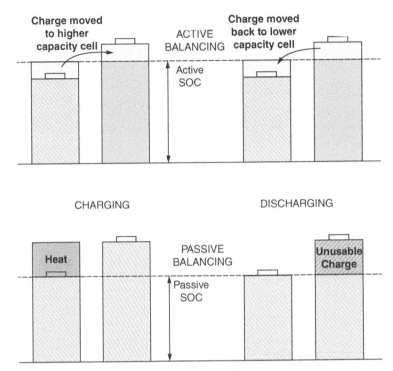

Figure 15.2 Passive and active charge balancing.

excess energy to the stronger cells during the charging process. When unloading, the energy from the stronger cells is then transmitted back to the weaker cells. Thus, all the cells can be completely discharged without any charge remaining in them. In summary, active charge compensation gives a higher effective capacity to a battery pack than passive charge compensation.

15.2 Battery Pack Design for an E-Bike

The battery of the investigated E-bike relies on a storage solution with Li-ion technology. The battery is built with 50 Samsung (18650-22P) cells. Ten packs are connected in series. Each pack has five parallel cells to achieve a higher pack voltage. The battery management system is responsible for monitoring each individual cell. Since, in this case study, measurements focus exclusively on the discharge characteristic, the shutdown criteria are of central importance to the battery management system. To examine the performance of active and passive balancing, it is necessary to keep the same switch-off criteria in both types of BMS (with active or passive balancing). This would maintain the same capacity level in both types of BMS at the initial stage without balancing. The BMS stops the discharge process when a cell voltage reaches below 3.0 V and stays at a voltage lower than 3.0 V for a period of five seconds.

The Stromer E-bike battery pack is shown in Figure 15.3. To ensure an E-bike performance, a guarantee is given in terms of 70% of the design capacity of the battery will be available after 1000 cycles or 3 years, which ever happens first.

Figure 15.3 E-bike battery pack: a Stromer power battery.

15.2.1 E-Bike Battery Pack Design Specifications

Tables 15.1 and 15.2 show the battery pack design specifications of the E-bike, which include the battery and cell respectively.

15.2.2 Testing

For designing and developing an E-bike battery system, an engineering group has used the testing system Evaluator-B from FuelCon, as shown in Figure 15.4. It was used to record the discharge characteristics of the Stromer battery. The Evaluator-B FuelCon enables the testing of a wide variety of performance classes of battery systems to be tested as well as various battery technologies such as Li-ion, Li-polymer, NiCd, NiMH, or lead-acid. Within the test bench, application-specific options/modes can be selected. These options are constant current (CC), voltage, power, and resistance modes. The control is carried out via the user-friendly "TestWork" software. For the E-bike applications, the Stromer battery was discharged in a first step with a CC. While testing we have considered three CC values: 0.5 C, 1 C, and 2 C.

The voltage of each individual cell is tapped and recorded separately using the Agilent Data Logger 34972A. The Agilent data logger has a 16-channel multiplexer module (34902A), so that each cell voltage can be measured independently. The battery is discharged through FuelCon until the battery management system, which is attached to

Table 15.1 Battery.

Technology-type	Power/ Wh	Voltage/V	Capacity/Ah	Charge cut-off voltage	Discharge cut-off voltage	Weight/kg
Li-ion SDI 18650-22P	396	36	11, 10.75 (0.2 C discharge)	4.2	3.0	3.1

Table 15.2 Cell.

Manufacturer/type	Minimum capacity/Ah	Typical capacity/Ah	Nominal voltage /V	Charge cut-off voltage /V
Samsung/18650-22P	2.050 (0.2 C discharge)	2.150 (0.2 C discharge)	3.62 (1 C discharge)	4.2 V ± 0.05 V

Figure 15.4 Evaluator B FuelCon battery test bench.

Figure 15.5 Test setup.

Table 15.3 Test setup details.

Test setup details	
FuelCon Evaluator-B:	Agilent Data Logger 34972A:
Serial number: 70414	Serial number: MY49004365
Model number: 420104500	Resolution: 22 bits
Charging discharge unit: max.1000 A, 100 V, 5 kW	Sampling speed: 3 MS/s
Resolution: up to 24Bit	

the battery, issues the shutdown command according to the discharged cut-off voltage. The cell voltages stored in the Agilent data logger are then processed in a MATLAB script so that the differences between the individual cell capacities can be determined. Simultaneously, the cells are monitored with the BMS Monitor Tool from Stromer and the individual cell voltages are also recorded. All tests are carried out at room temperature. Figure 15.5 shows the test setup and Table 15.3 gives the test setup details.

The effect of temperature variations on the cell behavior has not been investigated in this project. It is expected that the differences between the individual cells are likely to be even more pronounced with temperature variations.

15.3 Methodology

For processing Agilent monitored data and analysis, a MATLAB script was created. The MATLAB script reads the measurement series from the desired battery, the Filename, the battery number, the discharge voltage, and the time. The graphics for the discharge curves, extrapolation, and residual capacity are generated [1]. The characteristics of the battery such as the design capacity, the capacity at full charge, and the number of cycles are read out and displayed.

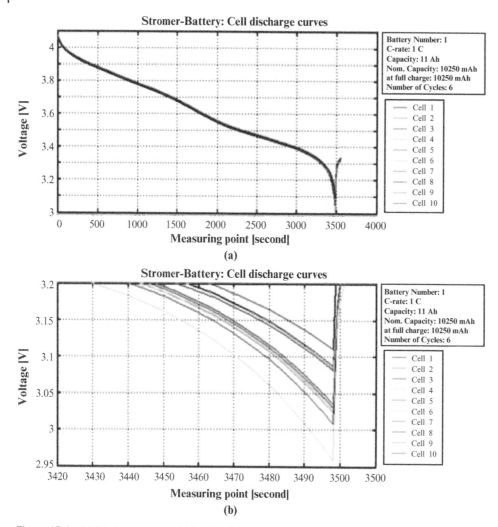

Figure 15.6 (a) Discharge curve of 10 cells of the Stromer battery; (b) zoom-in of the discharge curve at the end of the discharge when cells have voltage drift.

The GUI displays through the MATLAB script are given in Figure 15.6. Figure 15.6(a) shows the discharge curves of 10 cells of the Stromer battery. The battery information is displayed at the upper right corner of the graph. Figure 15.6(b) is a zoom-in of the cell voltages at the end of the discharge. The battery management system of the Stromer battery automatically turns off the discharge process when a cell has a voltage less than 3.0 V and it remains below 3.0 V for a period of five seconds. To simulate the test as a real-time operation, the shutdown condition is kept the same as described in BMS.

Figure 15.7(a) shows the weakest and strongest cells during the course discharge time as described in Figure 15.6(a). The strongest cell 10 is shown in red and the weakest cell 6 in blue. The plot clearly shows that the same cell (10) remained strongest throughout discharging and cell (6) remained the worst for the entire duration. Because the strongest and weakest cell did not change for the entire duration of the discharge, this suggests that balancing did not happen during discharging.

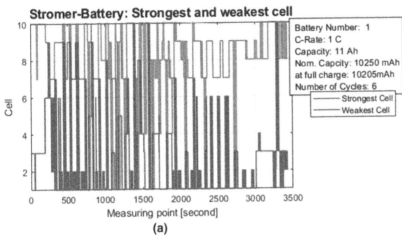

(a)

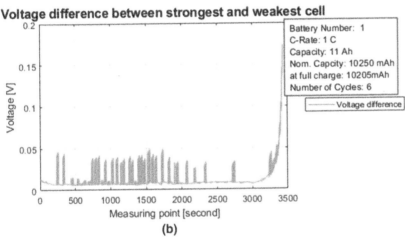

(b)

Figure 15.7 (a) Strong and weak cells during the discharge time and (b) difference of the voltages between the strong and the weak cells.

Figure 15.7(b) shows the voltage difference between the strongest and weakest cells. The voltage difference between the two cells remains minimal for most of the discharge time, but the voltage difference rises exponentially close to the discharge cut-off or to the end of the discharge.

The difference of the cell capacities can be determined by extrapolating the discharge curves of Figure 15.6(b). When the weakest cell reaches the discharge cut-off voltage the BMS stops the discharging process. Figure 15.8(a) shows the discharging curves of the cells at the discharge cut-off. Figure 15.8(b) provides the extrapolation of the discharge voltage curves beyond the discharge cut-off voltage. The extrapolation curves can determine the additional discharge time and energy that other strong cells could have been provided if the discharge process was not stopped due to the weakest cell voltage.

As specified in Figure 15.8(b), the extrapolation of the voltage profiles can determine the remaining capacities in each cell. Figure 15.9(a) shows the remaining capacity in each cell with blue bars, whereas the mean

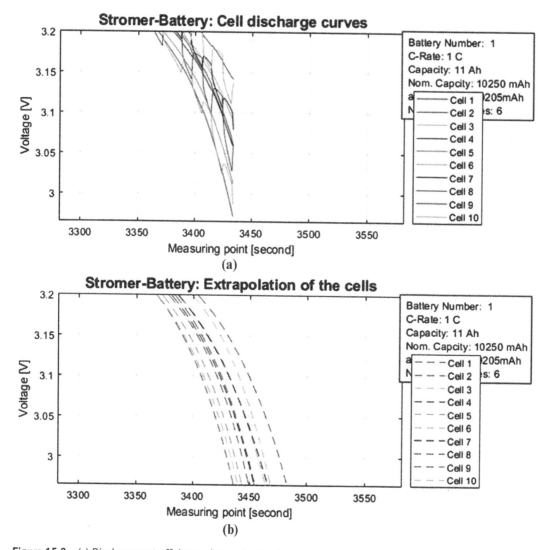

Figure 15.8 (a) Discharge cut-off due to the weakest cell and (b) extrapolation of the stronger cells to determine the additional discharge time they could have offered.

value over the 10 cells is given by the red bar. The weakest cell 6 has no remaining capacity while the strongest cell 10 has over 100 mAh remaining capacity. The mean of the remaining capacity in 10 cells is approximately 55 mAh.

Figure 15.9(b) shows the effect of active balancing on improving the discharge capacity of the Stromer battery. In active balancing, during charging and discharging processes, the strong cells support the weak cells in order to balance the differences of capacity among them. To determine the effect of active charge balancing, the mean value of the residual capacitances is compared with the output capacitance of the reference cell (the weakest cell).

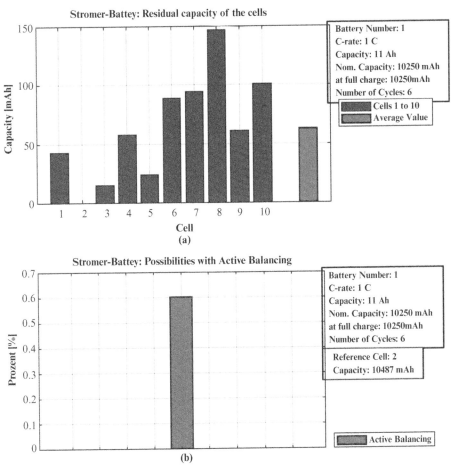

Figure 15.9 (a) Remaining capacity in each cell of the Stromer battery when the cut-off is at the end of the charge and (b) the effect of active balancing in recovering the remaining capacity from the Stromer battery.

15.4 Active Balancing Solutions

Among other companies, Linear Technology Corporation provides the active balancing solutions for the Li-ion battery. The LTC3300-1 balancing module from Linear Technology has a charge transfer efficiency of up to 92%. In the ideal case or in a MATLAB simulation, active balancing is able to extract 92% of the remaining capacity from the battery.

The LTC3300 from Linear Technology is a key element in a battery management system because it provides an active charge balancing feature. The charge balancing can reduce inequalities between cells, resulting in a longer battery life and, in the case of active balancing, a higher total capacity.

The LTC3300 is a bidirectional balancing device that can be used as a controller based on a synchronous fly-back topology and can manage up to six series-connected cells with a common mode voltage of up to 36 V. The charge may be shifted between a selected cell and 12 or more adjacent cells. All balancers can work independently.

The individual balancers are controlled via an SPI-compatible interface with level shifters, which makes it possible to connect several controllers of the LTC3300 in series.

15.4.1 Structure of LTC3300

The active charge balancing is performed by a bidirectional synchronous fly-back converter controlled by the LTC3300. Each LTC3300 controller has six independent fly-back controllers that allow individual loading or unloading of a cell. The six control signals within the LTC3300 can control the gates of the NMOS switches on the primary and secondary sides. The primary winding of the fly-back converter is connected to the battery potential while the secondary winding is connected to the potential of the respective cell. Figure 15.10 shows the basic structure of LTC3300.

15.4.2 Discharging Procedure

If a cell needs to be discharged in a battery, the MOSFET is switched on at the primary side, as shown in Figure 15.11. A current flows through the primary-side inductance, which rises ramp-shaped until a voltage of 50 mV is detected at the current measurement input of the LTC3300. The NMOS-FET is then switched off and the energy stored in the transformer is transmitted to the secondary side, whereby a current is generated in the secondary-side inductor of the transformer. In synchronism with switching off the primary MOSFET, the secondary side is switched on until the current returns to zero. Once the current has reset to zero, the entire cycle is repeated until the LTC3300 controller issues a command to stop the discharge mode. The process transfers the charge from one cell to all other cells.

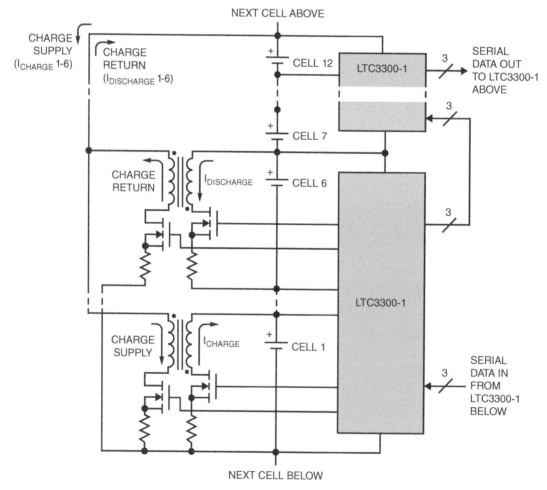

Figure 15.10 Structure of LTC3300.

15.4.3 Charging Process

If a cell needs to be charged, the secondary-side MOSFET for the selected cell is switched on, so that the current flows from the cells through the transformer, as shown in Figure 15.12. The MOSFET remains

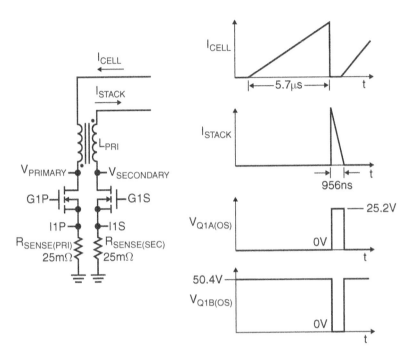

Figure 15.11 Discharging with LTC3300.

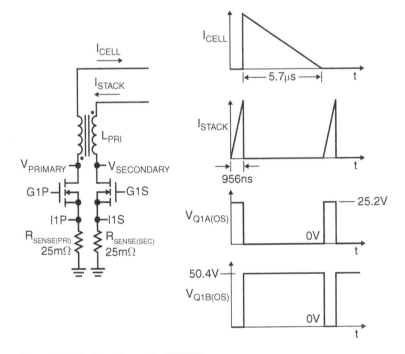

Figure 15.12 Charging with LTC3300.

switched on until the current measurement input is at a voltage of 50 mV; then it is switched off. Synchronously, the primary-side MOSFET is turned on, whereby the current flows through the inductor and charges the selected cell from the other cells in the stack. This cycle is repeated until the LTC3300 controller terminates the load mode.

Figure 15.13 shows a typical curve with active balancing. The stronger cells support the weaker ones by charge balancing. Active balancing results in a small voltage drop in the stronger cells and a small voltage boost in the weaker ones.

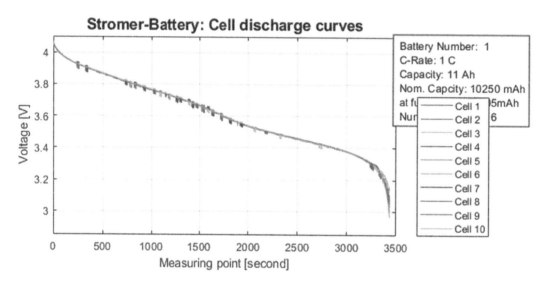

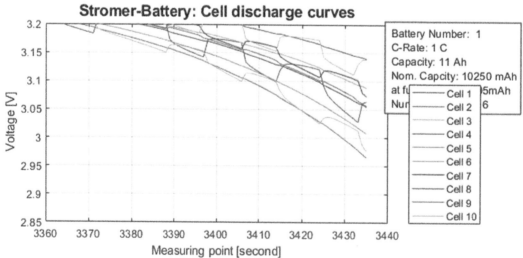

Figure 15.13 Test results of active balancing during the discharge process: (a) test results of active balancing and (b) enlarged view of test results of active balancing from 3360 s to 3440 s during the discharge process.

15.5 Test Results

The company Stromer has provided eight battery packs for the measurements: three new batteries, three batteries of medium age, and two batteries of advanced/old age. All eight batteries were discharged with the FuelCon battery test stand at the Bern University of Applied Sciences with three CCs, 0.5 C, 1 C, and 2 C, and the measured values of each individual cell were recorded with the Agilent data logger.

The testing was carried out with only three samples with a new age, a middle age, and an advanced/old age battery (see Table 15.4). All three aging battery pack samples were discharges with 1 C and measurements with and without BMS were taken. The BMS measurements were specifically examined to find the active balancing solution from Linear Technology.

Without BMS, testing is done with three samples of batteries that were discharged with CC of 0.5 C, 2 C, and 1 C. With BMS (active and passive), the test battery was discharged with 1 C. With passive balancing, the discharged was repeated five times and the measurement result was obtained by averaging five results. Active balancing was performed manually. In the process, only the best cell was unloaded and the worst cell was loaded. Three discharging processes were recorded for each battery, and only the best result for the evaluation was selected. Manual balancing is difficult to be effective because cells start diverging only at the end of the discharge and the time for compensation is very short for effective manual operation.

15.5.1 Measurements with Different Discharges

Battery NEW

Figure 15.14(a) and (b) shows the discharge curves of the battery as well as the voltage difference between the best and the worst cell with different discharge currents.

In the new battery, the voltage differences between the best and worst cells are between 0.08 V and 0.22 V. Figure 15.14(b) clearly shows that the greatest voltage difference is achieved by active balancing while discharging. However, the cell's voltage drift is smallest without BMS while discharging with 2 C.

Without active balancing, during discharge the voltage differential curve initially runs very flat and then rises exponentially shortly before the discharge cut-off voltage. In contrast, with active balancing, during discharge the curve remains flat, except that the voltage peaks can be detected, which indicates an active balancing for the entire period.

Table 15.4 History of the three test batteries.

	Battery 1 new	Battery 6 middle	Battery 4 old
Capacity (Ah)	11	10	10
Design capacity (Ah)	10.250	9.500	9.500
Capacity full charge (Ah)	10.205	8.916	6.172
Number of cycles	6	73	416
Date of production (week/year)	51/2012	35/2010	8/2011

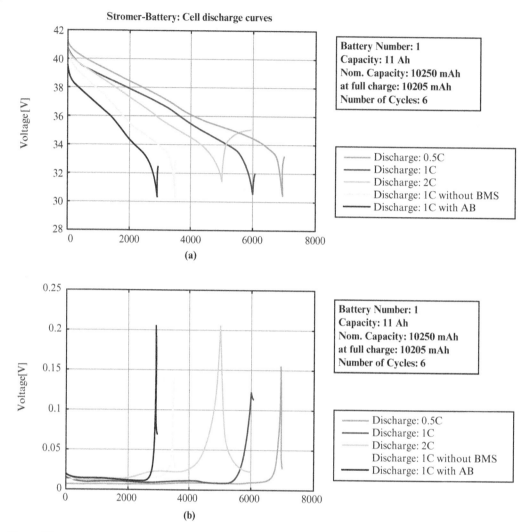

Figure 15.14 (a) Battery NEW – discharge curves and (b) voltage drift among cells while discharging with various discharge currents.

Figure 15.15 shows that the remaining capacity in the battery have different C-rate discharges. The active balancing effect can be derived from the remaining capacities. With a slow rate discharge of 0.5 C, the remaining capacity is very small and thus the effect of active balancing is relatively low at approximately 0.25%.

The new battery has the largest remaining capacity during discharge, even with active balancing. This indicates that the manual charge balancing has not been optimal and the result is worse than when the battery was tested without active balancing. The difficulty with manual charge balancing is that the cells do not drift until the end of the discharging process and a sudden drift occurs at the discharge cut-off voltage. The length of time for the determination of the weakest and strongest cell as well as starting the charge compensation is very short. Consequently, it is not possible to compensate energy optimally between the individual cells.

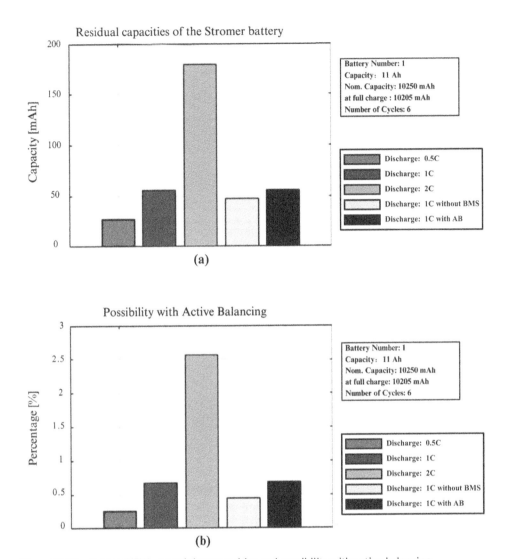

Figure 15.15 Battery NEW – remaining capacities and possibility with active balancing.

Battery MIDDLE Age

Figure 15.16 shows battery MIDDLE age – discharge curves and voltage differences. In the middle age battery, the voltage differences between the best and the worst cell range from 0.08 V to 0.26 V over various C-rate discharges. The 2 C-rate discharge shows the highest voltage difference in a middle age battery. However, the voltage drift gets smaller when using active balancing while discharging.

In the middle age battery, active balancing improves extracting the remaining capacity. The effect of active balancing can be derived from the remaining capacities average.

Most of the analysis and Figure 15.17 show that the residual capacities approximately remain equal to the capacity extracted with active balancing. With manual balancing, the mean remaining capacity is a little lower than 40 mAh. The remaining amount that would theoretically be extracted is 0.5% of the total designed capacity.

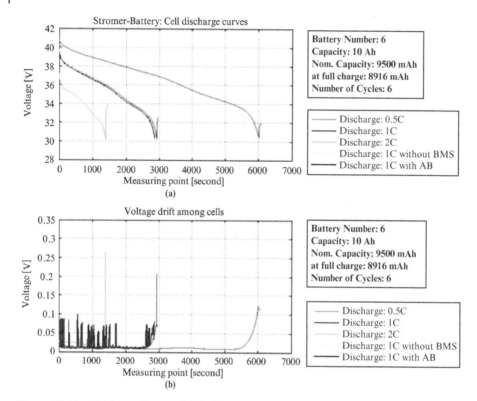

Figure 15.16 Middle age battery: (a) discharge curves, (b) voltage drift among cells while discharging with various discharge currents.

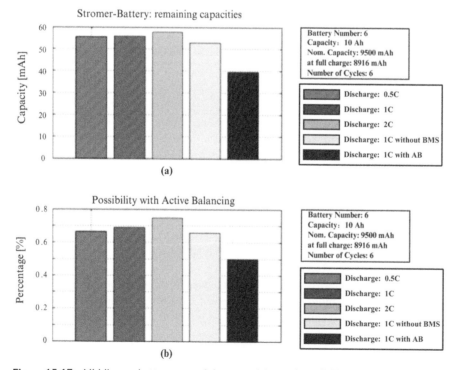

Figure 15.17 Middle age battery – remaining capacities and possibility with active balancing.

Battery OLD Age

The voltage drop across the internal resistance in the old battery was too high while discharging with 2 C and the discharge curve is not usable. Therefore, the discharge curve with 2 C is not included in this section for analysis.

The voltage difference between the best and the worst battery ranges from 0.15 V to 0.2 V, with various Crate discharges. Unlike others, the highest voltage difference was found with a discharge of 0.5 C. Figure 15.18(b) shows the voltage differences that remain are approximately equal when discharging with and without a battery management system and when discharging with active balancing.

Figure 15.19 shows, on discharging without BMS, that the old battery performs equally poorly for both cases of (a) remaining capacity and (b) possibility with active balancing. This is an indication that balancing is not effectively performed. The result was not verified on multisamples with various C-rate discharges. Therefore, the results are not universally applicable.

In Figure 15.19(b) active balancing shows a much higher 2.2–3.9% of capacity extraction in an old battery compared to other cases. However, this applies only to the ideal condition. The active charge transfer, otherwise, is also affected by losses.

The theoretical residual capacity that would have to be extracted with ideal active balancing has worsened here due to manual balancing and has a value of 3.9%. With this limited testing, we can conclude that manual charge compensation has not been optimal and the battery works better with the original BMS. As with the other measurements, this result has to be viewed subjectively, since it is very difficult to guarantee an optimal compensation of the cells manually.

In summary, it can be stated that the Stromer batteries under test work better with passive balancing (with BMS) as compared to active balancing. However, cells drift away when they age.

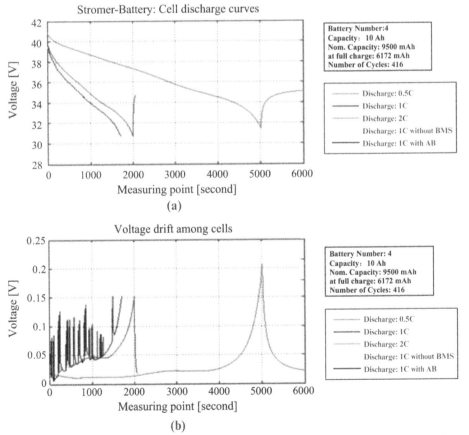

Figure 15.18 Old age battery: (a) discharge curves and (b) voltage drift among cells while discharging with various discharge currents.

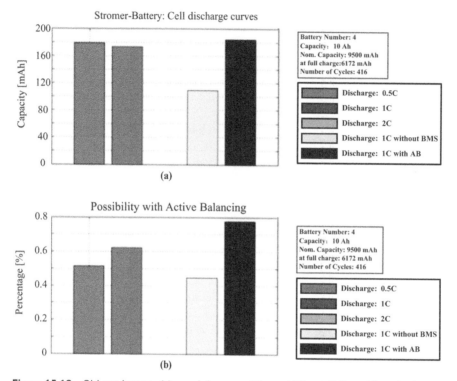

Figure 15.19 Old age battery: (a) remaining capacities and (b) possibility with active balancing.

15.5.2 Comparison Between the Batteries

15.5.2.1 Voltage Drift Between the Best and Worst Cells

Figure 15.20 shows the voltage drift of new, middle age, and old batteries at different rates of discharge. The voltage difference between the best and the worst cell ranges from 0.07 V to 0.26 V for all age batteries and including all discharge currents.

Figure 15.20 does not suggest conclusively that an old battery has a larger voltage drift, but the voltage drift is mostly a function of the discharging C-rate.

The oldest battery (battery 4) shows a higher voltage drift in the beginning; however, it matches other batteries eventually on discharging.

Nothing remarkable is found with the measurements on discharging the battery without a BMS.

It is very clear with Figure 15.21 that the battery loses capacity with increasing age. The oldest battery (green curve) reaches the discharge cut-off voltage much earlier than the other batteries. Discharging the battery with active balancing gives intermediate voltage peaks during the entire period of discharging. These intermediate peaks certainly indicate charge balancing among cells. In addition, the voltage differences are of the same order of magnitude as the original Stromer–BMS with passive balancing. Thus, we did not see a significant improvement by adding a complex active balancing mechanism.

15.5.2.2 Residual Capacity

The available residual capacity increases with the aging process of the battery. This can be seen very clearly in Figure 15.22, where the oldest battery has the most residual capacity in contrast to the new battery, which has the least residual capacity. The results suggest that voltages of the cells drift apart more quickly in older batteries and force the discharging to stop. Hence stronger cells in an old battery have been left with more residual capacity. Note that the analysis is based on the extrapolation discharge curve up to the discharge cut-off voltage.

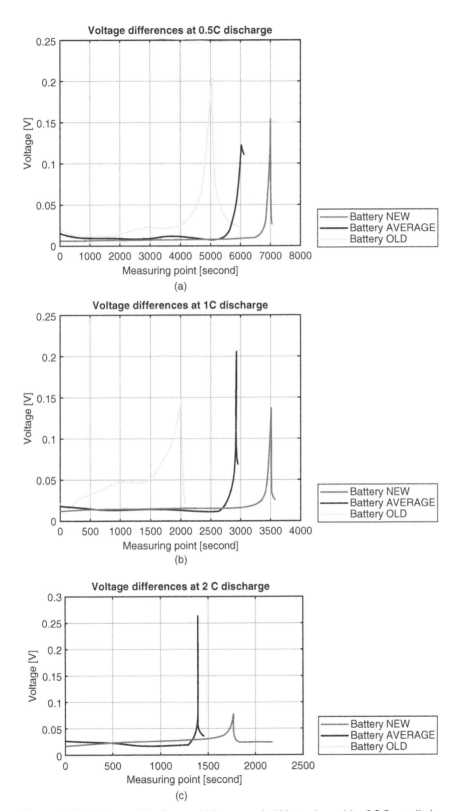

Figure 15.20 Voltage drift of new, middle age, and old batteries at (a) a 0.5 C rate discharge, (b) a 1 C rate discharge, and (c) a 2 C rate discharge.

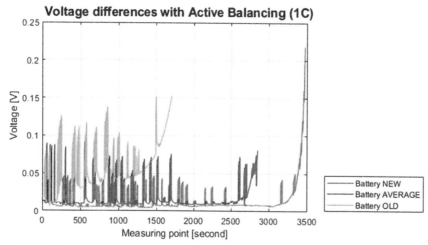

Figure 15.21 Voltage drift between the cells with active balancing for new, middle age, and old batteries.

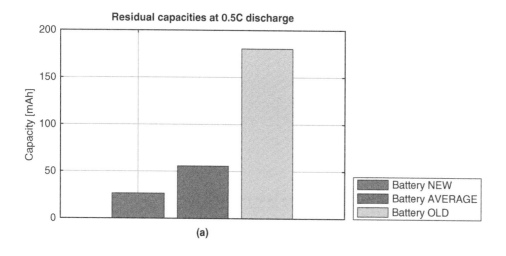

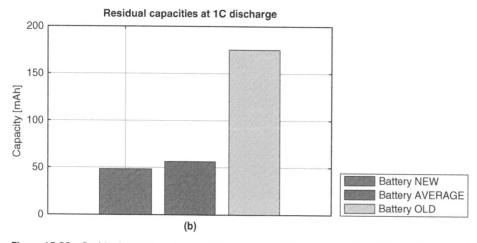

Figure 15.22 Residual capacity of new, middle age, and old batteries at (a) 0.5 C and (b) 1 C discharges.

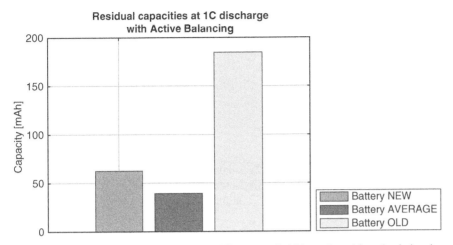

Figure 15.23 Residual capacity of new, middle age, and old batteries with active balancing.

Figure 15.22(a) also shows the discharge with 0.5 C, the oldest battery has a maximum residual capacity that is approximately 180 mAh, and the new battery has a residual capacity of approximately 25 mAh.

An exception occurs in Figure 15.23 when batteries are discharged with active balancing in the cells. In the figure the middle battery has less residual capacity than the new battery, which suggests that active charge balancing works best with the middle age battery. However, there were not enough measurements to validate the single test result.

15.6 Possibility with Active Balancing

With active charge balancing, when cells transfer energy among themselves, it is possible to utilize the total average capacity of a battery.

Figure 15.24 shows that the old battery can ideally get the highest capacity out as 3% more on using active balancing. Moreover, with the new battery, less than 0.5% more capacity can be extracted. The figure also suggests that the effect of active balancing is more pronounced with a higher discharge current.

When discharging with active and manual balancing, it is found that the possibility of extracting the remaining capacity has deteriorated in a new and an old battery. In Figure 15.25, it is clear that the middle age battery performance (the usable residual capacity) has improved with the manual compensation. It would be interesting to repeat the measurements with an implemented balancing algorithm and to compare the measured values.

15.7 Results and Evaluation

When designing the Stromer battery, the following test cases were considered.

Three battery packets were subjected to the following measurements, as shown in Table 15.5.

Final results have been generated by calculating the mean values of a higher number of measurement samples, such as the case using five or three of two samples. Only the best result was included in the evaluation for manual load balancing. The testing results clearly showed that the battery cells drift apart with age. Active charge balancing, if employed, extracts more capacity from a battery. In general, the battery packs are good and only a

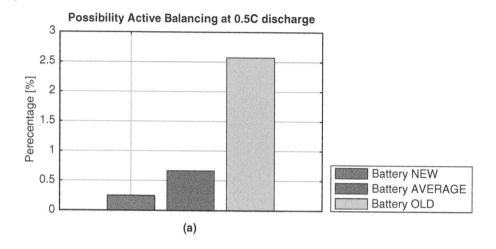

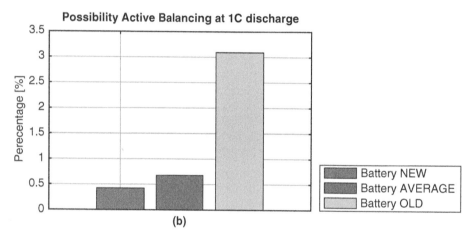

Figure 15.24 Possibility of extracting the remaining capacity with active balancing: (a) 0.5 C and (b) 1 C rate discharges.

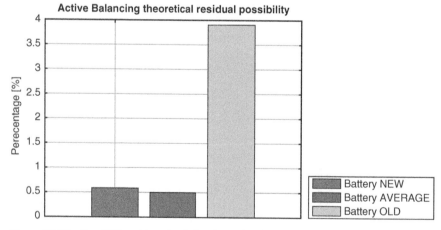

Figure 15.25 Possibility with active balancing using different discharges.

Table 15.5 Three test battery packets.

Discharge C rate	Electronics	Balancing	Sample measurement cycles
0.5 C	With BMS	Passive	1
1 C	With BMS	Passive	5
2 C	With BMS	Passive	1
1 C	Without BMS	No balancing	2
1 C	Without BMS	Manual active balancing	3

little more capacity could be extracted with active balancing. It makes more sense to develop a new balancing system with active balancing only when the new module can offer 92% or higher efficiency.

Since time was limited for the realization of a balancing algorithm in this project work, only the active balancing strongest cell was unloaded manually to load the weakest cell. In residual capacities, a middle age battery shows an improvement, but old and new batteries were deteriorating in a residual capacity.

In summary, it may be concluded that the capacity of the Stromer battery can be increased with active balancing. However, the scope of the gain in capacity is fairly low. For example, in an old battery only approximately 3% more capacity can be extracted with active balancing. A good balancing algorithm should be implemented in an LTC3300 demo board to achieve better gains in the capacity.

Reference

1 Honkura, K., Takahashi, K., and Horiba, T., "Capacity-fading prediction of lithium-ion batteries based on discharge curves analysis," *Journal of Power Sources*, 2011, 196: 10141–10147.

16

BMS for a Fork-Lift

In Europe, about half a million commercial vehicles are sold every year. Around half of these vehicles are now powered by lead-acid batteries. Lead-acid batteries are cost-effective, but they are heavy, show poor performance, and offer a shorter life. In addition, lead-acid batteries require a relatively large maintenance effort and have poor energy efficiency, particularly when partial discharge (shallow cycles) occurs. Lithium-iron-phosphate batteries behave differently. They produce higher efficiency, lighter weight, and a longer life, which make them increasingly attractive as an alternative. Lithium-ion (Li-ion) battery systems have been developed specifically for the conveyor belt area, with integrated battery management and adapted charging infrastructure. Use of Li-ion batteries has an economic advantage in the case of a precise lifetime cost analysis.

16.1 Lithium-Iron-Phosphate Batteries for Fork-Lifts

The lithium-iron-phosphate accumulator is the most suitable development of the Li-ion battery for fork-lift applications. Iron phosphate or $LiFePO_4$ is used as the cathode material. Iron phosphate can be produced less expensively than cobalt oxide, which is commonly used for Li-ion batteries. Even more problematic, cobalt is supplied by only a few countries, which makes this raw material more vulnerable to tremendous price fluctuations due to commodity speculation. Such imponderables are less likely to happen for iron phosphate as the raw materials for this substance are readily available worldwide.

The technology also has a number of key advantages to be used in industrial truck applications:

- Lithium-iron-phosphate technology is currently the most powerful, commercially available, rechargeable battery system with cell sizes above 100 Ah.
- Compared to conventional lead systems, a three to five times longer operating time is possible with the same weight.
- Due to the high thermal stability and firm bonding of oxygen, it is absolutely safe, maintenance-free, and a long-lasting operation.
- Li-ion batteries can be discharged at high currents, which makes it possible to use lower battery capacities for short-term systems.
- Li-ion batteries can be recharged quickly, which means that a battery can be re-used after use.
- The constant voltage levels across the wide range of discharge poses little demands on the converter technology (see Figure 16.1).
- The relatively low capacity reduction at low temperatures allows lower demands on temperature management.
- Lithium-iron-phosphate batteries have a very small self-discharge (<3% per month).
- Lithium-iron-phosphate batteries do not need a floating charge to maintain the SoC level while not in use.
- The charging factor and energy efficiency are close to one (see Figure 16.2). Higher energy efficiency together with a low self-discharge rate results in low losses.

Battery Management System and its Applications, First Edition. Xiaojun Tan, Andrea Vezzini, Yuqian Fan, Neeta Khare, You Xu, and Liangliang Wei.

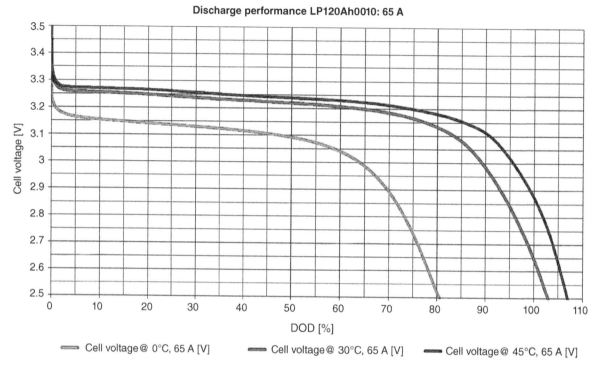

Figure 16.1 Discharge behavior of the LiFePO$_4$ cell at 2 C with various temperatures.

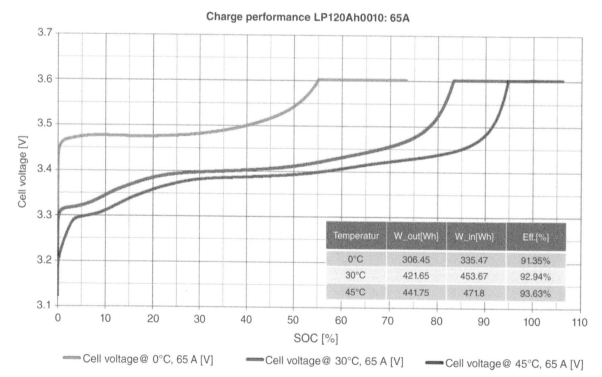

Figure 16.2 Charge behavior of the LiFePO$_4$ cell at 2 C with various temperatures (*Source:* Berner Fachhochschule/IPS AG http://www.integratedpower.ch).

- The proportion of gray energy in Li-ion batteries is very small and is not significant during the operating time as measured by the energy savings.
- Li-ion batteries use safe and recyclable materials.

16.2 Battery Management Systems for Fork-Lifts

The battery management system (BMS) plays a crucial role in the safe operation of a fork-lift. Various functions performed by the BMS are shown in Figure 16.3. There are major differences between protective and monitoring functions. While in the normal case only the monitoring functions play a role, the protective functions must intervene in the case of malfunctions. The following tasks are included in the category of monitoring functions:

- Voltage, current, and temperature monitoring: These values are frequently referred to as primary values and other variables are then derived from them. A directly derived value is, for example, the internal resistance, which allows conclusions to be made about the loading and health states.
- Charge balancing: Since charging and discharging due to the parameter dispersion and, for example, different temperature distributions in the battery pack can lead to an uneven load on the cells, they could drift over time and the weakest cell would then limit the capacity of the battery. The charge compensation is applied with the aid of parallel-connected discharge resistors or actively controlled charging circuits.

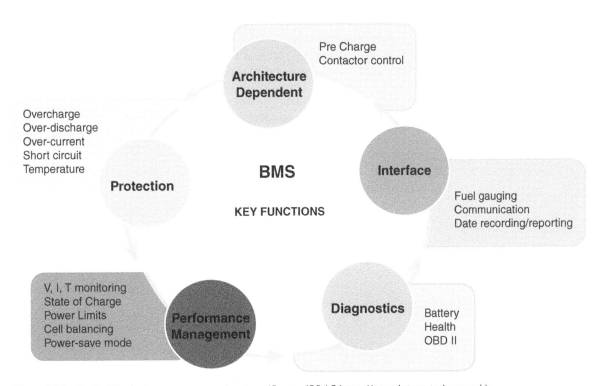

Figure 16.3 Task of the battery management system (*Source:* IPS AG http://www.integratedpower.ch).

- State of charge (SoC): The charge state plays a very important role, both for managing the battery and for displaying the information for the driver. While the battery management system monitors the operation of the battery within certain charge state limits, the system can also calculate the remaining available service life from the charge state using a superordinate consumption calculator.
- Health status (state of life/SoL or state of health/SoH): The BMS calculates the available capacity as a function of the number of charge and discharge cycles. Most commonly, a mathematical model is used to calculate the life time, which also includes calendar aging as well as depth of discharge, the current intensity, and the operating temperatures [1]. Intelligent systems periodically correct the calculated model values by means of so-called reference charges.

Other tasks of the BMS include the protection and control over external peripherals. External peripherals are charging circuits, cooling units and protective switches, and diagnosis and communication with the driver and the vehicle.

16.3   The LIONIC® Battery System for Truck Applications

Under a collaborative project, Integrated Power Solutions AG and Bern University of Applied Sciences have developed a complete battery system for industrial truck applications using lithium-iron-phosphate batteries. Under the project, an integrated protective circuit has been developed for an optimal reaction using high capacity (50/70/100/120 Ah) lithium-iron-phosphate cells. A "Smart Header" in the system stores digital DNA of a single cell. Digital DNA keeps the single cell voltage and temperature. This is the basic unit for the LIONIC® battery, under the name LIONIC® Smart Cell Battery System, as shown in Figure 16.4.

The Power Protection Unit (PPU) in an extended battery management system controls all the Smart Cells and continuously collects their data. Each of the cells and the battery are checked for a safe operating range (monitoring). In the event of a fault, switches completely disconnect the battery from the load and prevent any damage to the material and the human operator.

A LIONIC®-certified charger has an interface between the battery and the charger, which allows the PPU to intelligently influence the charging process. PPU controls the charging current as a function of the charge state.

The most important parameters, such as the charge status and the real-time operating status, can be read directly on the battery or via the external charge status indicator, which is an optional feature.

Figure 16.4   LIONIC® Smart Cell and LIONIC® Power Protection Unit (*Source:* IPS AG).

For several years, the LIONIC® lithium-iron-phosphate battery systems have been tested in various applications. A number of clear advantages have emerged for the customer:

- Maintenance-free: Li-ion battery systems are absolutely maintenance-free.
- Energy efficiency: Li-ion battery systems have a very high energy efficiency compared to conventional solid-state battery systems. They offer high efficiency up to 95% and vary with the application. A high efficiency provides approximately 20 to 40% of energy savings.
- Emission-free: Li-ion battery systems are absolutely emission-free and can therefore be used everywhere.
- Protection class: LIONIC® Li-ion battery systems meet IP54 protection. This means that the system has dust and splash protection. A higher degree of protection is also possible on request.
- Intrinsic safety and operational safety: LIONIC® Li-ion battery systems have subsystems that ensure intrinsic safety. During an emergency these subsystems lead to a shutdown of the battery system.
- Temperature-resistant: Li-ion battery systems can be used very well at large temperature ranges, especially at very low temperatures.
- Lifetime: According to the latest findings, Li-ion battery systems can be expected to have at least twice the life expectancy compared to conventional lead-acid batteries.
- Intermediate charge: Li-ion battery systems tolerate intermittent charging, as opposed to lead-acid batteries; however, the service life is shortened by intermediate charges.
- Fast charging: Li-ion battery systems accept fast-charging, which can provide more than 80% of the capacity in less than 1 h.
- Higher energy density: Li-ion battery systems are characterized by a high energy density (three times higher than conventional lead-acid batteries). Therefore, Li-ion battery systems offer greater performance at a smaller volume.
- Recuperation current: Li-ion battery systems are particularly capable of recuperation due to a relatively low internal resistance throughout the SoC range.
- New functionalities: LIONIC® Li-ion battery systems have a battery management system that provides accurate data, charge state, number of charging and discharging cycles, energy throughput, lifetime, and possibility of upgrading.
- Li-ion battery systems last longer, despite the initial high cost compared to a conventional lead-acid battery, but Li-ion battery mitigates the acquisition costs in the long run.
- Low self-discharge: The self-discharge rate is very small in comparison to the conventional lead-acid solutions.

16.4 Application

LIONIC® energy systems are capable of fast charging and can also be intermediate charged. Charging is carried out at a constant current. In a two-shift operation, a replacement battery is not required if intermediate charging is carried out during breaks (1 × 15 min and 1 × 30 min per shift). As an example, Figure 16.5 shows the capacity curve of a LIONIC® 24 V/9 kWh (360 Ah) energy system for a two-shift operation with intermediate charging.

LIONIC® with a higher energy efficiency reduces the costs and protects the environment!

Figure 16.6 shows that the conversion from an electrochemical process to electrical energy is only 70% efficient in the lead-acid battery. The losses arise due to the charge factor, the large voltage swing between the charge and discharge, and the temperature rise in the battery during the charging/discharging process. Only 64% of the fully charged battery energy is available for the operation of electric vehicles where lead-acid batteries are in use.

If LIONIC® energy systems are used instead of lead-acid batteries, the usable energy for the electric vehicle increases by up to 85.5%, which is significantly greater than 64%. This is because the substantially higher efficiency of the LIONIC® energy systems compared to lead-acid batteries, which is approximately 93%. The high efficiency is the result of a reduced charge factor, lower voltage swing, and lower temperature rise while charging and discharging. Thus, the LIONIC® energy systems provide much better energy efficiency. When compared with lead-acid batteries, the LIONIC® energy system requires 30% less electrical energy for every charging operation.

Figure 16.6 shows that, on using LIONIC® energy systems, the electrical energy cost saving is achieved by 30% and CO_2 emissions are reduced by the same factor.

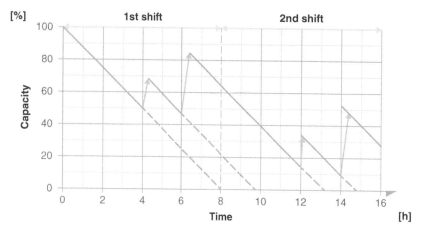

Figure 16.5 Capacity curve of a LIONIC® 24 V/9 kWh (360 Ah) energy system for a two-shift operation with intermediate charging.

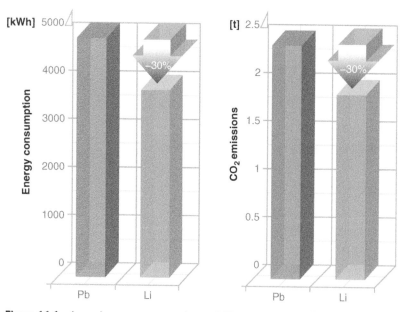

Figure 16.6 Annual energy consumption and CO_2 emissions for charging traction batteries – lead-acid battery (Pb)/ lithium-ion battery (Li).

16.5 The Usable Energy Li-Ion Traction Batteries

The energy contained in a traction battery is the product of the rated capacity (Ah) and the rated voltage (V). A fully-charged 24 V lead-acid battery with a capacity of 375 Ah (five hours) has an energy content of 24 V × 375 Ah = 9.0 kWh. In comparison, a 24 V – 240 Ah Li-ion battery comprising 2 × 8 cells, each with a cell voltage of 3.2 V, has a rated voltage of 25.6 V and an energy content of 6.1 kWh.

The cell voltage of an E/PzS lead-acid battery drops significantly during the discharge process. This voltage drop is further increased on discharging with higher rates (see Figure 16.7).

In contrast, Figure 16.8 shows the discharge voltages of Li-ion batteries. The discharge curves are very constant up to one hour of discharge time, so nearly 100% of the initial energy is available over the whole discharge range.

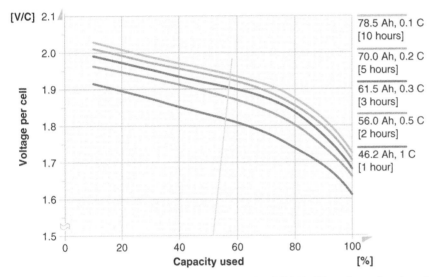

Figure 16.7 Discharge curve for an E/PzS lead-acid cell (70 Ah, 5 hours) as a function of capacity used.

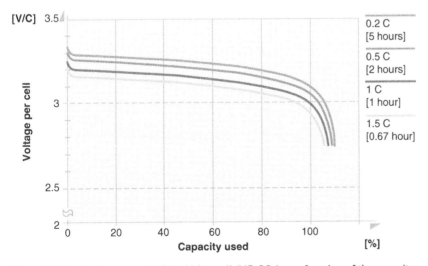

Figure 16.8 Discharge curve of the Li-ion cell (LiFePO$_4$) as a function of the capacity used.

To determine the nominal capacity of the E/PzS traction battery for a particular industrial truck and load profile, one needs to compensate for the reduced energy deficiency of the lead-acid battery. The compensation is mostly achieved by over-sizing the battery and by selecting a larger rated capacity (Ah) battery. A large rated capacity ensures that the energy supply to the truck is sufficient for an entire period of the load profile when the battery discharge increases. However, with a highly efficient Li-ion battery a higher compensation is not required and so the system does not need to over-size the battery. From previous experience, L-ion batteries with a 35% smaller capacity than E/PzS lead-acid batteries can be chosen for the same application. This is due to the excellent voltage stability of the Li-ion battery.

For example, Figure 16.9 clearly shows that a 240 Ah E/PzS lead-acid battery can be replaced by a 150 Ah Li-ion battery.

Figure 16.10 shows the voltage and current characteristics when charging the Li-ion cell with a charge current of 0.5 C.

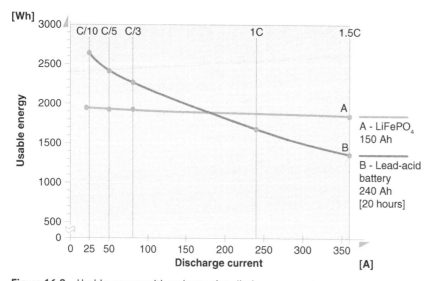

Figure 16.9 Usable energy with an increasing discharge current.

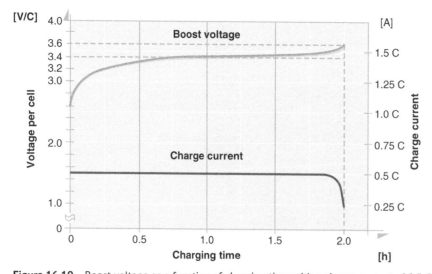

Figure 16.10 Boost voltage as a function of charging time with a charge current of 0.5 C.

With this charge current of 0.5 C, a fully discharged LIONIC® energy system requires two hours of charging time.

Li-ion technology is currently the most powerful, commercially available rechargeable battery system. Compared to conventional lead systems, a three to five times longer operating time is possible with the same weight. With a suitable choice of Li-ion technology, an absolutely safe, maintenance-free, and long-lasting operation is also possible. Li-ion batteries therefore fulfil all the requirements for a modern heavy E-vehicle system.

There has been consistent development of the battery system for applications like a heavy electric vehicle and case study results have early shown the economically viability of the LIONIC® Li-ion system. This is a success story of replacing a conventional battery by Li-ion technologies.

Reference

1 Kassem, M., Bernard, J., Revel, R., et al., "Calendar aging of a graphite/LiFePO$_4$ cell," *Journal of Power Sources*, 2012, 208: 296–305.

17

BMS for a Minibus

Compared with light-duty electric vehicles and special vehicles, comprehensive functions are required for the battery management system in the application of new energy vehicles. For example, for the minibus, the battery management system not only monitors and controls relevant features of the cells, but also provides an optimization strategy for the energy output and input of the whole bus. Its main functions include (1) a battery system discharging strategy management, (2) a battery system charging strategy management, (3) a consistent performance evaluation, prediction, and management of the battery system, and (4) battery safety management and protection. These four functions are described in this chapter.

17.1 Internal Resistance Analysis of a Power Battery System and Discharging Strategy Research of Vehicles

The battery management system used in the new energy vehicles must be developed according to the V model, as shown in Figure 17.1.

All items shown in Figure 17.1 are described below:

1) Definition requirements: According to the design concept of vehicle products, the total output voltage, output power, and functions of the battery system, such as the equilibrium mode, system charging mode, performance evaluation method, safety protection requirements, etc., are determined.
2) System-level specifications: According to the national specifications and the industrial requirements, the requirements that defined functions and performance must meet are established, including charging cut-off voltage, discharging cut-off voltage, charging current, maximum allowable discharge current and time, system temperature threshold, etc.
3) Subsystem design: According to the definitions and specifications of the battery system, the topological structure of the battery system is constructed and divided into multiple subsystems, and corresponding software, hardware, and supporting schemes are designed according to functions of various subsystems.
4) Subsystem realization: The designed subsystem hardware and software are debugged, modified, improved, and verified to finally complete the subsystems meeting the product requirements.
5) Subsystem integration and test: The completed subsystems are integrated, tested, and improved to ensure that the functions and performance of the subsystems meet the product requirements.
6) System-level integration and test: The whole system is completed after integration of the subsystems and the functions and indicators of the system are tested and improved.
7) Completion of integration and debugging: after completion of the above-mentioned steps, the product prototype is completed, tested, and modified to launch the final product.

Battery Management System and its Applications, First Edition. Xiaojun Tan, Andrea Vezzini, Yuqian Fan, Neeta Khare, You Xu, and Liangliang Wei.
© 2023 China Machine Press. All rights reserved. Published 2023 by John Wiley & Sons Singapore Pte. Ltd.

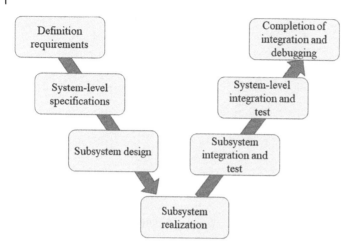

Figure 17.1 V model of the battery management system.

After the vehicle structure and relevant power parameters are determined, to meet the driving behavior of different users and complicated working conditions, the vehicle control mode is generally classified into (1) ordinary mode, (2) power mode, and (3) energy-saving mode. In the ordinary mode, the normal energy output is provided according to the linear/non-linear relationship between the accelerator and the motor, and the battery management system is only used to monitor the voltage, temperature, and current of the cells, and to give an early warning. In the power mode, on the basis of the ordinary mode, the relationship between the accelerator and the motor is properly changed so that the motor requests a higher output power from the energy system with the same stepping depth, so the battery management system provides the same functions as those provided in the ordinary mode. In the energy-saving mode, according to the power requirement of the motor and the battery characteristics, a set energy output optimization model is established by the battery management system to reduce energy consumption and increase the driving range of the vehicle. The energy-saving mode is the main one described in this chapter.

17.1.1 Internal Resistance Change Characteristic Research of a Power Battery

To research the energy-saving mode of the vehicle, it is first required to research the energy output performance of the battery. The internal resistance change of the battery is the essential reason for impact on the output of the battery system. Generally, the battery internal resistance includes ohm resistance and polarization resistance. Ohm resistance consists of contact resistance of anode and cathode materials, diaphragm, electrolyte, and parts, and relates to the production process, size, and structure of the battery, and is generally regarded as a constant. However, the ohm resistance value may vary with the aging and temperature of the battery. The polarization resistance refers to the departure of the electrode potential from the equilibrium potential for complicated causes such as adsorption and desorption of lithium ions, charge transfer, electrochemical reduction, and oxidizing reactions during the chemical reaction at the anode and cathode of the battery. Due to the active materials, the battery internal resistance presents a non-linear change characteristic while the battery is in use.

In conclusion, the equivalent internal resistance of the battery can be expressed by the following equation:

$$R_i = R_\Omega + R_f \tag{17.1}$$

where R_Ω is the ohm battery resistance and R_f is the polarization battery resistance.

The external characteristic of the battery internal resistance is shown in Figure 17.2, in which section B-D may be regarded as the characteristic of the ohm battery resistance and section C-D shows the characteristics of the polarization internal resistance.

Note the testing conditions: the 110 Ah lithium-iron-phosphate battery is discharged at 1 C and SoC = 97% for 10 s, and is kept in the non-charge-discharge state for 6000 s in order to observe the voltage change of the battery anode and the battery cathode.

For a certain battery system, the temperature, charge current, discharge current, and aging degree of the battery affect the change of the battery internal resistance.

The following method may be used to calculate the equivalent internal resistance of the battery by considering the impact at different temperatures and different currents:

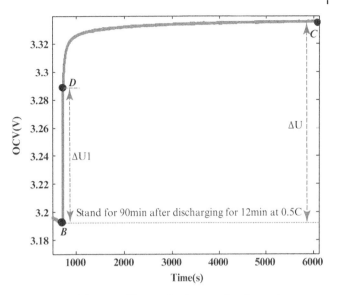

Figure 17.2 Voltage resilience after battery charging and discharging.

$$
\begin{cases}
r_{ic}(\mathrm{SoC}, I_c, T) = \dfrac{U_c(\mathrm{SoC}, I_c, T) - U_{oc}(\mathrm{SoC}, I_c, T)}{I_c} \\[2mm]
r_{id}(\mathrm{SoC}, I_d, T) = \dfrac{U_d(\mathrm{SoC}, I_d, T) - U_{od}(\mathrm{SoC}, I_d, T)}{I_d}
\end{cases}
$$

$$(17.2)$$

where $r_{ic}(\mathrm{SoC}, I_c, T)$ is the internal resistance at temperature T, charge current I_c, and SoC, U_c (SoC, I_c, T) is the working voltage at temperature T, charge current I_c, and SoC, $U_{oc}(\mathrm{SoC}, I_c, T)$ is the balanced electromotive force at temperature T, charge current I_c, and SoC, $r_{id}(\mathrm{SoC}, I_d, T)$ is the internal resistance at temperature T, discharge current I_d, and SoC, $U_d(\mathrm{SoC}, I_d, T)$ is the working voltage at temperature T, discharge current I_d, and SoC, and $U_{od}(\mathrm{SoC}, I_d, T)$ is the balanced electromotive force at temperature T, discharge current I_d, and SoC.

The battery is tested by the following steps to verify the impact of the charge and discharge currents I_c and I_d on the battery resistance at the same temperature.

1) Place a 110 AH lithium-iron-phosphate battery in a 20°C thermostat for about 2 hours.
2) Charge the battery at current I until a 3.8 V voltage is reached and keep it at a non-charge-discharge state until the temperature of the battery electrode is close to that of the thermostat.
3) Discharge the battery for 10 s at current I and keep the non-charge-discharge state until the temperature of the battery electrode is close to that of the thermostat.
4) During discharging, if the voltage of the battery is less than 2.7 V, perform step 5; otherwise repeat step 3.
5) Keep the non-charge-discharge state until the temperature of the battery electrode is close to that of the thermostat.
6) Charge the battery for 10 s at current I and keep the non-charge-discharge state until the temperature of the battery electrode is close to that of the thermostat.
7) During charging, check whether the voltage of the battery is more than 3.8 V. If yes, perform step 8 and if no, repeat step 6.
8) Keep the non-charge–discharge state until the temperature of the battery electrode is close to that of the thermostat.
9) Repeat steps 1 to 8 at current I of 0.1 C, 0.2 C, 0.3 C, 0.4 C, 0.5 C, 0.7 C, and 1 C.
10) Calculate the SoC of the battery and the internal resistance change of the battery during charging and discharging by using Equation (17.3) and draw the internal resistance change of the SoC curve.

Figures 17.3 and 17.4 show the internal resistance change curves of the battery during charging and discharging and Table 17.1 provides the charge–discharge capacity during the test. As shown in the figures, the charge and discharge currents have a certain impact on the battery internal resistance. At the end of charging and discharging, the charge and discharge currents have a more obvious impact on the change in the internal resistance: the larger the current, the larger the resistance change will be and the earlier the inflection point of resistance change will be. In the case of a small current, the resistance change at the end of the charge and discharge is much smaller

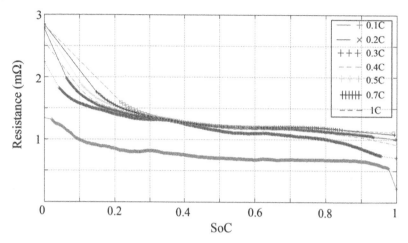

Figure 17.3 Battery internal resistance–SoC curve at the same temperature and different discharge currents.

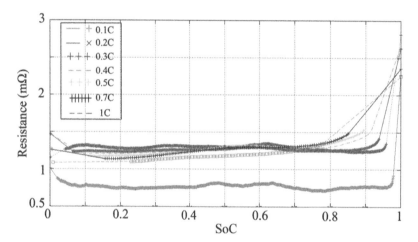

Figure 17.4 Battery internal resistance–SoC curve at the same temperature and different charge currents.

Table 17.1 Comparison of charge and discharge capacities of the battery at different currents.

Current	Charge capacity (Ah)	Discharge capacity (Ah)
0.1 C	110.7333	108.6250
0.2 C	110.5500	107.6333
0.3 C	110.5500	107.2530
0.4 C	109.6333	106.6820
0.5 C	108.6250	106.0083
0.7 C	107.1583	103.8000
1 C	103.5833	102.4167

than that of a large current. According to Table 17.1, if charged and discharged at the larger current, the battery may more rapidly reach the cut-off condition due to the increase in the battery resistance at the end of charging and discharging, when the battery cannot release all its energy. If the charge and discharge strategy is properly adjusted according to the resistance variation characteristic, the battery pack can be charged with more energy and discharge more energy.

A 110AH battery is tested by the following testing steps to verify the change of the battery internal resistance at different temperatures.

1) Place the battery in the thermostat with temperature T for about 2 h.
2) Charge the battery at a current of 0.5 C until a voltage of 3.8 V is reached. Keep the non-charge–discharge state until the temperature of the battery electrode is close to that of the thermostat.
3) Discharge the battery at a current of 0.5 C for 10 s. Keep the non-charge–discharge state until the temperature of the battery electrode is close to that of the thermostat.
4) During discharging, if the battery voltage is less than 2.7 V, perform step 5, otherwise repeat step 3.
5) Keep the non-charge–discharge state until the temperature of the battery electrode is close to that of the thermostat.
6) Charge the battery at a current of 0.5 C for 10 s. Keep the non-charge–discharge state until the temperature of the battery electrode is close to that of the thermostat.
7) During charging, check whether the battery voltage is more than 3.8 V. If yes, perform step 8 and if no, repeat step 6.
8) Keep the non-charge–discharge state until the temperature of the battery electrode is close to that of the thermostat.
9) Repeat steps 1 to 8) at temperatures T of –20°C, –10°C, 0°C, 10°C, 20°C, 30°C, and 40°C.
10) Compare the charge and discharge capacities at different temperatures. Calculate the SoC of the battery and the internal resistance change of the battery during charging and discharging using Equation (17.2) and draw the internal resistance change–SoC curve.

The charge and discharge capacity of the battery during the test is shown in Table 17.2. The charge and discharge capacity of the battery reduces with each temperature reduction, which means that the charge and discharge capacity of the battery significantly reduces with the temperature reduction when the battery is used in winter in the North. If no insulation measure is provided, the driving range of the electric vehicles is reduced at low temperatures. Therefore, during evaluation of the battery performance, full consideration should be given to the impact of the temperature on the charge and discharge capacity of the battery. During use of the battery, different temperature control strategies should be used for the battery pack according to the local temperature; for example, thermal insulation equipment is provided for the battery pack in cold regions and a heat dissipation strategy is used in warm regions.

Figures 17.5 and 17.6 show the battery internal resistance change curve. As shown in the figures, at less than 10°C, the battery internal resistance doubles and the charge or discharge capacity decreases correspondingly, mainly because the increase in the internal resistance leads to the battery voltage closing the cut-off voltage, which

Table 17.2 Comparison of charge and discharge capacity of the battery at different temperatures.

Temperature	Charge capacity (Ah)	Discharge capacity (Ah)
40°C	112.7539	111.8554
30°C	110.2051	110.1766
20°C	107.4965	111.7681
10°C	99.9545	107.9014
0°C	87.7082	92.0571
–10°C	70.8562	74.8113
–20°C	49.4409	54.543

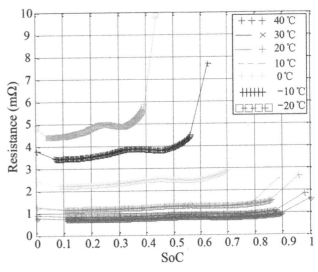

Figure 17.5 Battery internal resistance change during charging at different temperatures.

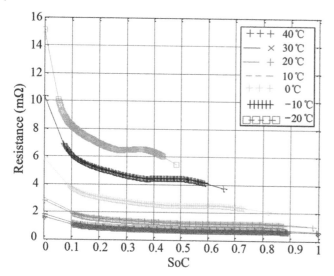

Figure 17.6 Battery internal resistance change during discharging at different temperatures.

makes the battery unable to be further charged or discharged. Therefore, if the battery is used at less than 10°C, a heating system should be installed to maintain the battery temperature to more than 10°C in order to ensure normal operation of the vehicle at low temperatures.

In conclusion, the charge and discharge currents and external temperatures of the battery have a direct impact on the battery internal resistance change, which results in the battery being unable to be charged by discharged to real energy during use of the electric vehicle, thus affecting the driving range of the vehicle. Therefore, a battery internal resistance diagram is drawn by testing the battery internal resistance at different temperatures and different charge and discharge currents in order to understand the change rule of the battery internal resistance at different temperatures and different currents and to provide a strategy for use of the battery in the vehicle. The battery may be tested and analyzed by the steps shown in Figure 17.7 to draw the battery internal resistance diagram.

In Figure 17.7, *I* is set as 0.1 C, 0.2 C, 0.3 C, 0.4 C, 0.5 C, 0.7 C, and 1 C respectively and *T* is set as –20°C, –10°C, 0°C, 10°C, 20°C, 30°C, and 40°C respectively. A 110 Ah power battery is used for the test. Figure 17.8 shows the resistance during discharging and Figure 17.9 shows the resistance during charging.

According to the battery internal resistance diagrams, as shown in Figures 17.8 and 17.9, the external temperature and the charge and discharge current have a direct impact on the charge and discharge performance of the battery system. A set of reference data can be provided for the charge and discharge mode of the vehicle by the

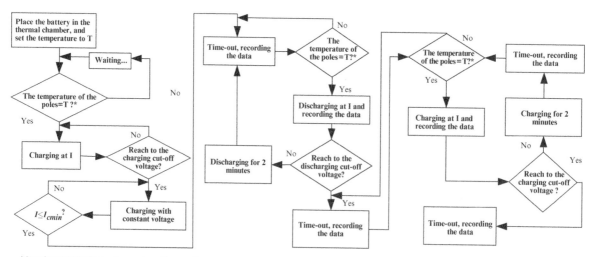

* In order to obtain the core temperature of battery, the temperature of poles, which connect the positive and negative sides of the battery, can be approximated to the core temperature of battery so that the sensors was fixed on the poles of battery and it was called "the temperature of poles" in the test.

Figure 17.7 Test steps to obtain a battery internal resistance diagram.

battery internal resistance diagrams during charging and discharging to guide the charge and discharge strategies of the vehicle.

17.1.2 Internal Resistance Characteristic-Based Discharge Strategy

The energy consumption of the vehicle is optimized by two aspects: (1) high efficiency drive: the drive efficiency is improved by improving the operation efficiency of the motor and reducing the energy consumption of the drive system and (2) more energy is released from the battery by optimal management and efficient utilization of energy during use of the vehicle. According to the battery internal resistance, temperature, and current curve, the larger the discharge current, the further the inflection point of battery internal resistance battery is from 0% SoC, and vice versa. This could be used to explain the condition that often occurs when the electric vehicle is used: SoC is very close to 0% during rapid acceleration (discharge at a large current), but the vehicle can be run for a long mileage during low acceleration (properly stepping on the accelerator). It can be seen from this that, according to the internal resistance change characteristics at different SoC ranges and different currents, if the discharge current of the battery is properly controlled, the service efficiency of the battery can be improved and the driving range of the vehicle can increase.

According to the impact of the current on the battery internal resistance, a sustainable and maximum ideal discharge current of the vehicle can be established during different SoCs, as shown by

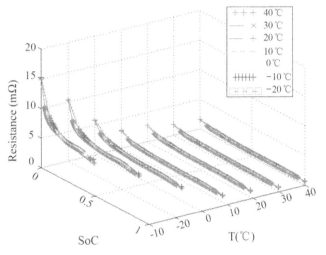

Figure 17.8 Battery internal resistance diagram during discharging.

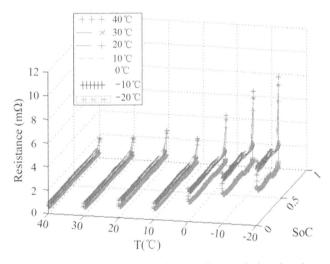

Figure 17.9 Battery internal resistance diagram during charging.

$$f(x) = \begin{cases} I_{d1}(\text{SoC}_1 < x < 1) \\ \dfrac{(I_{d1} - I_{d2})x + I_{d2}\text{SoC}_1 - I_{d1}\text{SoC}_2}{\text{SoC}_1 - \text{SoC}_2}(\text{SoC}_2 < x < \text{SoC}_1) \\ \cdots\cdots \\ \dfrac{(I_{cn-1} - I_{cn})x + I_{cn}\text{SoC}_{n-1} - I_{cn-1}\text{SoC}_n}{\text{SoC}_{n-1} - \text{SoC}_n}(\text{SoC}_n < x < \text{SoC}_{n-1}) \end{cases} \quad (17.3)$$

In Equation (17.3), I_{dn} is the sustainable and maximum ideal discharge current and SoC_n is the SoC corresponding to the inflection point of the resistance change at the discharge current I_{dn}.

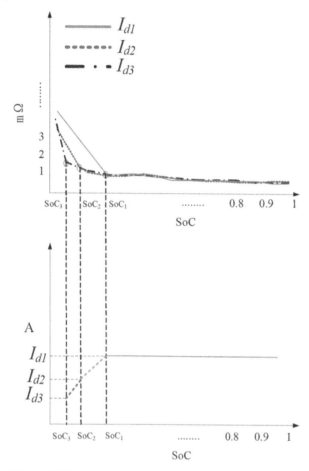

Figure 17.10 Ideal discharge current curve of the vehicle.

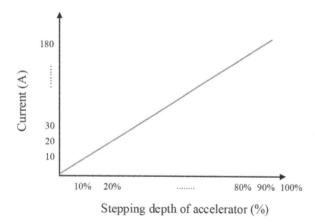

Figure 17.11 Stepping depth of the accelerator and the discharge current relation curve.

The output current of the vehicle is properly controlled by the ideal discharge current curve (as shown in Figure 17.10) in order to reduce the output power of the vehicle, slow down the increase of the battery internal resistance, and increase the discharge capacity of the battery and the driving range of the vehicle.

During actual use, the power performance of the vehicle relates to the discharge current. If the current is limited excessively, the power performance of the vehicle is compromised; for example, as the large current cannot be output from the battery, the vehicle may have a poor grade ability and fails to run during climbing and heavy-duty use.

During use, the output current of the vehicle relates to the stepping depth of the accelerator. Compared with the motor, the power of air conditioning and DC/DC is less, so it can be approximately considered that the output current is directly proportional to the stepping depth of the accelerator (see Figure 17.11):

$$I(\delta) = k\delta \quad (0 \leq \delta \leq 100\%) \tag{17.4}$$

where δ is the stepping depth of the accelerator, k is the scale factor of the stepping depth of the accelerator and current, and $I(\delta)$ is the output current at the stepping depth of the accelerator δ.

The stepping depth of the accelerator relates to road conditions and driving habits. During actual use, it is impossible to step on the accelerator at the 100% level, especially for an electric bus. Due to basically the same road conditions, the stepping depth is fixed to a certain level, except for when climbing and during acceleration. It may be considered that during driving, the driver basically controls the accelerator to a specific stepping depth. When the current limit is required due to SoC reduction and the current is limited by the stepping depth, if the current provided by the stepping depth can meet the use requirement of the vehicle, the driver does not deliberately step on the accelerator according to his/her driving habits. If the current cannot meet the use requirement, the driver will step on the accelerator while a large output current is increased by the control system to ensure the power performance of the vehicle.

Therefore, a discharge strategy can be established at different SoCs and a certain stepping depth range to limit the current without loss of the vehicle's power performance, according to the change in the battery internal resistance with the discharge current and the relation between the stepping depth and the current.

As shown in Figure 17.12, the maximum ideal discharging current is allowed to pass within a certain range of stepping depth of the accelerator. Therefore, the relation between the stepping depth of the accelerator δ and the current I is modified according to the following equation:

$$
f(x) = \begin{cases}
I_{d1} \, (\text{SoC}_1 < x < 1) \\
\dfrac{(I_{d1} - I_{d2})x + I_{d2}\text{SoC}_1 - I_{d1}\text{SoC}_2}{\text{SoC}_1 - \text{SoC}_2} \, (\text{SoC}_2 < x < \text{SoC}_1) \\
\quad\quad\quad \ldots\ldots \\
\dfrac{(I_{cn-1} - I_{cn})x + I_{cn}\text{SoC}_{n-1} - I_{cn-1}\text{SoC}_n}{\text{SoC}_{n-1} - \text{SoC}_n} \, (\text{SoC}_n < x < \text{SoC}_{n-1})
\end{cases}
$$

$$
I(\delta) = \begin{cases}
\dfrac{f(x)}{\delta_0}\delta & (0 \le \delta \le \delta_0) \\
f(x) + \left(\dfrac{I_{max} - f(x)}{1 - \delta_0} \right)(\delta - \delta_0) & (\delta_0 \le \delta \le 100\%)
\end{cases}
$$

(17.5)

where $f(x)$ is the sustainable and maximum ideal discharge current of the battery pack at the SoC value x, $I(\delta)$ is the output current of the battery pack at the stepping depth of the accelerator δ, δ_0 is the general stepping depth of the accelerator at the sustainable and maximum ideal discharge current of the battery pack, and I_{max} is the maximum discharge current of the battery pack.

Within a certain stepping depth range of the accelerator, although the increase of the battery internal resistance can be prevented and the driving range can be increased, this mode may reduce the driving comfort level. As shown in Figure 17.13A, under normal conditions, the stepping depth range of the accelerator is basically in direct proportion to the output current, that is to say, the output current of the battery varies with the stepping depth range of the accelerator to provide different power for the vehicle. When the relation between the stepping depth of the accelerator and the output current is changed (Figure 17.13B), the relation between the stepping depth of the accelerator and the output current becomes a piecewise function with a critical point δ_0. When the stepping depth of the accelerator is less than δ_0, the current change is less and the power performance of the vehicle is lower. When the stepping depth of the accelerator is more than δ_0, the current change is larger and the power performance of the vehicle is higher. Therefore, a condition may appear at this time that will affect the performance: when the accelerator is controlled to close to the critical value δ_0, the power performance of the vehicle may be sharply improved or reduced with the change of the stepping depth, especially at the end of a discharge, due to the current value corresponding to δ_0 becoming small, so when the stepping depth of the accelerator exceeds δ_0, the current change is quite large, the power performance of the vehicle will be improved instantaneously, and this will cause a feeling of tension to the driver, affecting the safety and comfort of the vehicle.

To avoid the sharp current change when the accelerator is operated to δ_0, the delay control is performed to the output current. Generally, the output current does not immediately match the current corresponding to the stepping depth, but shows a certain delay for a short time.

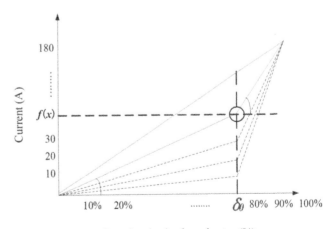

Figure 17.12 Relationship between the optimized stepping depth of the accelerator and the discharge current.

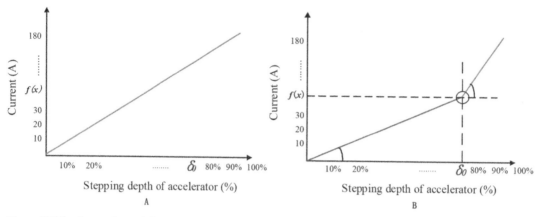

Figure 17.13 Comparison of the relationship between different stepping depths of the accelerator and discharge currents.

As shown in Figure 17.14A, at a certain stepping depth, the time Δt (very short, generally a few milliseconds) is required for change of the output current from 0 to $I(\delta)$. If the current is not slowly output, Δt relates to the cycle time of the procedure, that is to say, regardless of the stepping depth, Δt is required for the output current to reach a corresponding value. The current output is changed to the slow change mode by optimizing the slope k_0 of the current rise curve (the current response factor) in order to control the current response. When the stepping depth is higher than the stepping depth at the sustainable and maximum ideal discharge current δ_0, the current response factor is modified to k_1 (Figure 17.14B) according to

$$k_0 = \frac{f(x)}{\Delta t}$$
$$k_1 = k_0 \frac{f(x)}{\delta_0} \times \frac{1 - \delta_0}{I_{max} - f(x)}$$

(17.6)

The current is controlled to slowly rise to the output value by extending the response time of the current output so that tension is not caused to the drive due to the current change at different stepping depths in order to improve driving comfort and smoothness.

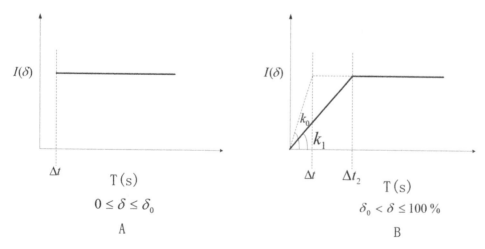

Figure 17.14 Current output delay.

A   B

Figure 17.15   Electric minibus.

To verify the effectiveness of the above-mentioned discharging strategy, the battery (Figure 17.15B, 110 Ah cell) used in an electric minibus (Figure 17.15A) is tested.

The test procedures are as follows:

1) Test the internal resistance of the 110 Ah lithium-iron-phosphate battery at the discharge current I_d of 1 C, 0.7 C, 0.5 C, 0.4 C, and 0.3 C, which are selected according to the characteristics of the lithium-iron-phosphate battery and the discharge current of the vehicle, and the temperature T of −20°C, −10°C, 0°C, 10°C, 20°C, 30°C, and 40°C.

2) Draw the average battery internal resistance diagram during discharging and calculate the inflection point of the battery internal resistance at different currents.

3) Draw the sustainable maximum discharge current curve according to Equation (17.3).

Obtain the inflection point of the battery internal resistance change at different temperatures and different currents, as shown in Table 17.3.

According to Table 17.3, the sustainable maximum discharge current curve of the vehicle is shown in Figure 17.16.

To verify the effectiveness of the method proposed here, a comparison test is performed for the pure electric bus by the following test procedures:

1) Fully charge the battery and test the no-load driving range of the vehicle using a conventional method until there is a full discharge of the battery.

2) Fully charge the battery and test the no-load driving range of the vehicle using the method proposed here and using same road conditions and the same driver until there is a full discharge of the battery.

3) Record the mileage and the energy output during testing using the conventional method and the method proposed here.

Their comparison results are given in Table 17.4. It can be seen that the discharge energy and the driving mileage can be improved by using the discharge strategy proposed here.

Table 17.3   Inflection point of the battery internal resistance change at different discharge currents.

Temperature	0.3 C	0.4 C	0.5 C	0.7 C	1 C
40°C	6.82%	9.08%	11.57%	16.38%	24.42%
30°C	6.82%	9.07%	11.56%	16.33%	24.37%
20°C	6.76%	8.98%	11.45%	16.01%	23.43%
10°C	6.62%	8.55%	10.83%	14.92%	22.14%
0°C	5.40%	7.15%	9.01%	12.53%	18.46%
−10°C	4.93%	6.54%	8.23%	11.53%	16.92%
−20°C	2.19%	2.91%	3.71%	5.24%	7.62%

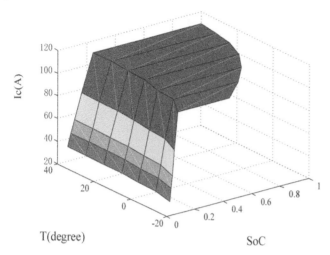

Figure 17.16 Sustainable maximum discharge current curve of the vehicle.

Table 17.4 Effect comparison of discharge strategies.

Control strategy	Discharge energy	Driving mileage
Conventional discharge strategy	31.69 kWh	189 km
Discharge strategy proposed here	32.75 kWh	193 km

17.1.3 Research of a Charging Method for a Power Battery System Based on an Internal Resistance Characteristic

According to the analysis results of the battery internal resistance change characteristic provided in Section 17.1.1, the charging time relates to the charge current: the larger the current, the shorter the charging time is, and vice versa. The charge current may be increased to shorten the charging time of the battery. Therefore, the charge or discharge current is described as "C" or the "multiplying power" in the industry. When the current is equal to the rated capacity, if the battery is charged to the rated capacity within 1 h, the charge current is expressed by the multiplying power 1 or 1 C, and the multiplying power 2 or 2 C shows that the current is twice the rated capacity. It is impossible to increase the charge current without a limit as it relates to the current accommodation ability of the battery. At present, a current of about 1 C can be used by the lithium-iron-phosphate battery. To ensure the battery life, however, it is generally recommended to use 0.2 C and 0.3 C as the operating current of the battery. Due to the current capacity of the battery, it is impossible for the electric vehicle to solve the problem of energy supply in a short time as in traditional cars. The problem has also become one of the reasons for restricting the promotion and use of electric vehicles.

In this chapter, an internal resistance characteristic-based variable current charging method is proposed in order to shorten the charging time and to ensure that more energy is charged to the battery. The basic charging principle is shown in Figure 17.17.

The charging procedures are as follows:

1) Perform the charging test at the same temperature and a different current $I_{cn}(n = 1, 2, 3, 4, \dots)$ and draw the battery internal resistance change curve at different charge currents. Multiple cells made of same material can be selected from the same brand and the same batch to perform the test or after a series connection to ensure that the representativeness of the data and the average resistance of the samples are taken for reference.

2) Determine the SoC_n corresponding to the inflection point of the battery internal resistance increase at different charge currents and use SoC_n as the cut-off position of the charge current I_{cn}, that is to say, when the battery is charged to SoC_n at the current I_{cn}, the charging is stopped.

3) Establish the coordinate system of the charge current I_{cn} and SoC, draw (SoC_n, I_{cn}) on the coordinate and draw the variable current charging curve by linear fitting.

It is assumed that there are $m + 2$ sample points on the battery internal resistance curve $r_{ni} = f(SoC_{ni})$. At the charge current I_{cn}, the battery internal resistance characteristic can be calculated by the resistance slope change:

$$k_{ni}(SoC_{ni}) = \frac{r_{ni+1} - r_{ni}}{SoC_{ni+1} - SoC_{ni}}$$

$$\Delta k_{ni}(SoC_{ni}) = k_{ni+1}(SoC_{ni+1}) - k_{ni}(SoC_{ni})$$

$$(17.7)$$

The inflection point of the battery internal resistance change SoC_n is determined by the maximum slope change:

for $(i = 1 : m)$
$if(\Delta k_{ni}(SoC_{ni}) = \max(\Delta k_{nj}(SoC_{nj}))$ $(j = 1, \ldots, m)$
$SoC_n = SoC_{ni}$
end

$$(17.8)$$

The internal resistance characteristic-based variable current charging curve may be expressed by the following piecewise function:

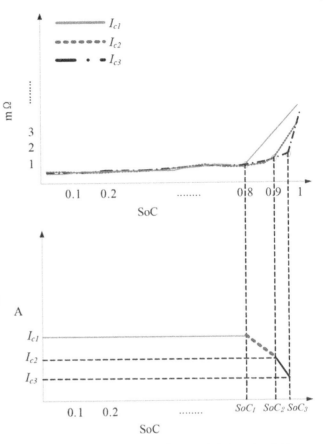

Figure 17.17 Internal resistance characteristic-based variable current charging method.

$$f(x) = \begin{cases} I_{c1}(0 < x < SoC_1) \\ \dfrac{(I_{c1} - I_{c2})x + I_{c2}SoC_1 - I_{c1}SoC_2}{SoC_1 - SoC_2}(SoC_1 < x < SoC_2) \\ \ldots \\ \dfrac{(I_{cn-1} - I_{cn})x + I_{cn}SoC_{n-1} - I_{cn-1}SoC_n}{SoC_{n-1} - SoC_n}(SoC_{n-1} < x < SoC_n) \end{cases}$$

$$(17.9)$$

The pure electric minibuses are tested in the following test procedures to obtain the variable current charging curve:

1) Test the charge internal resistance of the battery used in the pure electric minibuses at charge current I_C of 0.5 C, 0.4 C, 0.3 C, 0.2 C, and 0.1 C and temperature T of −20°C, −10°C, 0°C, 10°C, 20°C, 30°C, and 40°C.
2) Draw the average battery internal resistance diagram during charging and calculate the inflection points at different currents.
3) Draw the variable current charging curve according to Equations (17.7), (17.8), and (17.9).

After the charge internal resistance of the battery pack has been tested, the inflection point of the resistance change at different temperatures and different currents can be obtained, as shown in Table 17.5.
Using Table 17.5, draw the variable current charging model, as shown in Figure 17.18.

Table 17.5 Inflection point of resistance change at different currents.

Temperature\Current	0.3 C	0.4 C	0.5 C	0.7 C	1 C
40°C	98.84%	96.68%	94.20%	91.98%	89.60%
30°C	98.74%	96.49%	94.11%	91.89%	89.51%
20°C	97.77%	95.53%	93.27%	90.98%	88.62%
10°C	94.83%	92.95%	91.40%	86.70%	83.84%
0°C	78.41%	76.08%	74.52%	72.42%	69.75%
–10°C	72.64%	70.03%	68.09%	66.23%	63.72%
–20°C	31.68%	30.95%	30.22%	29.48%	28.71%

The Li-ion battery for the electric vehicle is charged generally using two charging methods: (1) first the constant current method and then the constant voltage charging method and (2) first the constant current method and then the variable current charging method.

The first constant current and then constant voltage charging method means that the battery is charged at the constant current in the initial stage and at the constant voltage when the total voltage of the battery pack reaches the set constant voltage. However, as the cells are connected in series, if the consistency of the battery pack is poor, it is possible that at the end of charging, a certain cell may reach the cut-off voltage first due to the more rapid voltage rise than other cells, causing some cells to be unable to fully charge and the failure of other cells to obtain more energy.

The first constant current and then variable current charging method means that the battery is charged at the constant current at the initial stage, and by reducing the charge current when the battery voltage reaches the set value, the cells are checked to determine whether the current is continuously decreased repeatedly until the current is decreased to zero. In this method, the performance of the cells is considered and the cells are prevented from meeting the cut-off condition first, due to the battery internal resistance increase resulting from the large current, thus protecting the battery while it is charged with more energy. Therefore, the first constant current charging method and then the variable current charging method are more ideal for the electric vehicle at present.

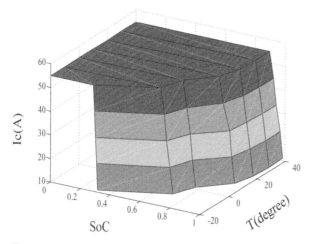

Figure 17.18 Variable current charging curve at different temperatures.

The method proposed here is compared with the first constant current method and then the variable current charging method using the following testing procedure to verify the effectiveness of the algorithm provided here:

1) Fully charge the battery in the first constant current and then variable current charging method, test the non-load driving range of the vehicle, and record the energy required to fully charge the battery, and the driving mileage of the vehicle upon full discharge of the battery.
2) Fully charge the battery in the method proposed here, test the non-load driving range of the vehicle, and record the energy required to fully charge the battery and the driving mileage of the vehicle upon a full discharge of the battery.
3) Compare the impact of the method proposed here and the first constant current and then variable current charging method on the energy level and driving mileage of the vehicle.

The basic procedures of the first constant current and then variable current charging methods are given below:

1) Charge the battery pack at 0.5 C. If the voltage of the cell is 3.55 V, continuously charge the battery pack after reducing the current by 5 A.

Table 17.6 Comparison of results achieved by different charging methods.

Charging method	Energy level	Driving mileage	Charging time
First constant current and then variable current charging method	33.10 kWh	189 km	2 h and 30 min
Internal resistance detection-based variable current charging method	34.84 kWh	192 km	2 h and 10 min

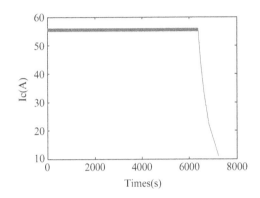
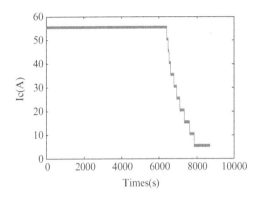

(a) Current change curve of variable current charging method   (b) Current change curve of traditional charging method

Figure 17.19 Comparison of results achieved by different charging methods.

2) Determine whether the voltage of the cells is more than 3.5 V. If yes, continuously charge the battery pack after reducing the current by 5 A. If no, continuously charge the battery pack at the same charge current.
3) After the current is reduced to 0 A, stop charging and consider that the battery pack has been fully charged.

The external temperature is about 23°C during the test and the variable current curve at 20°C is selected as the charging curve. Table 17.6 shows the energy level and driving mileage of the vehicle achieved by the traditional method and the charging method provided here.

Figure 17.19 (a and b) shows the current change during charging in two charging methods. It can be seen that during use of the first constant current and then the variable current charging method, a longer charging time is required and the energy level is lower, and during use of the proposed variable current charging method, a short charging time is required and the energy level is higher, so the driving mileage of the vehicle is increased.

17.2 Consistency Evaluation Research of a Power Battery System

17.2.1 Analysis of a Battery Pack Maintenance Strategy and Performance Evaluation Index

It is required to regularly maintain the battery pack in order to ensure normal use of the electric vehicle for a long time. The battery pack maintenance may be divided into the following stages according to the inconsistency degree of the battery pack: (1) when the consistency of the battery pack is good, there is only the need to charge and discharge the battery pack; (2) when the consistency of the battery pack is reduced but the battery pack has a good performance, there is only a requirement to balance the charge or discharge of the battery pack by the battery management system in order to ensure the consistency of the battery pack; (3) when the consistency of the battery pack is reduced further, the performance of some cells becomes poor, so these cells need to be checked and replaced; and (4) when the consistency of the battery pack is very poor, the battery pack needs to be replaced.

During the maintenance stage, the performance of the battery pack can be evaluated by the index σ. As shown in Figure 17.20, the different range of the index σ shows the different performances of the battery pack: (1) if $\sigma < d_1$, the consistency of the battery pack is better and the battery pack is normally used; (2) if $d_1 \leq \sigma < d_2$, the performance of the battery pack meets the requirement of the battery maintenance stage 2; (3) if $d_2 \leq \sigma < d_3$, the performance of the battery pack meets the requirement of the battery maintenance stage 3; (4) if $\sigma > d_3$, the performance of the battery pack meets the requirement of the battery maintenance stage 4.

17.2.2 Comparison of the Battery Pack Performance Evaluation Methods

The maintenance of the battery pack using a suitable method depends on the method used for accurately describing the consistency index σ of the battery pack. Generally, the consistency of the battery pack is classified into three types: voltage consistency, capacity consistency, and internal resistance consistency.

Voltage consistency means that the voltage deviation of the cells is determined by checking the real-time voltage of the battery pack and comparing the voltage difference of the cells to determine the consistency of the battery pack. This method is used to check the consistency when the battery pack is not charged and discharged, and can obtain a more accurate consistency. However, as the current change is large during the use of the electric vehicle, the voltage change of the battery is also large, a certain deviation may be caused by the voltage consistency, and it is difficult to accurately describe the performance of the battery pack.

The capacity consistency means that the capacity of the cell/single module is tested to determine the consistency of the battery pack, which, in theory, can be descried more accurately by this method. However, the cells or the modules need to be removed and their capacity during testing needs to be evaluated, but it is difficult to evaluate the consistency using this method due to a higher operation complexity.

Therefore, this method has a poor practical significance during the evaluation. The internal resistance consistency means that the difference in the internal resistance between the cells is determined in order to obtain consistency of the battery pack. The battery performance is described essentially in this method. The internal resistance of the cells can be calculated by the equivalent internal resistance to evaluate the consistency of the battery pack. However, this method requires time to place the battery pack, so it cannot be used to check the battery pack in a real-time manner when the vehicle is being used. As the consistency of the battery pack is determined together with the maintenance of the vehicle, the testing procedure is added for the equivalent internal resistance to determine the internal resistance consistency during regular maintenance of the vehicle (for example, the charging process).

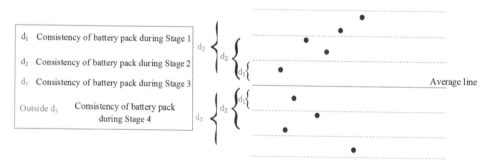

Figure 17.20 Division of the battery pack performance.

17.2.3 Internal Resistance Characteristic-Based Consistency Evaluation Theory of the Battery Pack

It is assumed that the internal resistance of the cells can be detected and the consistency of the battery pack can be described using the standard deviation calculation method:

$$R_{av} = \frac{\sum_{i=1}^{n} R_i}{n} \tag{17.10}$$

$$\sigma = \sqrt{\frac{\sum_{i=1}^{n}(R_i - R_{av})^2}{n-1}} \tag{17.11}$$

where σ is the standard deviation index of the battery pack, n is the number of cells, R_i is the internal resistance of cell I, and R_{av} is the average resistance of the battery pack.

According to Figure 17.20, the consistency of the battery pack is divided into four stages and there is a degree difference d_e ($e = 1, 2,$ and 3) in the battery performance index between each stage and the next. To establish the relationship between the d_e and the staged internal resistance standard deviation index σ_e, it may be considered that the distribution of the battery internal resistance meets the normal distribution rule. The following equation is used to calculate whether the number n of cells is given.

1) Given the normal distribution function,

$$f(x) = \frac{1}{\sqrt{2\pi}} e^{(-x^2/2)} \tag{17.12}$$

2) If d_e is used as the maximum difference between the internal resistance of the cells and the average resistance of the battery pack during each stage, the resistance distribution of each cell meets rule 3σ, that is to say, at the internal resistance difference d_e, this corresponds to the position in Equation (17.12) of $x = \pm 3$, and the occurrence probability of the internal resistance difference of n cells is distributed within $[-3, 3]$. Then

$$x_j = -3 + (j-1) \times \frac{6}{n-1}$$
$$d_j = \left(\frac{1}{\sqrt{2\pi}} e^{(-x_j^2/2)} - \frac{1}{\sqrt{2\pi}} \right) \times d_e \bigg/ \left(\frac{1}{\sqrt{2\pi}} e^{(-9/2)} - \frac{1}{\sqrt{2\pi}} \right) \tag{17.13}$$

where d_j is the difference between the internal resistance of the cell j and the average internal resistance.

3) When the internal resistance difference of the cells is distributed within d_e, the standard deviation σ_e, which is calculated in Equation (17.13), is given:

$$\sigma_e = \sqrt{\frac{\sum_{i=1}^{n} d_j^2}{n-1}} \; (j=1,\ldots,n) \tag{17.14}$$

4) During the consistency evaluation of the battery pack, the standard deviation σ is calculated for the cells in Equations (17.10) and (17.11), and compared with the staged standard deviation σ_e to determine the stage of the standard deviation of the resistance of the battery pack and to adopt the corresponding maintenance strategy.

As the internal resistance change of the battery pack varies with the SoC and charge and discharge current of the battery, especially at the end of charging and discharging, it is possible to cause further polarization and more obvious inconsistency in the battery. The SoC and the charge–discharge process should be comprehensively considered during the consistency evaluation of the battery pack for the electric vehicle.

17.2.4 Internal Resistance Characteristic-Based Consistency Evaluation of the Battery Pack

To verify the effectiveness of the above-mentioned method, the charge and discharge test is performed for a new 110 Ah 10 Li-ion battery pack by the following test procedures in order to obtain the internal resistance consistency evaluation curve of the new battery pack:

1) Place the battery pack in a 20°C thermostat and keep the non-charge–discharge state until the temperature of the battery electrode is close to that of the thermostat.
2) Charge the battery pack at the current I. When the voltage of the cell is 3.8 V, consider that the battery pack is fully charged and keep the non-charge–discharge state until the temperature of the battery electrode is close to that of the thermostat.
3) Discharge the battery pack at the current I for 30 s, keep the non-charge–discharge state until the temperature of the battery electrode is close to that of the thermostat, and record the voltage $U_{idischarge}$ of the cell 1 s before stopping the discharge and the voltage $U_{idischarge}$ of the cell 1 s before the standing state.
4) If the voltage of the cell is less than 2.7 V during discharging, perform step 5; otherwise repeat step 3.
5) Keep the non-charge–discharge state until the temperature of the battery electrode is close to that of the thermostat.
6) Charge the battery pack at the current I for 30 s, keep the non-charge–discharge state until the temperature of the battery electrode is close to that of the thermostat, and record the voltage $U_{icharge}$ of the cell 1 s before stopping the charge and the voltage $U_{icharge}$ of the cell 1 s before the standing state.
7) Check whether the voltage of the cell is more than 3.8 V during charging. If yes, perform step 8; otherwise repeat step 6.
8) Keep the non-charge–discharge state until the temperature of the battery electrode is close to that of the thermostat.
9) Set the current I respectively to 0.3 C, 0.5 C, 0.7 C, and 1 C, calculate the equivalent internal resistance of each cells in Equation (17.2), and calculate the consistency curve of the battery pack.

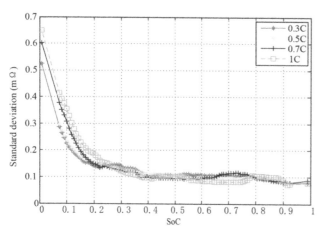

Figure 17.21 Consistency curve of the battery pack during discharging.

Figures 17.21 and 17.22 show the charge and discharge internal resistance consistency curve of the battery pack. It can be seen that at the same temperature, 20–90% SoC, and different charge and discharge currents, the standard deviation curves of the battery internal resistance basically coincide, that is to say, the standard deviation curves of the battery internal resistance are rarely affected by the charge and discharge currents. In order to ensure that the battery is not damaged, the battery internal resistance consistency test is generally performed at charge and discharge currents of 0.5 C.

Generally, as a series of screening and matching tests are performed for the battery pack upon delivery to ensure its good consistency, the consistency evaluation results of the new battery pack may be used as the basic curve for the normal evaluation of the battery pack.

17.2.5 Internal Resistance Characteristic-Based Staged Consistency Evaluation Method for the Battery Pack

The staged evaluation may be performed for the battery pack by the following equation:

$$g_{dn}(\lambda(1-x)) = r_{dn}(1-x) + f_{d1}(1-x)$$

$$(17.15)$$

$$g_{cn}(\lambda x) = r_{cn}(x) + f_{c1}(x) \qquad (17.16)$$

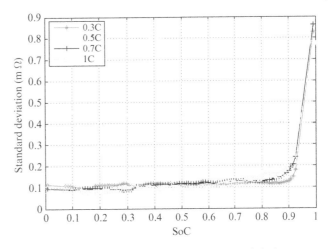

Figure 17.22 Consistency curve of the battery pack during charging.

where x is the SoC of the battery pack, $g_{dn}(x)$ is the function of SoC related to the maximum average internal resistance difference of the battery pack during stage n of the discharging process, $f_{d1}(x)$ is the function of SoC related to the maximum average internal resistance difference of the battery pack during stage 1 of the discharging process, $g_{cn}(x)$ is the function of SoC related to the maximum average internal resistance difference of the battery pack during stage n of the charging process, $f_{c1}(x)$ is the function of SoC related to the maximum average internal resistance difference of the battery pack during the stage 1 of the charging process, λ is the attenuation coefficient of the battery pack, which is generally considered to be the performance attenuation of the battery pack, $r_{dn}(x)$ is the relation function of the maximum average internal resistance difference of the battery pack during stage 1 of the discharging process related to SoC, and $r_{cn}(x)$ is the relation function of the maximum average internal resistance difference of the battery pack during stage 1 of the charging process related to SoC; $r_{dn}(x)$ and $r_{cn}(x)$ can be considered as the resistance increase resulting from the performance attenuation of the battery.

A cycle aging test is performed for the cell by the following procedure to determinate λ, $r_{dn}(x)$, and $r_{cn}(x)$:

1) Connect 10 110 Ah cell samples in series, place them in a 20°C thermostat, and keep a non-charge–discharge state until the temperature of the battery electrode is close to that of the thermostat.
2) Charge the battery pack at 0.5 C until the voltage of the cell is 3.8 V and keep the non-charge–discharge state until the temperature of the battery electrode is close to that of the thermostat.
3) Calculate the charge and discharge internal resistance and charged and discharged energy of the battery pack in steps 3 to 9 of the battery internal resistance evaluation method provided in Section 17.2.4.
4) Discharge the battery pack at 0.5 C until the voltage of the cell is less than 2.7 V and keep the non-charge–discharge state until the temperature of the battery electrode is close to that of the thermostat.
5) Charge the battery pack at 0.5 C until the voltage of the cell is 3.8 V and keep the non-charge–discharge state until the temperature of the battery electrode is close to that of the thermostat.
6) Repeat steps 4 and 5 for nine times and then evaluate the battery internal resistance in step 3.
7) Repeat steps 2 to 6 for 150 times, namely 1500 charge and discharge cycles, and test the internal resistance and calculate the charged and discharged energy of the battery pack every 10 times.
8) Record the energy attenuation and the change of average resistance and the SoC curve.

Then calculate the energy attenuation using the following equation:

$$\eta_i = \frac{A_i}{A_1} \times 100\%$$

(17.17)

where A_i is the energy (A_h) discharged from the battery pack during discharging for time i, A_1 is the energy (A_h) discharged from the battery pack during discharging for time 1, and η_i is the ratio of energy discharged from the battery pack during discharging for time i to the initial energy.

Figure 17.23 shows the energy attenuation with the increase in the charge and discharge cycles of the battery. It can be seen that when more than 400 charge and discharge cycles are performed for the battery pack, the energy attenuation is about 5%, and when more than 800 charge and discharge cycles are performed for the battery pack, the energy attenuation is about 10%.

As many cells are connected in series for the electric vehicle, the performance attenuation of the cells has a direct impact on the energy released from the battery pack. Due to the impact of the production process, the use process, and the current change of the battery, during actual use the attenuation of the battery is more rapid than the attenuation tested in the laboratory. It can be considered that when the released energy is 95% of the initial energy, there is a requirement to balance the battery, namely stage 2 of the battery performance evaluation; when the released energy is less than 90%, there is an urgent requirement to maintain the battery, namely stage 3 of the battery performance evaluation. When the released energy is further reduced, the maintenance strategy of stage 4 provided here is used.

The internal resistance standard deviation is calculated at 95% and 90% energy attenuation of the battery pack in the resistance standard deviation curve of the battery described in Section 17.2.3, as shown in Figures 17.24 and 17.25.

The internal resistance standard deviation of the battery pack during the initial energy and the energy attenuation of 95% and 90% are compared, as shown in Figures 17.26 and 17.27, in which $r_{cn}(x)$ and $r_{dn}(x)$ can be calculated by fitting the different resistance standard deviations at the same SoC and resistance standard deviations during the initial energy.

As many cells are used for the vehicle, it is impossible to perform the cycle test for each cell to obtain the consistency evaluation curve after installation of the battery pack in the vehicle. This test should be performed for the same batch of the battery pack made in the same process and the same material in order to obtain the initial

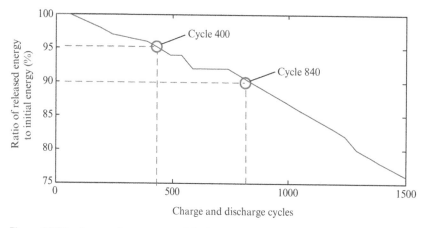

Figure 17.23 Energy change curve of the battery during the charge and discharge cycle.

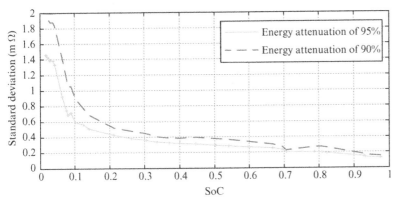

Figure 17.24 Discharge internal resistance standard deviation curve of the battery pack during energy attenuation.

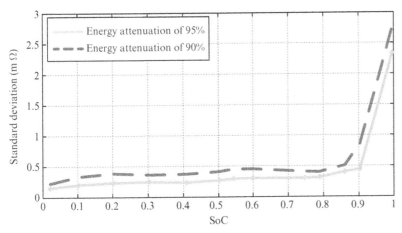

Figure 17.25 Charge internal resistance standard deviation curve of the battery pack during energy attenuation.

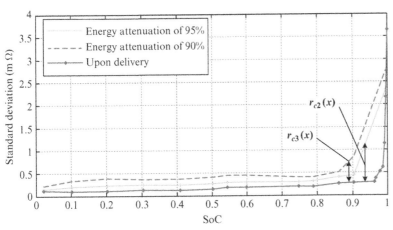

Figure 17.26 Charge internal resistance standard deviation curve of the battery pack during different energy attenuation stages.

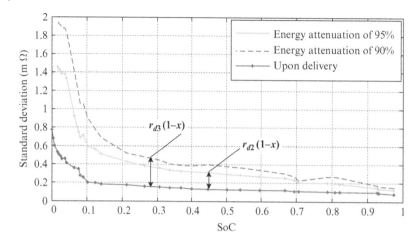

Figure 17.27 Discharge internal resistance standard deviation curve of the battery pack during different energy attenuation stages.

resistance standard deviation curve of the battery pack before installation in the vehicle. The functions $r_{cn}(x)$ and $r_{dn}(x)$ are found using the aging test of a few battery packs in the method described to determine the internal resistance consistency curve during different stages.

17.2.6 Internal Resistance Consistency Evaluation Test of the Battery Pack for a Pure Electric Vehicle

A pure electric minibus FDG6601 independently developed by the Dongguan Institute, Sun Yat-sen University was tested to verify the effectiveness of the method proposed herein.

The test procedures are as follows:

1) Perform the charge and discharge test for the battery pack of the vehicle in the method provided in Section 17.2.4 to obtain the staged internal resistance consistency curve of the battery pack.
2) Perform the charge and discharge test for the new battery pack, calculate the internal resistance characteristic of each cell during charge and discharge, and draw the internal resistance standard deviation curve of the battery.
3) Perform the charge and discharge test for the battery pack used for one month, calculate the internal resistance characteristic of each cell during the charge and discharge, and draw the internal resistance standard deviation curve of the battery.
4) Compare the change between the internal resistance standard deviation curve of the new battery pack and the battery pack used for one month, and compare this with the staged internal resistance consistency evaluation curve of the battery pack to check the performance of the battery pack.

Figures 17.28 to 17.31 show the internal resistance standard deviation curve of the battery pack. It can be seen that the internal resistance of the battery pack is basically under the normal curve. After using for one month, the internal resistance of the battery pack begins to fluctuate during charging and discharging, and sometimes changes from stage 1 to stage 2 of the evaluation curve. Equilibrium charge–discharge maintenance is required for the battery pack in the staged evaluation and maintenance method proposed here after use of one month.

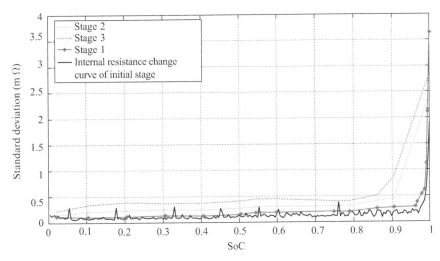

Figure 17.28 Charge internal resistance change curve of the battery for a minibus during the initial stage.

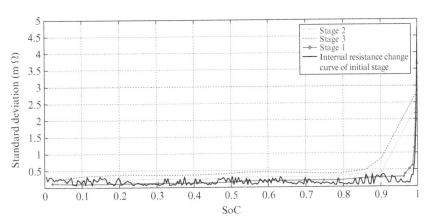

Figure 17.29 Discharge internal resistance change curve of the battery for a minibus during the initial stage.

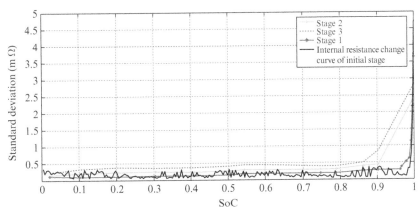

Figure 17.30 Charge internal resistance change curve of the battery for a minibus after use of one month.

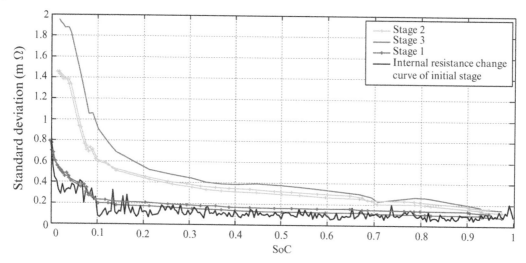

Figure 17.31 Discharge internal resistance change curve of the battery for a minibus after use of one month.

17.3 Safety Management and Protection of a Power Battery System

Battery safety is a primary problem for normal use of the new energy vehicles. Therefore, the safety protection of the battery management system is the most important thing in battery management. At present, the common safety protection functions include "over-current protection", "over-voltage protection," and "over-temperature protection."

1. Over-current protection

Over-current protection means that the safety protection measure must be taken if the operating current exceeds the safety range of the battery during charging and discharging. Popularly, if the current is too large during use, the power supply is disconnected. At present, common power batteries support continuous charge and discharge at less than 1 C. If the current is more than 1 C, the impact on battery performance depends on the over-current duration. As the battery characteristics vary with the manufacturers, the effective duration of the battery is generally tested at different over-currents to establish the over-current duration database. The over-current duration is obtained by the database to provide power-off protection or alarming during use.

2. Over-voltage protection

This function is used to provide over-charge and over-discharge protection of the battery. In principle, the over-charge and over-discharge of the battery means that the charged or discharged energy exceeds the maximum capacity of the battery.

In the case of over-charge or over-discharge, it is possible to cause different degrees of damage to the battery, and even a safety accident. Therefore, when the battery is close to an over-charge or over-discharge state, current input and output are actively disconnected by the battery management system so that the battery is not charged or discharged.

Considering the statistical error of SoC, the voltage change of the battery during use is generally used to determine the over-charge or the over-discharge. Therefore, charging cut-off voltage and discharging cut-off voltage are set as the basic condition for over-voltage protection in the battery management system. Considering the characteristics of the cells, its protection conditions include the cut-off voltage of the cells and the battery pack. The cut-off voltage of the cells is set for the safety of the cells and the cut-off voltage of the battery pack is set for the safety of the power battery system.

It is worth noting that due to the battery inconsistency, the cut-off voltage of the battery pack is not the multiple of the cut-off voltage of the cells connected in series. The cut-off voltage of the cells is determined by the battery manufacture after the safety test of the same batch of cells. The cut-off voltage of the battery pack is determined according to the use condition of the vehicle and the voltage and performance of the cells. Therefore, the over-voltage protection threshold must be set for the battery management system according to the type and model of the power battery and installment of the battery pack.

3. Over-temperature protection

Over-temperature protection means that when the battery temperature or the ambient temperature exceeds a certain threshold, certain protective measures are taken to protect the power battery from being damaged.

For most batteries, the ideal operating temperature is 20°C–45°C. Due to the external environment and the storage space of the battery, the actual operating temperature of the battery cannot always be within the range. In order to ensure normal use of the battery and the safety of the vehicle, the temperature protection strategy is generally divided into two parts: over-temperature protection and low-temperature management.

1. Over-temperature protection strategy

If the battery temperature is higher than the ideal operating temperature, it is difficult to control chemical reactions, resulting in battery damage and even fire accidents. Therefore, in the case of battery over-temperature, a series of stepped protection measures, such as starting the cooling system, early warning, and finally a forced stop, are generally taken according to the relationship between the actual temperature and the boundary temperature.

Figure 17.32 shows the basic battery temperature management strategy of a new energy vehicle. At more than 30°C temperature, the cooling system is started. If the temperature is 40°C, the temperature alarm is activated. In the case of a continuous temperature rise to 44°C, the system is forced to stop. During alarming or cooling, if the temperature descends, to avoid the frequent starting or stopping of the system at a critical temperature, the temperature used to close the cooling system or the alarming system should not be different from the temperature used to open them. Therefore, 28°C and 38°C were respectively selected as the temperatures for closing the cooling system and the alarming system. At the same time, considering the hysteresis of the temperature change, when the protection temperature is set, the temperature lead is often considered. For example, 44°C is used as the temperature for stopping the vehicle, that is to say, the whole system does not operate to prevent the battery from damage caused by a temperature rise.

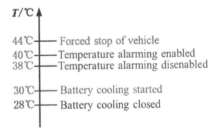

Figure 17.32 Over-temperature control strategy of a new energy vehicle.

2. Low-temperature management strategy

If the battery temperature is lower than the temperature, the battery internal resistance will increase and its output and input current will decrease. Therefore, when the battery temperature is too low, a heating module is generally added inside the battery box or around the cells. If the temperature is too low, the battery management system gives an instruction to control the operation of the heating module, in order to ensure that the battery works within the appropriate temperature range.

Figure 17.33 shows the heating system control strategy of a new energy vehicle. When the temperature is below 10°C, the battery heating system is turned on. If the temperature is lower than -10°C,

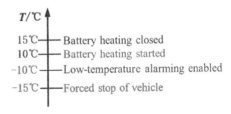

Figure 17.33 Lower-temperature control strategy of a new energy vehicle.

most of the batteries can output less than 40% of the normal output. The low-temperature warning function is enabled to indicate when the battery is no longer suitable for long-term use at this temperature. If the temperature continues to decrease to –15°C, the system will be forced to stop. Similarly, in order to avoid frequent starts or stops of the heating system at the critical temperature, 15°C is selected as the temperature used for stopping the heating system.

Usually in Northern China, a heating module is added to the battery systems of most new energy vehicles to adapt to the low temperature in winter; In Southern China, especially the south-east coastal areas, the temperature is basically below 10°C in winter, so few battery systems are equipped with the heating module.

Index

a

AC: Alternating Current 8, 161–163, 194–198, 205–211, 264, 289
A/D: Analog-to-Digital Converter 171, 186, 189
ADC: Analog-to-Digital Converter 189
AhC: Ampere-hours Counter Unit 6
APU: Auxiliary Power Unit 8
AQW214: photomos 139–140
AUTOSAR: AUTomotive Open System Architecture 22

b

BCU: Battery Control Unit 16–18, 135–138, 149, 166
BMC: Battery Monitoring Circuit 16–18, 135–136, 166
BMS: Battery Management System 3–10

c

CAN: Controller Area Network 4, 135–136
CC: Constant Current 196–198, 201–206
CCCS: Current Controlled Current Source 111, 125

CC method: Coulomb Counting method 137, 146, 158–161, 163–164, 170, 176
CCS: Combined Charging System 208
CCVS: Current Controlled Voltage Source 111
CD4051: general electronic analog switch 140
CF: Compact Flash card 317
CID: Circuit Interrupting Device 211
CP: Constant Power 198, 202
CTO: Chief Technology Officer 9
CV: Constant Voltage 196–199, 201–206
CV: Cell or mono blocks Voltage measuring units 6

d

DA: Data Acquisition unit 6–7
DC: Direct Current 7–8, 12, 73–80, 87–91, 117, 140, 147, 149, 161–162, 194–198, 202–208, 211
DC/DC: Direct Current to Direct Current 370
DOD: Depth Of Discharge 80–86, 90–91, 93–95

Battery Management System and its Applications, First Edition. Xiaojun Tan, Andrea Vezzini, Yuqian Fan, Neeta Khare, You Xu, and Liangliang Wei.
© 2023 China Machine Press. All rights reserved. Published 2023 by John Wiley & Sons Singapore Pte. Ltd.

DS2782: integrated chip for the battery management 151

DST: Dynamic Stress Test 122–124

e

EBC: Energy Bus Controller 299–300

E-Bike: Electric Bike 329–332

ECE: Economic Commission of Europe 53

ECI: Electronic Concepts Industrial 8

EEROM: Electrically Erasable Read-Only Memory 15

EICAS: Engine Indicating and Crew Alerting System 8

EKF: Extended Kalman Filter 170–172, 174–177, 179, 181, 183–191

EMC: Electro-Magnetic Compatibility 21

EMF: Electromotive Force 20, 23, 38, 105–107, 109–115, 125, 143–144, 157–162, 164–165, 169, 172, 174, 177–186, 191

EOL: End Of Life 87, 95

EPROM: Electrical Programmable Read Only Memory 6

E-R Diagram: Entity Relationship Diagram 311–312

EUDC: Extra Urban Driving Cycle 53

EV: Electric Vehicle 7, 194–196, 206–211

EVSE: Electric Vehicle Supple Equipment 195–198, 209

f

FET: Field Effect Transistor 139

FIFO: First Input First Output queue 307

FLASH: Flash memory 317–318

FUDS: Federal Urban Dynamic Schedule 122–124

g

GA: Genetic Algorithm 122–124

GB: China national standards 38, 53–55, 58

GUI: Graphical User Interface 292–293

h

HEV: Hybrid Electric Vehicles 34, 196

HV: High-Voltage 292–293

i

I^2C: Inter-Integrated Circuit 4, 295–296

IC: Integrated Circuit 3–4, 17, 141, 294, 296–297

IEC: International Electrotechnical Commission 39, 54–55, 65

INEEL: Idaho National Engineering and Environmental Laboratory 39

IO: Input/Output ports 306–307

ISO: International Organization for Standardization 22, 37, 296

isoSPI: Isolated SPI(Serial Peripheral Interface) 142

j

JARI: Japan Automobile Research Institute 210

k

KB: Kilobyte 6

l

LAN: Local Area Network 142

LCD: Liquid Crystal Display 304–305

LEV: Low Emission Vehicle 299–300

LFP: $LiFePO_4$ battery 204

LLB: Low-Level Balancing 234–238, 240–244, 248, 250, 253–254

LV: Low-Voltage 292–293

m

MAE: Mean Absolute Error 122–123

MAX: Maxim Integrated Products 140, 141

MAXIM: Maxim Integrated Products 141, 151

MCU: Micro Controller Unit 139–142, 151

MOSFET: Metal-Oxide-Semiconductor Field-Effect Transistor 141, 338–340

n

NCA: Lithium Nickel-Cobalt-Aluminum Oxide 30

NCM: Lithium Nickel Cobalt Manganese Oxide 31, 143

NEDC: New European Driving Cycle 182, 189

NEMA: National Electrical Manufacturers Association 196

NHTSA: National Highway Traffic Safety Administrative 8

NP: Number of Population 121

NTSB: National Transportation Safety Board 8

o

OCV: Open Circuit Voltage 107, 109–116, 125–126, 128

OSI: Open System Interconnection Model 298

p

PC: Personal Computer 5–6, 315–319

PCS: Power Control Software 292–293

PHEV: Plug-in Hybrid Electric Vehicles 34, 198

PLC: Programmable Logic Controller 6

PMBus: Power Management Bus 294

PNGV: Partnership for a New Generation of Vehicles 39, 65, 104

PPU: Power Protection Unit 356

PV: Photovoltaic 4

r

RAM: Random Access Memory 15

RC: Resistor-Capacitance circuit 103–105, 110–117, 172, 174, 282

RF: Radio Frequency 142

RMSE: Root Mean Squared Error 89–91, 95, 122–123, 274, 276, 288

RTU: Remote Terminal Unit 296

s

SAE: Society of Automotive Engineers 38–39, 65, 207, 211

SBS: Smart Battery Systems 294

SD: Secure Digital memory card 307

SDIO: Secure Digital Input and Output 307

SEI: Solid Electrolyte Interface 29–30, 35, 214

SMBus: System Management Bus 293–294

SoC: State of Charge 11

SoE: State of Energy 11

SoF: State of Function 11

SoH: State of Health 11

SoL: State of Life 11

SoP: State of Power 11

SoR: State of Range 11

SPI: Serial Peripheral Interface 293–297, 337

t

TCP/IP: Transmission Control Protocol/ Internet Protocol 292–293

TDMA: Time Division Multiple Access 297

TI: Texas Instruments 141, 222–223

TSA: Tree Seed Algorithm 121–124, 131

u

UDDS: Urban Dynamometer Driving Schedule 125, 127–129

UFCS: Ultra-Fast Charging Station
208–209
UPS: Uninterruptible Power Supply   194
USABC: United States Advanced Battery
Consortium   122, 262
USB: Universal Serial Bus   295–297
USCI: Universal Serial Communication
Interface   294

v

VCVS: Voltage-Controlled Voltage Source   111
VRLA: Valve-Regulated Lead-Acid
6, 10

w

WIFI/Wi-Fi: a wireless networking
technology   307